Synthesis Lectures on Mechanical Engineering

This series publishes short books in mechanical engineering (ME), the engineering branch that combines engineering, physics and mathematics principles with materials science to design, analyze, manufacture, and maintain mechanical systems. It involves the production and usage of heat and mechanical power for the design, production and operation of machines and tools. This series publishes within all areas of ME and follows the ASME technical division categories.

Xiaobin Le

Simulation-Based Mechanical Design

 Springer

Xiaobin Le
Mechanical Engineering Program
School of Engineering
Wentworth Institute of Technology
Boston, MA, USA

ISSN 2573-3168 ISSN 2573-3176 (electronic)
Synthesis Lectures on Mechanical Engineering
ISBN 978-3-031-64134-3 ISBN 978-3-031-64132-9 (eBook)
https://doi.org/10.1007/978-3-031-64132-9

This Springer imprint is published by the registered company Springer Nature Switzerland AG
The registered company address is: Gewerbestrasse 11, 6330 Cham, Switzerland

If disposing of this product, please recycle the paper.

Preface

Objectives

This book is written primarily for senior mechanical engineering students who have learned the mechanics of materials and machine element designs and are now ready to conduct mechanical design. It can be used as a textbook for a course related to FEA (Finite Element Analysis) simulation for mechanical design. By using SolidWorks Simulation, the FEA simulation package used throughout this book, they can simulate the stress, strain, and factor of safety of any mechanical components experiencing any type of loading. These results are typically exceedingly difficult to calculate by hand using theoretical calculations. FEA Simulation provides an alternative approach for them to complete and validate mechanical designs.

This book can be also used as a good and practical reference book for practicing mechanical engineers who want to start using FEA simulation tools. Many simulation design examples and design cases are provided with step-by-step instructions, to guide them through implementing FEA simulation, to help in their daily mechanical design activities.

This book can also be used for self-study for those who want to become familiar with mechanical design in general. The first two chapters provide an overview of the fundamentals of mechanical design and the fundamentals of FEA theory, to prepare them for running mechanical design using FEA simulation.

Features of This Book

This book is user-centered. The contents of the book and the examples with step-by-step instructions are arranged in such a way that we can easily follow, understand, and implement the various skills of FEA simulation for mechanical design. The following are several features of this book.

This book is a good textbook for mechanical design. It is different from traditional textbooks where stress, strain, and factors of safety are mainly calculated by hand with theoretical calculations. This book is also different from some FEA analysis books where FEA software is the only topic covered, and the fundamentals of mechanical design are barely discussed. This book discusses and explains the fundamentals of mechanical design such as the engineering design process, typical stress/strain equations, and failure theory, but also uses SolidWorks Simulation to simulate the stress, strain, and the factor of safety of any mechanical components. So, they can complete their mechanical design even if the components' geometric shapes and loading are complicated.

To most of us, the FEA simulation program is a black box that generates some outputs when the necessary inputs are provided. This book provides a chapter: Fundamentals of FEA Theory, to discuss the fundamental concepts of FEA theory such as nodes, elements, meshing, deformation functions, and element stiffness matrix. The provided hand-calculation example of using the FEA method will help to create a better understanding of FEA theory. Understanding these fundamental concepts of FEA theory will help with properly setting up simulation projects for conducting FEA simulations.

The book contents are driven by a general procedure for FEA simulation, and many design examples are provided with accompanying step-by-step instructions. The book contents are arranged in such a way that significantly facilitates a better understanding of the topics and mastering them. For each chapter, the topics are arranged according to the steps of the general procedure for FEA simulation. Each step will be covered by a section of a chapter. A total of 33 design examples are provided with step-by-step instructions to address each key step and important topic introduced in this book.

The only way to accumulate experience with FEA simulation is by conducting many real-industrial simulation projects. This book also provides minor and major design projects to provide experience with implementing the skills of FEA simulation on real-industrial design projects.

This book covers most of the key topics for mechanical design. The main topics are stress, strain, and the factors of safety of components and assemblies through FEA simulation. This book also covers three special topics with several examples. These special topics are thermal analysis and thermal stress analysis, natural frequency analysis, and fatigue analysis by FEA simulation.

All of the models of examples and exercises can be downloaded from the publisher-provided links. The link is listed as "Supplementary Information" at the bottom of the first page of each chapter. This will let us focus on the key contents: mechanical design and FEA simulation.

Contents of This Book

This book consists of seven chapters. The following is a brief description of each chapter.

Chapter 1 "Fundamentals of Mechanical Design" is a concise summary of mechanical design. It covers the engineering design process; the definition of stress and strain; their typical equations; material mechanical properties; dimension tolerances; engineering materials and manufacturing processes; mechanical failures; factor of safety; principal stresses and maximum shear stress; and static failure theories. A knowledge of these concepts will prepare us for the rest of the chapters in this book.

Chapter 2 "Fundamentals of Finite Element Analysis (FEA)" concisely explores and explains the fundamental concepts of FEA theory. It covers the fundamental concepts of FEA theory; deformation functions and shape functions; types of typical elements; and the element stiffness matrix (element properties); and provides an example of a uniaxial stepped bar under axial load which is manually solved using FEA methods.

Chapter 3 "FEA Simulation on Components" explains and demonstrates how to use the "Static" module in SolidWorks Simulation to conduct FEA simulation to determine and display the stress/strain and the factors of safety of a component. The chapter starts with a general procedure and explains and demonstrates each step using an example. The key steps of this procedure, such as pre-processing, types of elements, fixtures, loads, meshing, and post-processing are each split into a section of this chapter and discussed in detail with supporting examples. The convergence condition is presented as a method to check whether the FEA simulation result is acceptable. At the end of this chapter, a minor design project is presented, which is a real-industrial design project and can be followed to implement all skills learned in the chapter.

Chapter 4 "FEA Simulation on Assemblies" explores and demonstrates how to run an FEA simulation on assemblies. The definitions and settings of types of typical contacts are discussed in detail and demonstrated by examples with step-by-step instructions. Several typical connectors such as blot connectors, pin connectors, and shrink fits are discussed in detail and demonstrated using examples. The mixture of different types of elements in an assembly is discussed, and an example is included demonstrating this mixing. At the end of this chapter, the major design project is presented. This project is based on a real-industrial design project and all skills learned so far are used to complete this major design project.

Chapter 5 "Natural Frequency Analysis" discusses how to use SolidWorks Simulation to estimate the first five natural frequencies. One of the main tasks in mechanical design is to prevent it from experiencing resonance. For this task, this chapter discusses and demonstrates how to determine the natural frequencies of components or mechanical systems using FEA simulation.

Chapter 6 "Thermal Analysis and Thermal Stress Analysis" discusses how to run a thermal analysis using FEA simulation to obtain temperature distributions, and then use these distributions as a thermal load to conduct thermal stress analysis by FEA simulation. Two examples are demonstrated for this special topic.

Chapter 7 "Fatigue Analysis by FEA Simulation" discusses how to use SolidWorks Simulation to estimate fatigue damage of a metal component. Fatigue failure is the single largest cause of failures in metal components, estimated to be the cause of 90% of all metallic failures. First, the fundamentals of fatigue theory are briefly discussed and explained. Then the procedures for fatigue analysis using FEA simulation are discussed and explained in detail. Three examples are demonstrated for this special topic.

Possible Syllabus

If this book is used as a textbook for an FEA simulation-related course, the course could be a 3-2-4 course (3-hour lectures, 2-hour labs with 4 credits) or a 2-4-4 course (2-hour lectures, 4-hour labs with 4 credits) in a 15-week semester. An example course syllabus could be:

Week #	Contents
1	Chapter 1: Fundamentals of Mechanical Design Review of SolidWorks models and drawings
2~3	Chapter 2: Fundamentals of Finite Element Analysis (FEA) Exam 1 for Chap. 1 and Chap. 2
4~9.5	Chapter 3: FEA Simulation on Components Minor design project Exam 2 for Chap. 3
9.5~12	Chapter 4: FEA Simulation on Assemblies Major design project
13~15	Major design project Chapter 5: Natural Frequency Analysis Chapter 6: Thermal Analysis and Thermal Stress Analysis Chapter 7: Fatigue Analysis by FEA Simulation Exam 3 for Chaps. 4–7

Boston, USA   Prof. Xiaobin Le, Ph.D., P.E.
December 2023

Acknowledgments

Over the last decade, I have worked with a group of the best colleagues I could ask for Profs. Ali Moazed, Anthony Duva, Richard Roberts, Illie Talpasanu, and John Voccio to teach the same FEA simulation-related course in our mechanical engineering program. We worked together to discuss course plans, course progress, and assessment data collection. I sincerely thank them! I also sincerely thank Associate Dean and Prof. Michael Jackson for his great support.

I also want to thank my two sons: Mr. Zelong Le and Mr. Linglong Le, who are my first readers of this book and provided lots of valuable suggestions.

Finally, I want to sincerely thank my wife, Ms. Suyan Zou. With her continued support, I could complete this book.

Boston, USA

Prof. Xiaobin Le, Ph.D., P.E.

December 2023

Contents

About the Author

Xiaobin Le Ph.D., P.E., received a B.S. in Mechanical Engineering in 1982 and an M.S. in Mechanical Engineering in 1987 from Jiangxi University of Science and Technology, Ganzhou Jiangxi. He received his first Ph.D. in Mechanical Design of Mechanical Engineering from Shanghai Jiao Tong University, Shanghai, in 1993, and his second Ph.D. in Solid Mechanics of Mechanical Engineering from Texas Tech University, Lubbock, Texas, in 2002. He is currently a professor in the Mechanical Engineering Program at Wentworth Institute of Technology, Boston, Massachusetts. His teaching and research interests are Computer-Aided Design, Mechanical Design, Finite Element Analysis, Fatigue Design, Solid Mechanics, Engineering Reliability, and Engineering Education Research.

Fundamentals of Mechanical Design

1

Abstract

This chapter covers all fundamental concepts, design theory, and formulas for mechanical design without any examples for theoretical hand calculation because all calculations will be done by FEA (Finite Element Analysis) simulation. First, it uses the five-phase design process to discuss and explain an engineering design process. Second, it discusses and explains the definitions of stress and strain; it provides typical stress formulas such as normal stress of a bar, direct shear stress, shear stress of a shaft under torsion, bending stress, and shear stress of a beam under a shear force, as well as it explores principal stresses and maximum shear stresses of a general state of stress. Third, selections of materials and manufacturing processes are explained and discussed. Fourth, it explains the definition of free dimension and mating dimension and shows how to determine the dimension tolerance for free dimensions and mating dimensions. Finally, it explains the classification of materials, material mechanical properties, definitions of mechanical failures, typical failure modes, and typical failure theories for ductile and brittle materials. Therefore, this chapter is a concise summary of mechanical design. It prepares us to be ready for conducting mechanical design while stress, strain, and factors of safety are determined by FEA simulation.

1.1 Introduction

This chapter is a concise summary and review of mechanical design including fundamental concepts, theories, and formulas. It will prepare us for the rest chapters of this book. The calculations of stress & strain of mechanical components will not be explored through theoretical approaches. They will be simulated through Finite Element Analysis (FEA) simulations, which will be the main contents of the rest chapters of this book.

© The Author(s), under exclusive license to Springer Nature Switzerland AG 2025

X. Le, *Simulation-Based Mechanical Design*, Synthesis Lectures on Mechanical Engineering, https://doi.org/10.1007/978-3-031-64132-9_1

A brief description of each section are listed here.

Ch1.2 Engineering design process will discuss a general procedure for conducting design projects. The five-phase design process will be discussed and explored.

Ch1.3 Stress and strain will discuss the definitions of stress and strain and their applications in several typical cases.

Ch1.4 Material mechanical properties will discuss and explain typical mechanical properties obtained from tensile tests and shear tests.

Ch1.5 Dimensions and tolerances will discuss dimensions, dimension tolerances, types of dimensions, types of fits, and how to determine dimension tolerances of free dimensions and mating dimensions.

Ch1.6 Materials and manufacturing processes will discuss typical classifications of materials, their typical properties, and typical manufacturing processes. It will also discuss how to select materials and select manufacturing processes for a mechanical component.

Ch1.7 Mechanical failures will discuss definitions of failures and three typical types of mechanical failures for mechanical design.

Ch1.8 Factors of safety will discuss the definition of the factor of safety and how to properly select it.

Ch1.9 Principal stresses and maximum shear stresses will discuss the definition of principal stress and how to calculate principal stresses and maximum shear stresses, which are used in static failure theories.

Ch1.10 Failure theories due to static loads will discuss typical failure theories due to static loads. Two failure theories for ductile materials and two failure theories for brittle materials will be discussed.

Ch1.11 Concepts of simulation-based mechanical design (SBMD) will explain the definition of simulation-based mechanical design and its main features. This section will also provide brief descriptions of the rest of the chapters.

1.2   Engineering Design Process

This section will briefly discuss and describe one of the engineering design processes, by which a design project is initiated and finally converted into a real product.

Humans have been designing things for as long as we have existed as a society to improve our abilities and our lives. For example, our earliest ancestors designed stone knives and other basic tools to meet their basic needs. Design or engineering design is one of our common activities. We have a lot of different definitions for it and we have accumulated a lot of successful experiences and lessons from engineering design.

The ABET (Accreditation Board for Engineering and Technology) definition of engineering design is: *"Engineering design is the process of devising a system, component, or process to meet desired needs. It is a decision-making process (often iterative), in which the basic science and mathematics, and engineering sciences are applied to convert resources optimally to meet a stated objective. Among the fundamental elements of the design process are the establishment of objectives and criteria, synthesis, analysis, construction, testing, and evaluation.....".*

This definition specifies the four key aspects of engineering design: (1) Engineering design is a process; (2) Engineering design utilizes many different skills with iterative, decision-making, and systematic approaches; (3) Engineering design will build an object, including a system, component, or process under some constraints; and (4) The purpose of the engineering design is to meet the project's needs.

The most important aspect of engineering design is that engineering design is a process. Engineering design is a process not because it consists of several steps and might take a long time to be completed, but mainly because the final approval of the design is by the actual testing of a prototype, not by theoretical calculations or numerical simulations. For an engineering design project, even though accurate theoretical calculations and complicated numerical simulation results can suggest that the design would be safe and could satisfy design requirements, the design product should not be released for mass production and customers without testing. Engineering design experiences have indicated that products without thorough testing of prototypes might cause significant problems for customers and companies.

Any engineering design project is an activity with a lot of loose ends at its beginning. During its design process, each loose end will be closed by corresponding good options. At the end of its design process, the design project is fully specified and can be duplicated according to its engineering documents. Lots of different theories about the design process have been proposed to guide engineers in conducting the engineering design [1~6]. The five-phase engineering design process proposed by Gerard Voland [1] is one of these good theories. It includes (1) Phase One: Needs assessment; (2) Phase Two: Design Specifications; (3) Phase Three: Conceptual designs, (4) Phase Four: Detailed design, and (5) Phase Five: Implementation. We will provide brief descriptions of the five-phase engineering design process as follows [7, 8].

- **Phase One: Needs assessment**

The main task in Phase One is to check whether a need is real and feasible. The "need" is a current problem or current unsatisfactory status. The need can come from personal experience, customers, and society. However, some claim demands might not be real needs and might fade quickly. Some needs might not be feasible for a project team because they might not be technically feasible or not financially viable. For example, constructing more advanced airplanes is always a real need, but this will not be financially viable for a small company because it has limited human resources and funding.

The outcomes of Phase One are: (1) The need is not a real need. No further action is needed. (2) The need is a real need, but it is not technically or financially feasible for a project team. The need might be stored for future use. No immediate action is needed. (3) The need is real and feasible for the project team. Only if the need assessment passes Phase One, it will proceed to Phase Two: Design Specifications. However, the project team is not officially established or formed yet.

- **Phase Two: Design Specifications**

The main tasks in Phase Two are (1) Gaining a true understanding of the need, (2) Searching for existing related solutions for the need, and (3) Determining design specifications for the need. The need, which is generally described in a general language in Phase One, must be rephrased by engineering language.

In Phase Two, the first step is to have a true understanding of the need and then set up some primary criteria. Before we start to search for existing or possible solutions, we must fully and truly understand the need, the real meanings of the need, or the real requirements of the customers. For example, if customers require us to design transportation for them. If we provide transportation that is driven by gasoline, diesel, or electricity, we make a mistake when the power source in that region for the customers is natural gas. The second step is to search for existing or possible solutions for the need by using the primary criteria established in the first step. Do not reinvent the wheel. The design project for the need should use up-to-date techniques. After lots of background investigations have been done, and lots of related up-to-date technical information has been collected, the third step is to establish design specifications for the need. Several common or general design goals are usually associated with design projects such as safety, environmental protection, public acceptance, reliability, performance, durability, ease of operation, use of standard parts, minimum cost, minimum maintenance, ease of maintenance, and manufacturability [2]. The design specifications do not refer to these common or general goals, which are certainly considered in detail. The design specifications properly established in the third step are (1) Any special statement for the design project, and (2) Any numerical value related to the design project. Two examples of design specifications are: "The device must work properly in a very moist environment." and "The factor of safety is 2.5."

The outcomes of Phase Two are: (1) The need could be satisfied by an existing product, or a profitable product could not be constructed due to the highly competitive market. So, no further action is required. The design project will not be set up. (2) Proper design specifications have been established and can be satisfied by a new design. The design project will be set up formally. Now the design project will move to Phase Three: Conceptual Design.

- **Phase Three: Conceptual Design**

The main task in Phase Three is to develop a pool of all possible alternative solutions, and then select several best alternative solutions for the design project from the pool.

Modern design projects are typically complicated with several functions or subsystems. The following four steps can be used to conduct a conceptual design.

Step 1: Decompose a design project into several sub-systems according to its functions. It is very difficult to generate solutions for a design project with several functions. However, it will be relatively easy to generate solutions for a sub-system with only one function. So, in this step, the design project will be decomposed into a lot of sub-systems, which have only one key function or performance.

Step 2: Develop a lot of possible options for each sub-system. A sub-system with a single function or performance can be abstracted into a general catalog, which will be backed up by lots of existing engineering techniques and knowledge. Then, the project team can generate lots of solutions for it. In this step, full information research should be conducted. The quality of these solutions will determine the quality of the final design solution because it will be formed from these solutions.

Step 3: Form a pool of alternatives for the design project. The combination of one option from each sub-system will form one alternative design. This combination will typically generate a large quantity of possible alternative designs for the project. For example, if there are six subsystems, and each subsystem has three solutions, the possible alternatives for a design project will be equal to $3^6=729$.

Step 4: Use personal experiences, judgment, or group discussions to choose several best prospective alternatives for the design project. The project team does not have enough time and money to build & test each possible alternative. It is also not necessary. So, the project team will pick several possible best alternatives for further investigations in the next phase.

The outcome of Phase Three- Conceptual Design is several possible best alternatives for the design project, from which the final design solution will be selected.

- **Phase Four: Detailed Design**

The main tasks in Phase Four are (1) For each of the several possible best alternatives selected in Phase Three, create virtual key sub-systems, or virtual products and run necessary virtual or physical experiments, (2) Choose the final design option for the design project, and (3) Test and modify the prototype and finalize the final design option.

In modern engineering society, lots of modern advanced tools are available for engineering design. For example, CAD (Computer-Aided Design) software can be used to build a virtual component (VC), virtual assembly (VA), or virtual product (VP). A virtual component has the same dimensions, the same material, and the same behaviors as a real component. So, it looks like and behaves like the real component but is stored in a digital form. A virtual assembly is the digital form of a real assembly. The virtual product is the digital form of a real product. Lots of engineering simulation software such as Solid-Works Simulation and ANSYS can be used to test the functions and stress/strain of a virtual component, assembly, or product under different loading conditions. A lot of different CAMs (Computer-Aided Manufacturing) can be used to animate the manufacturing process of a virtual component to check the feasibility of manufacturing.

The first task in Phase Four is to construct virtual components, virtual assemblies, or virtual products, and then use available computer software to conduct numerical simulations to check the performances and functions of different design options. Only some key components or key sub-assemblies might be manufactured. Prototypes of those are experimented on to check their performances and functions.

The second task is to determine the best design option after the performances and functions of several design options under consideration have been evaluated through virtual tests. Since a design project has several design specifications and functions, a systematic evaluation method such as Decision Matrix [1] can be implemented to choose the best option.

The last task is to build the prototype of the final design option and then thoroughly test and modify it until all required design specifications, performance, and functions are properly satisfied.

The outcome of Phase Four is a tested and approved final design option, which is ready to be released for production.

- **Phase Five: Implementation**

The last phase in an engineering design process is implementation, which mainly means transforming a design into reality. The main task in Phase Five is to prepare a complete set of engineering documentation, such as drawings, part lists, operation & maintenance manuals, quality control procedures, manufacturing procedures, and manufacturing tool & fixture design. The complete set of engineering documentation is the technical information by which the final design option can be manufactured and duplicated. The outcome of Phase Five is a complete set of engineering documentation.

Summary of the engineering design process:
An iterative-interactive five-phase engineering design process shown in Fig. 1.1 includes six items [7].

The center is a project team. The project team is the core of any engineering design process. The iterative-interactive process and all other design activities are carried out by and through the project team. The other five elements are needs assessment (Phase One), design specifications (Phase Two), conceptual design (Phase Three), detailed design

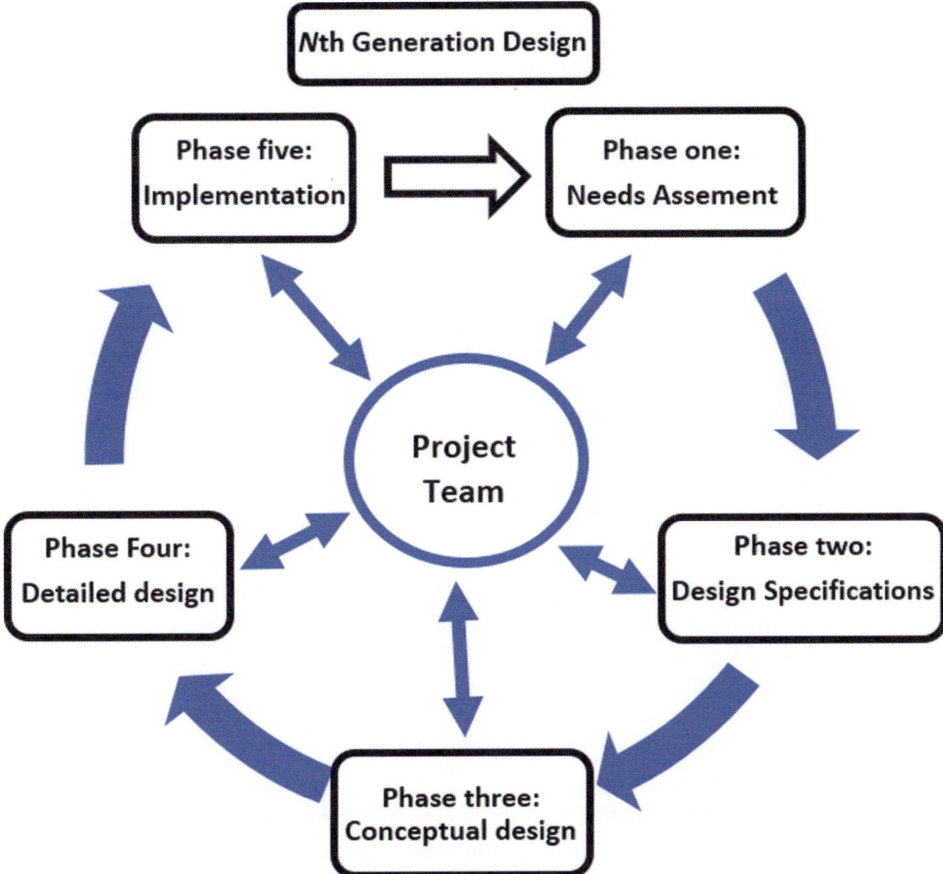

Fig. 1.1 The iterative -interactive five-phase design process

(Phase Four), and implementation (Phase Five). An engineering design project has its own life, starting with Phase One: needs assessment, and ending with Phase Five: implementation. The flow chart of the engineering design process flows naturally from Phase One to Phase Two, to Phase Three, then to Phase Four, and finally to Phase Five. During this design process, interactive activities could happen during any phase. For example, during Phase Four-detailed design, the project team could modify the final design option based on virtual numerical simulation and physical testing of the prototype. For another example, during Phase Three-conceptual design, the project team could go back to work on Phase Two- design specifications to make some necessary modifications.

Description and procedure of the engineering design process such as the five-phase engineering design process is the summary of past successful and unsuccessful design experiences. It is a piece of critical knowledge and design skills. The project team should follow it. It does not mean that following the five-phase design process will guarantee to have a successful design result for a design project. It only suggests that there is a very high possibility that the design project will end with a failure, high cost, or a long design period if the procedure of the engineering design process is not followed.

1.3   Stress and Strain

Component failure is due to external loads. Components with different dimensions will have different external loads at failure. Therefore, stress is a better variable to describe component failure because components of the same material will fail at the same stress level. This section will briefly describe concepts of stress and strain, and their formula under several typical loading conditions.

1.3.1   Normal and Shear Stresses

Normal stress is defined as the ratio of the normal force to the acting-on area.

In Fig. 1.2a, there is one normal force ΔF_z on the shaded area ΔA. The normal stress on the shaded area ΔA can be expressed as:

$$\sigma_z = \lim_{\Delta A \to 0} \frac{\Delta F_z}{\Delta A} \tag{1.1}$$

where ΔF_z is a resultant normal force component along the z axis. the symbol "σ" represents normal stress. The subscript "z" represents a stress direction. So, σ_z is a normal stress along the z-axis.

The sign convention for normal stress is: (1) When the normal stress "pulls" on the area ΔA, it is a tensile stress and treated as a positive stress; and (2) When the normal

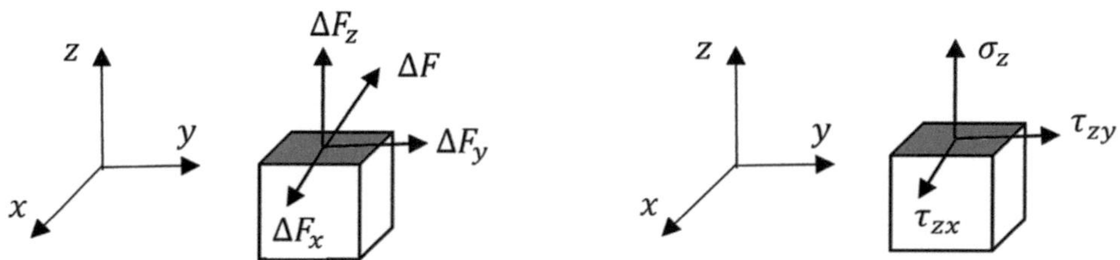

a) Three resultant force components   b) Three stress components

Fig. 1.2 The schematic of normal stress and shear stresses

stress "pushes" on the area ΔA, it is a compressive stress and is treated as a negative stress.

Shear stress is defined as the ratio of the shear force to the acting-on area ΔA.

In Fig. 1.2a, there are two shear force components ΔF_x and ΔF_y on the shaded area ΔA. The two shear stresses can be expressed as:

$$\tau_{zx} = \lim_{\Delta A \to 0} \frac{\Delta F_x}{\Delta A}; \quad \tau_{zy} = \lim_{\Delta A \to 0} \frac{\Delta F_y}{\Delta A} \tag{1.2}$$

where ΔF_x and ΔF_y are resultant shear force components on the shaded area along the x and y axes. The symbol "τ" represents shear stress. The first subscript of shear stress represents the normal direction of the area ΔA. The second subscript represents the shear stress direction. τ_{zx} is shear stress on the area ΔA which has a normal along the z-axis and its stress direction along the direction of the x-axis.

Stress is induced by forces. The normal force will induce only normal stress. The shear force will only generate the shear stress. No force will result in zero stress.

The sign convention for the shear stress is: When both the area's normal direction and the shear stress direction are in the same positive directions or the same negative directions of the Cartesian coordinate axes, the shear stress is positive. Otherwise, the shear stress is negative. The two shear stresses shown in Fig. 1.2b are all positive.

The unit for stress in the SI unit system is $\left(N/m^2\right) = Pa$, called a pascal. The unit of stress in the English unit system is $\left(lb/in^2\right) = Psi$, that is, pound per square inch.

The general state of stress: For a general case, the stresses on a point can be fully specified by three orthogonal cuttings through the same point as shown in Fig. 1.3. On each surface, there is one normal stress and two shear stresses. But since the forces at this point are still in equilibrium, the shear stresses have the following relationships:

$$\tau_{zy} = \tau_{yz}, \quad \tau_{xy} = \tau_{yx}, \quad \tau_{zx} = \tau_{xz} \tag{1.3}$$

Fig. 1.3 General state of stress

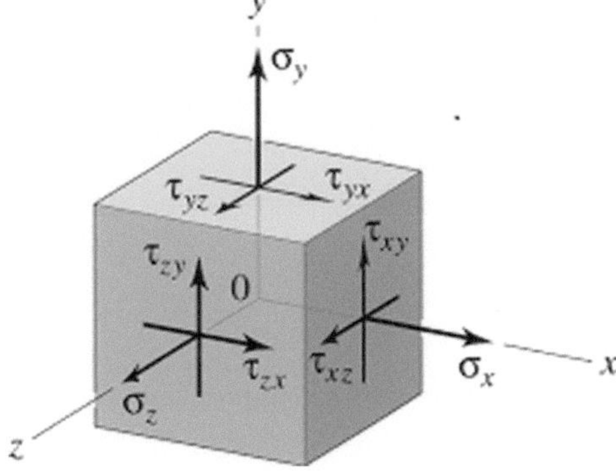

Therefore, there are only six independent stress components $\left(\sigma_x, \sigma_y, \sigma_z, \tau_{xy}, \tau_{yz}, \tau_{zx}\right)$ at a point. These will uniquely define the state of stress of a point. Lots of textbooks or literature [2, 3, 9] demonstrate how to calculate these six independent stresses.

The normal stress of a bar under axial loads

One of the simple normal stress examples is a bar under axial loading. The normal stress σ_z of a bar under an axial load P as shown in Fig. 1.4 can be calculated by:

$$\sigma_z = \frac{P}{A} \tag{1.4}$$

Where P is the resultant axial loading which is normal to the cross-section and along its center axis. A is a freely deformed cross-section area. z is the z-axis, which is the bar center axis. So, σ_z is the normal stress along the z-axis of the bar.

Direct shear stress

One of the simple shear stress examples is direct shear stress on direct shearing sections such as bolts in bolted joints and a pin in a pin connection as shown in Fig. 1.5a, where the bending moment is small and can be negligible. There will be shear stress on the potential shearing-off cross-section, which is the shaded area as shown in Fig. 1.5b.

The average shear stress τ of a direct shearing section can be determined by this equation:

$$\tau = \frac{V}{A} \tag{1.5}$$

Where V is the resultant shear force which is along the potential shearing-off cross-section and through the centroid of the cross-section. A is the area of the potential shearing-off cross-section under consideration.

Fig. 1.4 Normal stress of a bar under axial loading

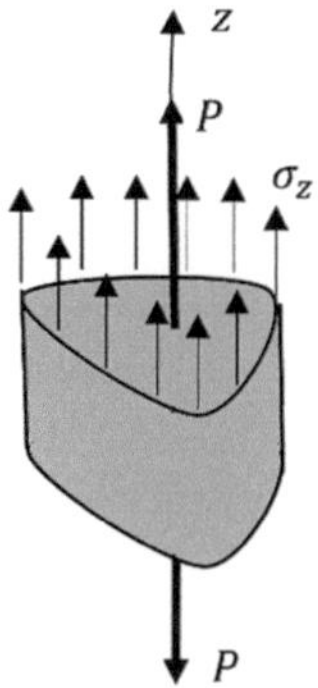

Fig. 1.5 Schematic of a pin connection

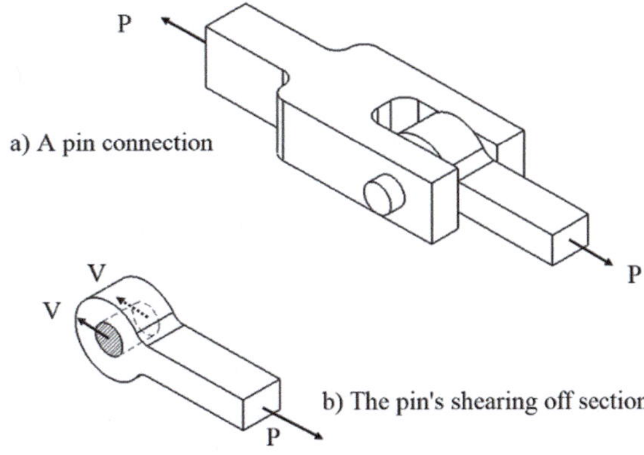

Shear stress of a shaft under torsion

A shaft is a typical mechanical component, the functions of which are to deliver power with a rotation motion and to provide a rotation axis for components such as gears, pulleys, flywheels, cranks, and sprockets. When a shaft is subjected to torsion, there is a resultant torque on a shaft cross-section. This torque will twist a shaft cross-section and induce shear stress on the cross-section. A schematic of shear stress on a cross-section of a solid circular shaft is shown in Fig. 1.6, where O is the center of the cross-section and T is an internal resultant torque about the shaft center axis.

The shear stress τ_A at any point A on a cross-section of a circular shaft can be determined by:

$$\tau_A = \frac{T\rho}{J} \tag{1.6}$$

Fig. 1.6 Schematic of shear stress on a cross-section of a solid circular shaft

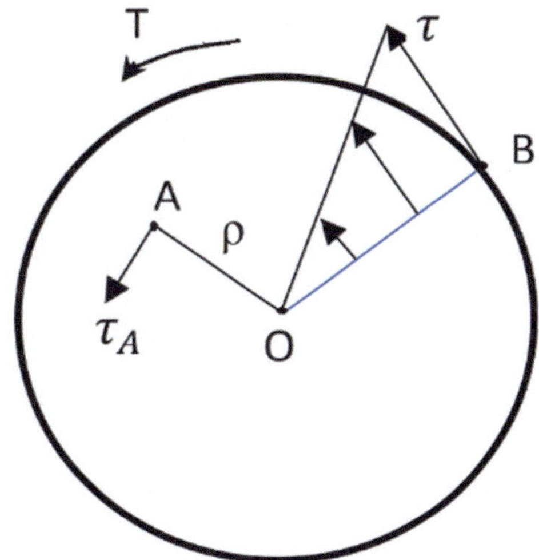

Where ρ is the radius of point A under consideration. J is the polar moment of inertia of the cross-section. The polar moment of inertia of a hollow circular shaft can be calculated by this equation.

$$J = \frac{\pi}{32}\left(d_o^4 - d_i^4\right)\tag{1.7}$$

Where d_i and d_o are the inner diameter and the outer diameter of the hollow circular shaft.

Bending stress of a beam under bending moment

A beam is a typical mechanical component that carries lateral loads and/or bending moments. For a beam, the ratio of the span between two supporting points to the largest dimension of a cross-section is typically more than 10. The lateral loads will generate the bending moment on the cross-section as shown in Fig. 1.7a. The bending moment will induce normal stress as shown in Fig. 1.7b. "O" is the centroid point of the cross-section. x-Axis is the beam center axis, which is also the neutral axis of the beam under bending. y-Axis is a symmetric axis of the cross-section. z-Axis is normal to the symmetric axis and through the centroid of the cross-section. z-Axis is also called the neutral axis on the cross-section. The bending moment M_z is the resultant bending moment on the cross-section, which is about the z-axis and applies on the xoy plane.

The bending stress (normal stress) σ_x can be calculated by the following equations:

$$\sigma_x = -\frac{M_z y}{I}\tag{1.8}$$

Where σ_x is the normal stress at the point with a coordinate value y along y-axis, which is the distance from the point to the neutral z-axis. I is the moment of inertia of the cross-section. The negative sign in Equation (1.8) is added to comply with the normal stress sign convention. The maximum bending stress on a cross-section of a beam under a bending moment is:

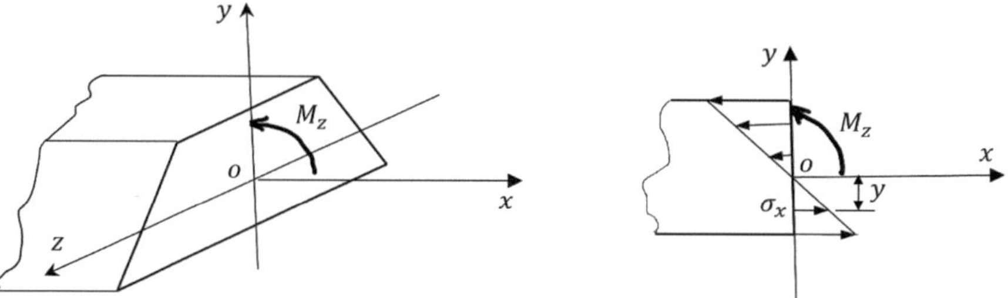

a) A beam cross-section with a bending moment   b) The bending stress on the cross-section

Fig. 1.7 Schematic of bending stress on a beam cross-section

$$\sigma_{max} = \frac{M_z C}{I} \tag{1.9}$$

Where C is the maximum distance from the neutral z-axis to the outermost point of the beam cross-section.

The moment of inertia of a cross-section with a rectangular shape and a circular shape can be calculated by the following equations.

$$I = \frac{1}{12}bh^3, \; rectangular \; shape \tag{1.10}$$

$$I = \frac{\pi}{64}\left(d_o^4 - d_i^4\right), \; solid \; circular \; shape \tag{1.11}$$

Where b is the width of the cross-section along the neutral axis, z-axis. h is the height of the rectangular cross-section which is along the direction normal to the neutral axis, z-axis. d_o and d_i are outer and inner diameters of a circular hollow cross-section shape.

Shear stress of a beam under a lateral load
There is typically a resultant shear force on a cross-section of a beam. The shear force will induce the shear stress. A schematic of shear stress on a beam cross-section is shown in Fig. 1.8. Fig. 1.8a is the schematic of a beam cross-section with a resultant shear force along the symmetric axis, y-axis, and through the centroid O of the cross-section. z-Axis is the neutral axis which is the same as the definition in Fig. 1.7. Fig. 1.8b is a schematic of shear stress on a beam cross-session. The shear stress is induced by the resultant shear force and has the same direction as the resultant shear force V.

Shear stress at point C on a cross-section of a beam due to the resultant shear force can be calculated by Equation (1.12).

$$\tau = \frac{VQ}{It} \tag{1.12}$$

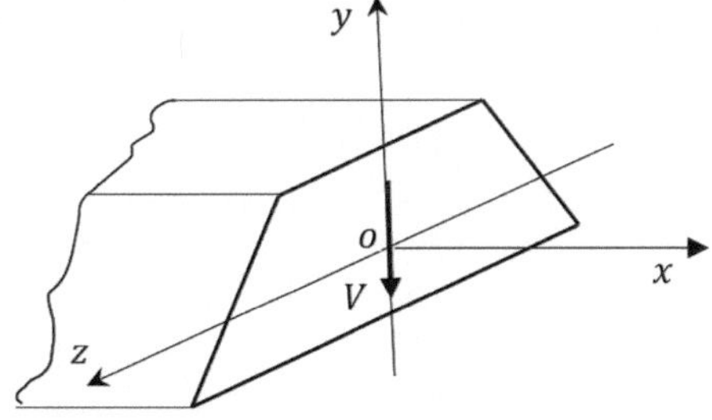
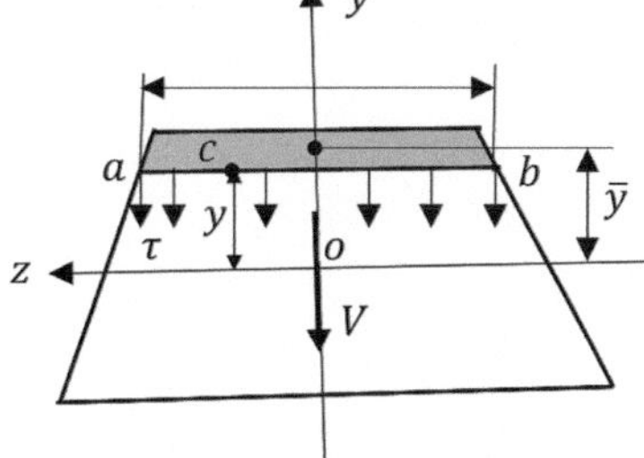

a) A beam cross-section with a shear force   b) The shear stress on the cross-section

Fig. 1.8 Schematics of shear stress on a beam cross-section

Where τ is the shear stress at point C with the coordinate value of y along the y-axis. t is the length of a line segment ab which is through point C and parallel to the neutral axis, z-axis. I is the moment of inertia of the cross-section of a beam, which is the same as the definition in Equation (1.8). $Q = \bar{y} \times A$, is the first moment of the shaded cross-section area about the neutral axis, z-axis. A is the area of the shaded area which is above the line segment ab and $\bar{y}$ is the distance from the centroid point of the shaded area to the neutral axis, z-axis.

1.3.2 Normal and Shear Strain

Mechanical components will deform when a force is applied to them. Deformation is a change in dimension or a change in shape. These two types of deformation can be described by normal strain and shear strain.

Normal Strain is defined as the change in the length of a line per unit length. A normal strain is induced by a normal force or normal stress.

A schematic of a normal strain is shown in Fig. 1.9. The dashed-line shape refers to the original small element without any deformation and the solid-line shape refers to the deformed element due to normal stress. The original length of the small element in the x-axis direction is L and the change in length is δ_x. Per the definition of the normal strain, the normal strain is:

$$\varepsilon_x = \lim_{L \to 0} \frac{\delta_x}{L} \tag{1.13}$$

Where ε refers to the normal strain. The subscript x represents the direction of the normal strain. So, ε_x is a normal strain in the x-axis direction.

Normal strain is a dimensionless physical parameter. It can be thought of as $\left(\frac{in}{in}\right)$ or $\left(\frac{m}{m}\right)$.

Shear strain: The shear strain is defined as the change in the angle of two lines that are originally perpendicular to each other. The shear strain is induced by a shear force or shear stress.

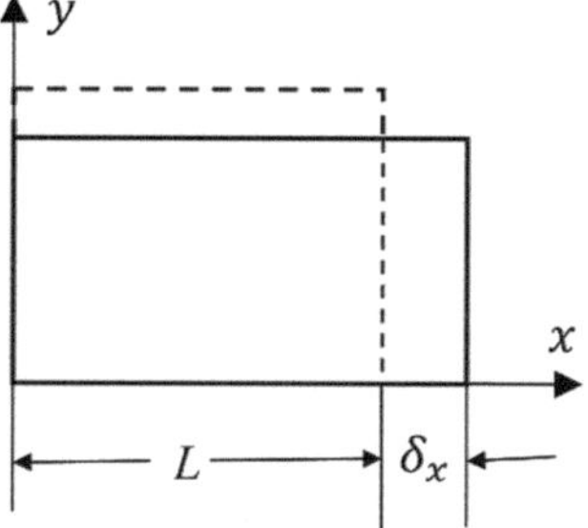

Fig. 1.9 The schematic of a normal strain

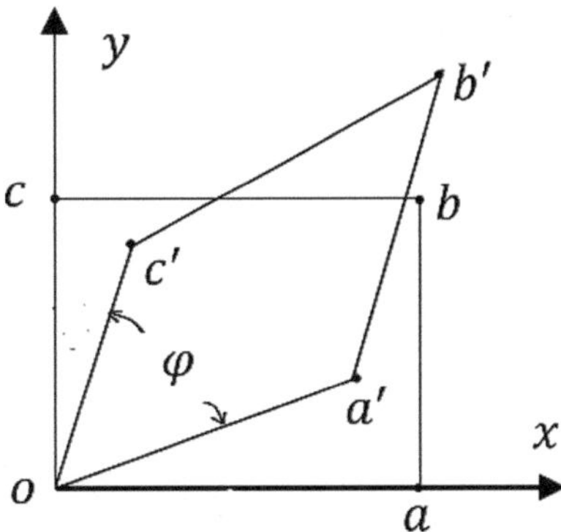

Fig. 1.10 A schematic of a shear strain

A schematic of a shear strain is displayed in Fig. 1.10. The rectangular shape $Oabc$ is the original shape of a small element without any deformation. The quadrilateral $Oa'b'c'$ is the deformed shape of the original rectangular shape $Oabc$. The lines Oa and Oc are perpendicular to each other. After deformed, they become Oa' and Oc'. φ is the angle between the lines Oa' and Oc'. Per the definition of a shear strain, the shear strain at the point O between the x-axis and the y-axis directions are:

$$\gamma_{xy} = \frac{\pi}{2} - \varphi \tag{1.14}$$

Where γ represents a shear strain. The subscripts x and y represent the directions of two original perpendicular lines. γ_{xy} is the shear strain at point O along the x and y axis. The unit of a shear strain is radian.

The best way to calculate normal strain and shear strain is through deformation functions. Point A in an original object without any deformation is deformed into point A' in the deformed object under external loading as shown in Fig. 1.11. The general deformation $\vec{\mathbf{d}}$ of the point A can be expressed as:

$$\overrightarrow{\mathbf{d}(x, y, z)} = u(x, y, z)\,\vec{\mathbf{i}} + v(x, y, z)\,\vec{\mathbf{j}} + w(x, y, z)\,\vec{\mathbf{k}} \tag{1.15}$$

Where, $\vec{\mathbf{i}}$, $\vec{\mathbf{j}}$ and $\vec{\mathbf{k}}$ are unit vectors along the x-, y-, and z-axes, respectively. $u(x, y, z)$, $v(x, y, z)$ and $w(x, y, z)$ are deformation function components of $\overrightarrow{\mathbf{d}(x, y, z)}$ along the x-, y-, and z-axes, respectively.

Per the definitions of normal strain and shear strain, we can calculate three normal strains and three shear strains as follows.

$$\varepsilon_x = \frac{\partial u(x, y, z)}{\partial x}, \quad \varepsilon_y = \frac{\partial v(x, y, z)}{\partial y}, \quad \varepsilon_z = \frac{\partial w(x, y, z)}{\partial z} \tag{1.16a}$$

$$\gamma_{xy} = \frac{\partial u(x, y, z)}{\partial y} + \frac{\partial v(x, y, z)}{\partial x}, \gamma_{yz} = \frac{\partial v(x, y, z)}{\partial z} + \frac{\partial w(x, y, z)}{\partial y},$$

Fig. 1.11 A schematic of a
general deformation

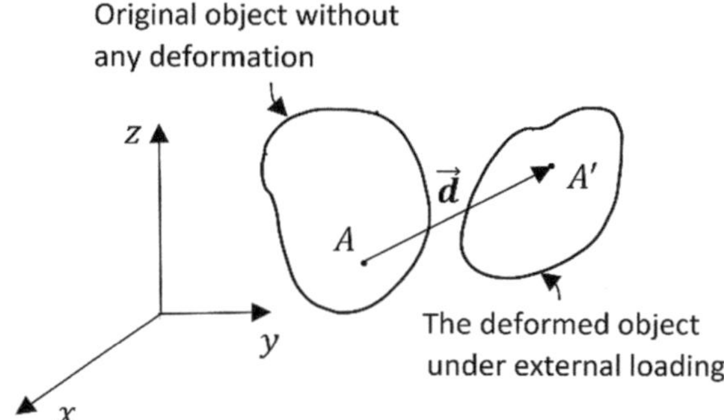

$$\gamma_{zx} = \frac{\partial w(x, y, z)}{\partial x} + \frac{\partial u(x, y, z)}{\partial z} \qquad (1.16b)$$

Where, ε_x, ε_y and ε_z are normal strains along the x-, y-, and z- axe, respectively. γ_{xy}, γ_{yz} and γ_{zx} are shear strains in the xy plane, the yz plane, and the zx plane, respectively.

1.4   Material Mechanical Properties

For mechanical design, we must use material mechanical properties that are solely obtained from experiments. This section will briefly describe and explain two typical experiments for obtaining material mechanical properties under static loads.

Tensile stress-strain diagram and material mechanical properties
One of the most important experiments to explore material mechanical behaviors is a tensile test. To make the experimental results useful, specimens for tensile tests are standardized per ASTM E8/E8M-22 Standard test methods of tension testing of metallic materials [10]. The standard specimen for a round tensile test specimen is shown in Fig. 1.12. It has a constant circular cross-section with enlarged threaded ends so that failure will not occur at the threaded section. Before tensile testing, two small punch marks separated by a gauge length of 2″ are placed along the specimen's uniform cross-section segment. This will be used to determine the percent elongation.

During a tensile test, the axial load P on the standard specimen is gradually increased until the specimen fractures. During testing, the cross-section of the standard specimen is

Fig. 1.12 The standard tensile
test specimen

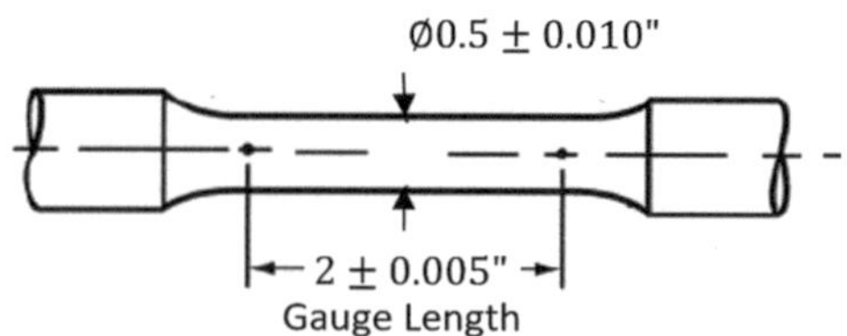

slightly reduced due to its stretch along the loading direction. However, our main concerns are the material behaviors before its fracture. Therefore, conventionally, the stress of the specimen is calculated by using its original cross-section. It is called engineering stress. per Equation (1.4), the engineering stress of the specimen is:

$$\sigma = \frac{P}{A_0} \tag{1.17}$$

Where P is the axial tension load; A_0 is the original cross-section area of the constant cross-section of the standard specimen.

During the tensile testing, the stain is also calculated by using the original gauge length and is called the engineering strain. Per Equation (1.13), the engineering strain of the specimen is

$$\varepsilon = \frac{\delta}{L_0} \tag{1.18}$$

Where, L_0 is the original gauge length and is typically 2". δ is the deformation of the gauge length.

The Stress-strain Diagram is the engineering stress of a specimen vs. the engineering strain of the specimen during tensile testing as shown in Fig. 1.13. Static material mechanical properties for mechanical design are mainly obtained from the engineering stress-strain diagram of tensile testing.

There are several special points on a typical stress-strain diagram, which are marked as a, b, c, d, and e in Fig. 1.13.

The proportional limit is the stress value at point a. Between the origin o and point a, there is a linear relationship between stress and strain, that is, Hooke's Law

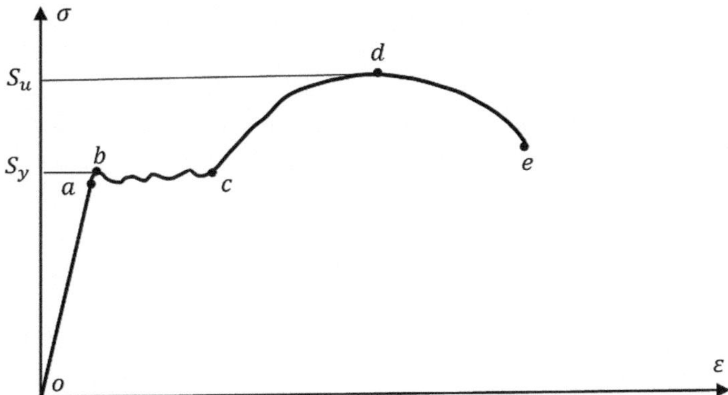

Fig. 1.13 A typical stress-strain diagram for a ductile material

$$\sigma = E\varepsilon \tag{1.19}$$

Where E is the slope of the straight line oa and is called the modulus of elasticity or Young's modulus.

Young's modulus is one of the most important material mechanical properties. It is the coefficient of a linear relationship between stress and strain and is inversely proportional to the deformation of a component. The bigger Young's modulus, the smaller the deformation for the same stress. For example, when a component's deformation is the main concern, the material with a bigger value of Young's modulus will be the choice of material for this situation.

The elastic region is the region where the strain is between point o and point a. In this region, the deformation is elastic. So, when the load is removed from the specimen, the deformation will become zero. For general engineering design, we do want the stress/stain of a component under the rated working loadings and conditions to be in an elastic region so that it can be reused again and again during the specified service life.

Yielding is such a phenomenon that when the stress reaches point b, the specimen will have significant deformation even if the stress remains the same value. This is the region between point b and point c.

The plastic region is the region after point b. When the strain passes the yielding point b, the component deformation will include elastic deformation and plastic deformation. When the load is removed from the specimen, its deformation will not become zero and will still have some plastic deformation. Therefore, for a general engineering design, we don't want a component under the rated service loadings to work in the plastic region.

Yield strength S_y is the stress value at the yielding point b. Where the symbol "S" means material strength and subscript "y" means yielding. Yield strength is one of the important material mechanical properties for component design. To make the component safe and work properly during the specified service life, yield strength is widely used as the allowable strength for components made of ductile materials.

For brittle materials, there is no visible yielding as shown in Fig. 1.13. There isn't a well-defined yielding point for brittle materials. For brittle materials, the material's allowable strength for component design is not a yield strength S_y. However, the yield strength for a brittle material can be determined by a 0.2%-offset graphical procedure as shown in Fig. 1.14. From the point with a value of 0.002 on the strain axis, a line parallel to the linear straight line of the stress-strain curve is drawn.

The value at the interaction of this line with the stress-strain curve will be the estimated yield strength.

Strain hardening refers to the region between point c and point d as shown in Fig. 1.13, where after yielding, the material restructures its internal structures due to the plastic deformation and gains more strength to resist the external loading.

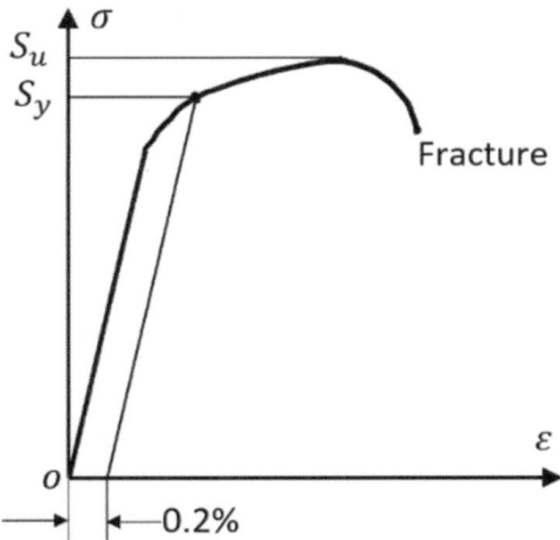

Fig. 1.14 Typical stress-strain diagram for a brittle material

Ultimate strength S_u is the stress at point d, which is the maximum stress point of the engineering stress-strain curve. The ultimate strength S_u is typically used as the material strength for a component design when the material of the component is brittle.

Necking refers to the phenomenon that after the stress of the specimen reaches ultimate strength S_u, its cross-section area will decrease as shown in Fig. 1.15. For ductile materials, there is visible necking before it fractures as shown in Fig. 1.15a. But for brittle materials, there is no visible necking when it fractures as shown in Fig. 1.15b.

Percent elongation is one of the material mechanical properties and is a ratio expressing how much a tensile specimen stretched during testing. It is defined as

$$Percent\ elongation = \frac{L_f - L_0}{L_0}\% \tag{1.20}$$

Where L_0 is the specimen's original gauge length between two marked dots as shown in Fig. 1.16. L_f is the specimen's fracture length between the two marked dots.

Ductile material is a material that can be subjected to large strains before it fractures. Mild steel is a ductile material. This type of material is widely used for components when there are some shock loadings. Another feature of this material is that the component can

Fig. 1.15 Pictures of fractured surfaces of ductile vs. brittle materials

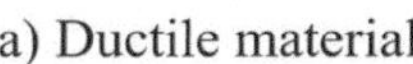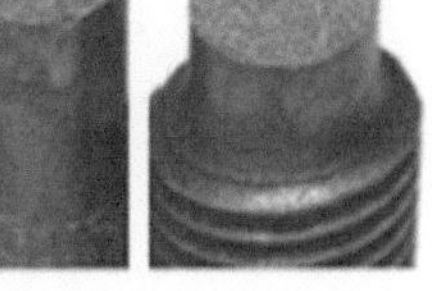

a) Ductile material   b) Brittle material

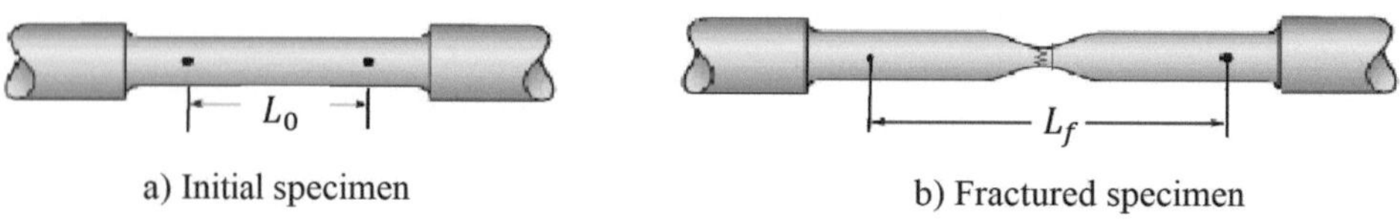

a) Initial specimen   b) Fractured specimen

Fig. 1.16 Schematic of specimen for a percent elongation

have visible deformation before it fractures. If the percent elongation of materials is more than 5%, the material is classified as a ductile material.

Brittle material is a material that exhibits little or no yielding before it fractures. Brittle material is not typically used for components that will be subjected to shock loadings because it will not show any visible deformation as a warning sign before it fractures. Gray cast iron is an example of brittle material. If the percent elongation is less than 5%, this material is classified as a brittle material.

Poisson's ratio v is one of the material mechanical properties and is defined as the ratio of the lateral strain to the strain along the axial load direction as shown in Fig. 1.17. The dashed line is the original shape of a component. The solid line is the current shape of the component under a tensile load P. The component will be stretched in the load direction and at the same time, the component will be shrunk in the lateral direction. The Poisson's ratio is:

$$v = -\frac{\varepsilon_{lateral}}{\varepsilon_{load}}$$ (1.21)

Where, ε_{load} is the normal strain along the load direction. $\varepsilon_{lateral}$ is the lateral normal strain in the direction perpendicular to the load direction. Poisson's ratio of most materials has a value between 0 to 0.5.

Shear stress-strain diagram and material mechanical properties
Material mechanical behaviors under shearing are very similar to those under tensile loading. Material mechanical properties under shearing can be obtained through a shearing

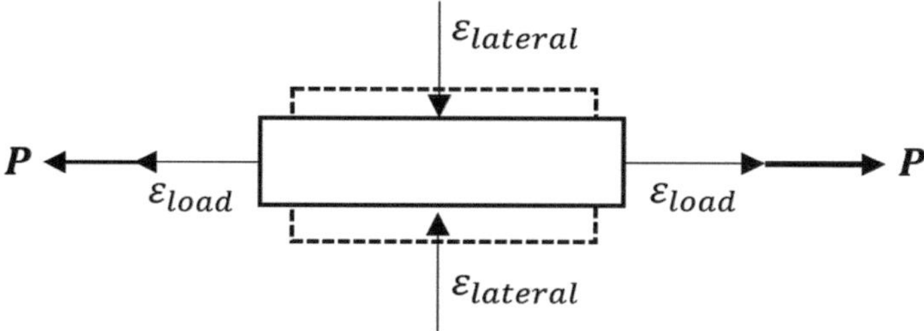

Fig. 1.17 A schematic for the Poisson ratio

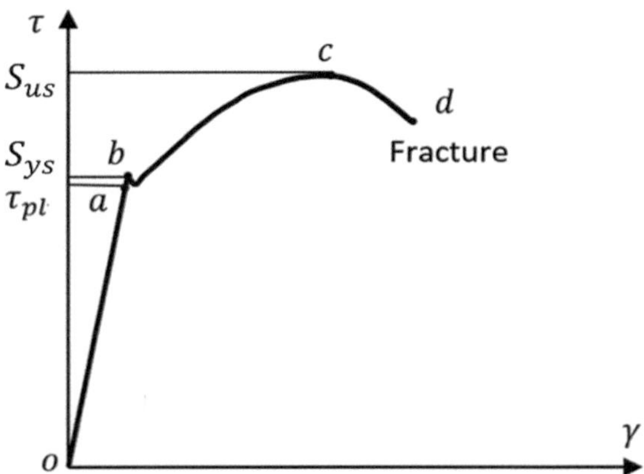

Fig. 1.18 A schematic of a shear stress-strain diagram

test. The shear stress and the strain of a round specimen under pure torsion can be calculated by using its original geometric dimensions. Two ends of the shearing test specimen are typically hexagon-shaped so that there will not be a slip between the shear specimen and the clamp jaws. A schematic of the shear stress-strain diagram is depicted in Fig. 1.18.

Hooke's Law and Shear Young's Modulus: Below the proportion limit, that is, point a, there is a linear relationship between the shear stress and shear strain. This is the Hook's Law.

$$\tau = G\gamma \tag{1.22}$$

Where G is the slope of the straight line oa and is called the shear modulus of elasticity or shear Young's modulus.

There is a relationship between Young's modulus E, shear Young's modulus G, and Poisson's ratio v.

$$G = \frac{E}{2(1 + v)} \tag{1.23}$$

The elastic region is the region between point O and point a. In this region, the shear strain or deformation will become zero when the external torque is removed.

Shear yield strength S_{ys} is the shear stress at the yielding point b. It is one of the material mechanical properties and is used as allowable material shear strength for ductile materials.

The plastic region refers to the region after the shear stress passes the yield point b. After material yields, the deformation will contain an elastic deformation and a plastic deformation.

Ultimate shear strength S_{us} is the shear stress at point d which is the utmost point on the shear stress-strain diagram. It is one of the material mechanical properties and can be used as allowable material shear strength for brittle materials.

In the absence of available shear mechanical properties, reasonable approximations can be obtained from tensile mechanical properties [2, 3].

$$S_{ys} = 0.577 S_y \tag{1.24}$$

$$S_{us} = \begin{cases} 0.80 S_{ut} & \text{for steels} \\ 0.75 S_{ut} & \text{for other ductile material} \end{cases} \tag{1.25}$$

1.5   Dimensions and Tolerances

Component and assembly dimensions are always associated with allowable variations, that is, tolerances. This section will briefly describe dimension tolerances and how to determine them.

1.5.1   Dimension Tolerances

Dimension tolerances of mechanical components and assembly are specified by Geometric Dimensioning and Tolerancing (GD&T).

Geometric Dimensioning and Tolerancing (GD&T) is a language of symbols and standards designed and used by engineers and manufacturers to specify a product and facilitates communication between entities working together to manufacture products. GD&T includes dimension tolerances and geometric tolerances of shapes such as location, orientation, profile, form, and runout. This book will only briefly discuss dimension tolerances. However, the envelope principle in GD&T states that the surface of a feature of size may not extend beyond an envelope of perfect form at MMC (Maximum Material Condition) [2, 11, 12]. In this way, dimension tolerances specify the maximum allowable geometric tolerances if geometric tolerance is not specified.

Dimension tolerance is the total allowable variation in a dimension, that is, the difference between the upper and low limits. For example, a diameter $\varnothing.750^{-0.001}_{-0.006}$ of a bar has an upper limit of $0.750 - 0.001 = 0.749''$ and a low limit of $0.750 - 0.006 = 0.744''$. Therefore, the dimension tolerance of this example will be 0.005".

Two key benefits of specified dimension tolerances are:

- Properly specified dimension tolerance will help to reduce manufacturing costs. Tight tolerance means a high cost. Therefore, generous dimension tolerance should be assigned to the dimensions of the part if it can satisfy the design functions.

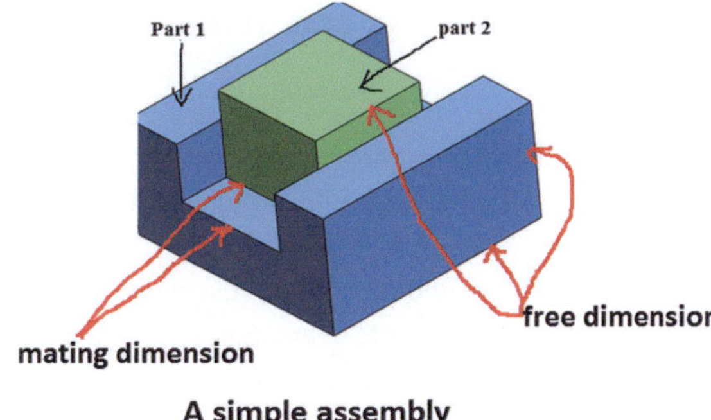

Fig. 1.19 A schematic used for explaining types of dimensions

- In the mass production of a product, the interchangeability of the same component is a must. Components with properly specified dimension tolerances can guarantee the interchangeability of the same components.

1.5.2 Types of Dimensions

All dimensions of a component can be grouped into two types of dimensions: (1) a free dimension and (2) a mating (assembly) dimension [13].

A free dimension is defined as a dimension the variation of which will not affect or interfere with other parts for the required functions in an assembly.

A mating (assembly) dimension is defined as a dimension the variation of which will affect other parts and the required functions in an assembly.

Let us assume that the design intent in this assembly shown in Fig. 1.19 is that part 2 can freely slide along the slot of part 1. In this case, the width of the slot of part 1 and the width of part 2 are mating dimensions. All other dimensions will be free dimensions.

1.5.3 Dimension Tolerances of Free Dimensions

For a mechanical part, generally, the maximum possible variation of the dimension could be 1/32", that is, 0.032". When the variation of a dimension is over 1/32" and this variation does not affect other components or the functions of the assembly, it is a free dimension. Since the tolerance of a free dimension will not affect other parts and assembly functions, its dimension tolerance will be determined by the manufacturing process of a company with the least manufacturing cost. This is typically the default tolerance of a company. In other words, the dimension tolerance of a free dimension is solely determined by a manufacturing process.

1.5.4 Types of Fits and Standard Fits

For a mating dimension, two components will be assembled or mated together through the design interface controlled by mating dimensions. Let us use a generic hole to be an empty shape of geometry and a generic shaft to be a solid shape of geometry as shown in Fig. 1.20.

Let us use $d_{h_L}^{h_U}$ and $d_{s_L}^{s_U}$ to represent the dimension tolerances of a generic hole and a generic shaft dimension tolerance. d is the nominal dimension of the mating dimension. h represents a hole and s a shaft. The subscript U represents the upper limit of the dimension tolerance and the subscript L the lower limit of the dimension tolerance. Let us use $d_{a_L}^{a_U}$ to represent the assembly tolerances. a represents assembly tolerance. Then, the governing equations for assembly tolerance of mating dimensions are:

$$a_U = h_U - s_L = \begin{cases} > 0 \ \text{Maximum clearance} \\ < 0 \ \text{Minimum interference} \end{cases} \tag{1.26}$$

$$a_L = h_L - s_U = \begin{cases} > 0 \ \text{Minimum clearance} \\ < 0 \ \text{Maximum interference} \end{cases} \tag{1.27}$$

a_U and a_L are numerical values. They can be positive or negative. The positive value means the gap or clearance on the interface. The negative value means an interference.

When both sides of Equations (1.26) and Equation (1.27) are subtracted, we have:

$$(a_U - a_L) = (h_U - h_L) + (s_U - s_L) \tag{1.28}$$

This equation indicates that the assembly tolerance on the interface is the sum of the shaft dimension tolerance and the hole dimension tolerance. Since the tolerance is

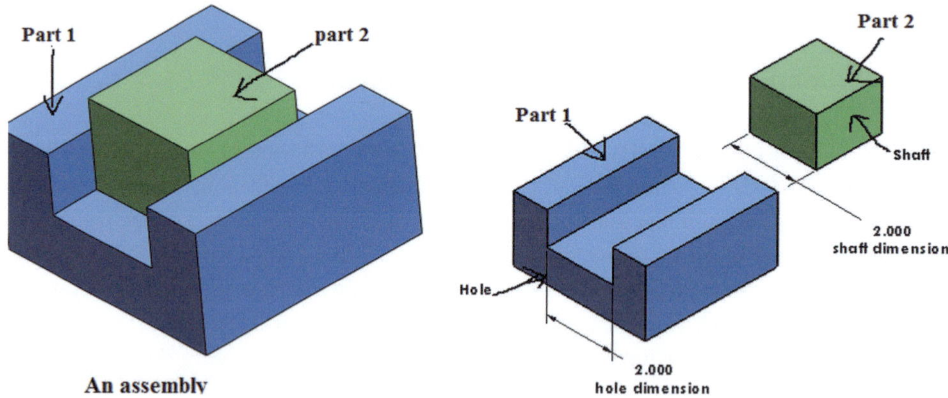

Fig. 1.20 A generic hole and a generic shaft

always a positive value, the assembly tolerance could not be less than the shaft dimension tolerance or the hole dimension tolerance.

Based on the values of a_U and a_L for mating dimensions, there are three possible types of fits: Clearance fit, Interference fit, and Transition fit.

Clearance fit means that there is always a gap between a shaft and a hole. The dimension of the hole is always larger than the dimension of the shaft in a clearance fit.

Interference fit means that there is always some interference between a shaft and a hole. The dimension of the shaft is always larger than the dimensions of the hole in an interference fit.

Transition fit means that there might be a gap or an interference between a shaft and a hole. In this case, sometimes the dimension of the hole is larger than the dimension of the shaft, but sometimes, the dimension of the shaft is larger than the dimension of the hole.

Engineers specify mating (assembly) tolerance or type of fits to satisfy design intents or required functions on the interface. The ANSI standard fits (ANSIB4.1-1967(R2009)) [14] is one of the standards for three different types of fits. It defines three groups of standard fits per possible practical applications or features of the mating tolerances. These three groups of fits are (1) Running and sliding clearance fits, (2) force or shrink fits, and (3) locational fits.

Application guidelines of these standard fits and all standard fits tables with a nominal dimension up to 19.69" are provided in ANSIB4.1-1967(R2009)) [14].

1.5.5 Dimension Tolerances of Mating Dimensions

Unlike free dimensions, mating dimension tolerances cannot be only determined by a manufacturing process and must satisfy the required assembly tolerance, which is specified per required functions. Mating dimension tolerances are governed by Equations (1.26) and (1.27). There are three approaches for determining mating dimension tolerances.

Approach One—Use the ANSI standard fit tables: When the assembly tolerance such as type of fits and corresponding class level is specified per required function, we could directly use the ANSI standard fits [14] to obtain corresponding dimension tolerances of the hole and shaft dimension.

Approach Two—Basic hole system: After the assembly tolerance per required functions is specified and if the hole dimension tolerance $d_{h_L}^{h_U}$ is predetermined, the dimension tolerance of the shaft can be calculated through the following two equations.

$$s_L = h_U - a_U \tag{1.29}$$

$$s_U = h_L - a_L \tag{1.30}$$

Approach Three- Basic shaft system: After the assembly tolerance per required functions is specified and if the shaft dimension tolerance $d_{s_L}^{s_U}$ is predetermined, the dimension tolerance of the hole can be calculated through the following two equations.

$$h_U = s_L + a_U \tag{1.31}$$

$$h_L = s_U + a_L \tag{1.32}$$

1.6   Materials and Manufacturing Processes

The cost of materials and manufacturing is always a big portion of products' prices. Design engineers should consider a selection of materials and manufacturing processes at the early stage of the engineering design process such as the phase three conceptual design and phase four detailed design for easier and more economical production of the design projects. This section will briefly discuss materials, material selection, and manufacturing processes.

1.6.1   Materials and Material Selection

There are thousands of materials available for mechanical components. There is a lot of literature that provides detailed descriptions and the applications of engineering materials such as references [2, 15, 16]. The following are some concise summaries of the classifications of materials. Engineering materials can be classified into three different categories: metals, non-metals, and composite as shown in Fig. 1.21.

Metals are composed of one or more metallic elements such as iron, copper, and nickel, and often also nonmetallic elements such as carbon in a relatively small amount. As for mechanical characteristics, these solid materials are typically hard, strong, malleable, and ductile with good electrical and thermal conductivity. Metals are widely used for lots of mechanical components.

Metals are typically further classified into two general subcategories: ferrous metals and nonferrous metals.

Ferrous metal is any metal that is primarily composed of iron and has magnetic properties. Some common ferrous metals include alloy steel, carbon steel, cast iron, and wrought iron. Due to its hardness, durability, and tensile strength, ferrous metals are widely used in almost all industries such as shipping containers, industrial piping, automobiles, railroad tracks, ships, and many commercial and domestic tools.

Non-ferrous metals refer the metals other than iron and alloys that do not contain an appreciable amount of ferrous (iron). Some commonly used non-ferrous metals are

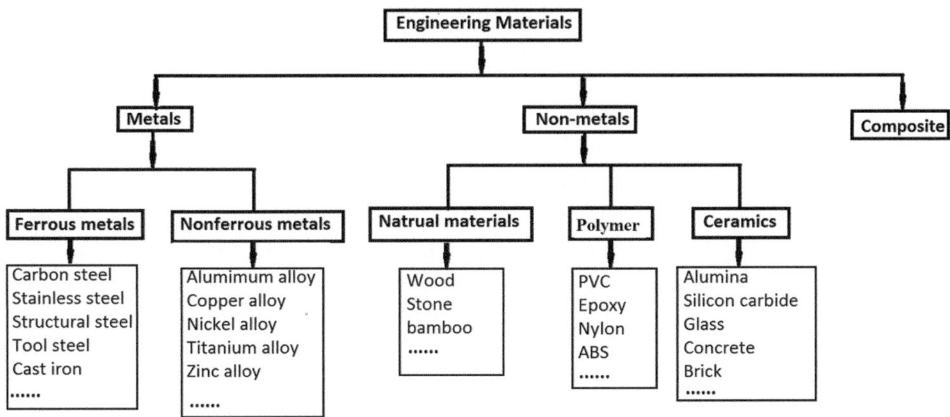

Fig. 1.21 A classification of engineering materials

copper, zinc, aluminum, lead, nickel, cobalt, chromium, gold, silver, and many others. A distinguishing feature of non-ferrous metals is that they are highly malleable. Another valuable advantage over ferrous metals is that they are highly corrosion-resistant and rust-resistant. Non-ferrous metals are also widely used for lots of mechanical components and are suitable for highly corrosive environments such as liquid, chemical, and sewage pipelines.

Non-metal materials are any materials, which do not contain metal and include three subcategories: natural materials, polymers, and ceramics.

Natural material is any product or physical matter that comes from plants, animals, or the ground. Some examples of natural materials are wood, bamboo, and stones. They are used sometimes for mechanical components but are typically used as building materials.

Polymers are a material consisting of very large molecules, composed of many repeating subunits. Polymers include plastic and rubber materials. Many of them are organic compounds that are chemically based on carbon, hydrogen, and other nonmetallic elements. These materials typically have low densities and are not as stiff or strong as these metals or ceramics. Many polymers are extremely ductile and pliable. In general, they are relatively inert chemically and unreactive in many environments. Some common and familiar polymers are polyethylene (PE), nylon, polyvinyl chloride (PVC), polycarbonate (PC), polystyrene (PS), and silicone rubber.

Ceramics are compounds made between metallic and nonmetallic elements. They are frequently oxides, nitrides, and carbides. For example, common ceramic materials include aluminum oxide, silicon dioxide, silicon carbide, silicon nitride, porcelain, cement, and glass. As for mechanical behavior, ceramic materials are relatively stiff and strong, but they are typically very hard and brittle. However, newer ceramics are being engineered to have improved resistance to fracture; these materials are used for cookware, cutlery, and even automobile engine parts. Furthermore, ceramic materials are typically insulative to the passage of heat and electricity and are more resistant to high temperatures and harsh environments than metals and polymers [15].

Composites are composed of two (or more) individual materials that come from the categories previously discussed: metals, ceramics, and polymers. The design goal of a composite is to achieve a combination of properties that are not displayed by any single material and also to incorporate the best characteristics of each of the component materials. Many composite types are represented by different combinations of metals, ceramics, and polymers. One of the most common and familiar composites is fiberglass, in which small glass fibers are embedded within a polymeric material. The glass fibers are relatively strong and stiff, whereas the polymer is more flexible. Thus, fiberglass is relatively stiff, strong, and flexible.

The selection of a material for a component is one of the most important decisions in mechanical design. We typically make this decision in Phase Three: Conceptual Design or Phase Four: Detailed Design. The decision is usually made before the dimensions of the part are established. We could follow the following four steps to select a material for a component.

Step one: Understand the functions of a product
The process of material selection must commence with a clear understanding of the functions and design requirements of a product. For example, for a component with the same functions but in different products such as an airbus and a school bus, the selected material might be different.

Step two: Understand the functions of a component
After we have a clear understanding of the functions of the product, we then need to have a clear understanding of the functions & requirements of a component. We need to collect information and fully understand the following: the nature of the forces applied to the component; the types of magnitudes of stresses; allowable deformation of the component; manufacturing process; interface with other components of the product; the working environment; and cost targets for the product as a whole and this component.

This information will help us to select a group of candidate materials.

Step three: List of candidate materials
We need to search for the key mechanical properties of materials that are needed to satisfy the component functions. Typical mechanical properties are Young's modulus; yield strength; ultimate strength; percent elongation; hardness; density; ductile or brittle; and others such as wearing resistance, fatigue strength, corrosion resistance, etc.

For example, if static strength is the main concern, the yield strength or ultimate strength will be the key index for choosing the candidate of materials. Metal with proper strength such as high-strength steel would be a good choice. If fluctuating loading is the main concern, fatigue strength will be the key index for choosing the candidate materials.

If there is dynamic or impact loading, ductile materials should be used. If defection is the main concern, Young's modulus or shear Young's modulus will be the index for choosing the candidate of materials. Metal with proper strength such as mild carbon steel would be a good choice. Since the Young's modulus of all steels is almost the same, it is not necessary to use high-strength steel. If corrosion is the main concern, we might choose stainless steel or some non-ferrous metals such as aluminum.

Step four: Select the final material

After a list of candidate materials has been determined, we can use the following three factors (materials availability, manufacturability, and raw materials price) to evaluate each candidate of materials.

Materials availability: Companies will only buy raw materials after the customer places the order for the product. So, material availability is very important.

Manufacturability: One of the main costs of the product is the manufacturing cost.

Raw material price: Another of the main costs of the product is the raw material cost.

Design engineers will combine all considerations and finalize the choice of material for the component.

1.6.2 Considerations of Manufacturing Processes for Components

Design engineers need to select proper manufacturing processes so that components and products can be produced more easily and economically. More detailed information about design for manufacturing can be found in a lot of literature [17, 18]. There are several typical types of manufacturing processes such as machining, casting, molding, forming, joining, and additive manufacturing. The following are their concise summaries.

Machining is a process in which a material is cut to a desired final shape and size by a controlled material-removal process. Three principal machining processes are turning, drilling, and milling. Much of modern-day machining is carried out by computer numerical control (CNC), in which computers are used to control the movement and operation of mills, lathes, and other cutting machines. Machining is used for many metal products, but it can also be used on other materials such as wood, plastic, ceramic, and composite materials.

Casting is one of the oldest manufacturing processes in which a molten liquid material is poured into a mold, which contains a hollow cavity of the desired shape, and then allowed to solidify. Casting can create complex or simple shapes for any kind of meltable metal with wide options of designs. However, casting is not suitable for a low-volume product due to the high upfront investments in molds. There are several different types of casting such as sand casting, centrifugal casting, die casting, permanent mold casting, continuous casting, etc.

Molding is a process of manufacturing by shaping liquid or pliable raw material with a rigid mold, which is a hollowed-out block. The liquid hardens or sets inside the mold, adopting its shape. There are lots of different moldings such as casting, blow molding, compression molding, extrusion molding, injection molding, etc. This manufacturing process is not suitable for low-volume production due to the huge upfront investments in molds.

Forming or metal forming is the manufacturing process of fashioning metal parts and objects through mechanical deformation. Forming operates on the materials science principle of plastic deformation, where the physical shape of a material is permanently deformed.

Joining processes include welding, brazing, soldering, mechanical fastening, and adhesive bonding. Mechanical fastening can be used to provide either temporary or permanent joints, while adhesive bonding, welding, brazing, and soldering processes are mainly used to provide permanent joints.

3D printing or **additive manufacturing** is the construction of a three-dimensional object from a CAD model or a digital 3D model. It can be done in a variety of processes in which material is deposited, joined, or solidified under computer control, with the material being added together (such as plastics, liquids, or powder grains being fused), typically layer by layer. While 3D printing can be expensive, it offers the potential to reduce financial capital, raw materials, and waste and lets companies create and test components, or products before committing to them on a larger scale.

At the early stage of the design process such as phase three: conceptual design and phase four detailed design, design engineers must work closely with manufacturing engineers to select proper manufacturing processes for their components and products. Design engineers typically need to consider three key factors for selecting proper manufacturing processes for components or assembly.

The first factor is the material of the components. The choice of materials has been selected to satisfy the required functions. The choice of materials for components typically significantly narrows down the choice of manufacturing processes. For example, if sheet metal is selected for manufacturing a component, the proper manufacturing process might be the metal forming. For another example, if steel is selected for a component, the injection molding manufacturing process will be excluded.

The second factor is the volume of production. If the volume of production is not high, manufacturing processes with expensive upfront costs should be excluded. For example, when the volume of a gearbox is not high, we might not use casting to manufacture the housing because of the expensive cost of the mold.

The third factor is component dimension tolerances. Component dimension tolerances are determined by the functions of components. Higher tolerance means higher manufacturing costs. So, we should use as generous tolerance as possible if it satisfies the required functions of components. Each manufacturing process typically has its common range of tolerances with the least manufacturing cost. Design engineers need to select a manufacturing process that can easily satisfy the required tolerance. For example, if the component has a tolerance of ± 0.0005, we might select a CNC machine for making this part.

1.7 Mechanical Failures

The purpose of mechanical systems or mechanical devices is to safely serve customers or society. Each component of mechanical systems or devices has its specified functions. When it fails to perform its required functions, it can be identified as a failure. This section will briefly discuss typical mechanical failures.

Mechanical failure means that a mechanical component or system cannot perform its required functions under normally rated loadings and working conditions within the service life.

This definition specifies three key aspects of failure. The first aspect is the required functions. If a pump with a rated flow rate of 100 liters per minute can only provide a flow rate of 50 liters per minute, this pump will be identified as a failure. The second aspect is the normally rated loadings and working conditions. The fracture or under-performance of a mechanical system or component caused by abnormal loading or working conditions such as misuse is a failure, but it is not the mechanical failure considered for mechanical design. The third aspect is the service life. Any possible failure beyond the service life will not be considered in mechanical design. For example, any metal component will eventually fail due to fatigue damage. But this is not an issue if it will not fail in the service life.

There are a lot of causes, such as loads, temperature, chemical environment, metallurgical environment, and wearing that might induce mechanical failure [19]. Mechanical failure due to corrosion, wearing, fracture, and creep are mechanical failures that should be properly considered for mechanical design. However, this book will only discuss mechanical failures due to external loads. Mechanical component designs of this book consider the following three typical modes of mechanical failures.

Mechanical Failure due to static stress is a typical mechanical failure when the component's maximum stress at a critical area is more than the allowable material strength.

The maximum stress could be maximum normal stress, maximum shear stress, or maximum equivalent stress which will be discussed in Chapter 1.10.

For example, a component of a ductile material under uniaxial stress will be identified as a mechanical failure due to stress when the maximum normal stress is more than material yield strength. For another example, a component of a brittle material under uniaxial stress will be identified as a mechanical failure due to stress when the maximum normal stress is more than the material ultimate strength.

Mechanical Failure due to excessive deformation is a typical mechanical failure when the maximum deformation is more than the allowable deformation. Excessive deformation will cause under-performance of mechanical systems or devices or damage to other components.

For example, if a shaft with gears in a gearbox has excessive deflection, it will cause big noise and big impact loads on the engaged gears and might even cut the gearbox case.

Mechanical failure due to fatigue damage is a typical mechanical failure when the component's cumulated fatigue damage is more than its material allowable fatigue damage.

Fatigue and fatigue damage will be discussed in Chapter 7: Fatigue analysis by FEA simulation of this book. For metal components, fatigue damage is inevitable due to cyclic stress. Fatigue damage is in-reversible, therefore metal components under fluctuating stress will eventually fail due to fatigue damage.

1.8   Factor of Safety

The primary objective of the mechanical design is to prevent failure under its intended service. Typically, mechanical engineers use the concept of a factor of safety to fulfill this objective. This section will discuss and explain the concept of the factor of safety.

A factor of safety is a measurement of component safety status and is a ratio of component allowable strength index S_S to component condition index L_C. To prevent component failure, the factor of safety n must be more than 1.

$$n = \frac{S_S}{L_C} \tag{1.33}$$

Component allowable strength index S_S is the component's capability for normal service and functions. It can be material yield strength or ultimate strength when we deal with mechanical failure due to static stress. It can be fatigue strength at the given fatigue life or the number of cyclic stress cycles to fatigue failure at a given cyclic stress when we deal with mechanical failure due to fatigue damage. It also can be the allowable deflection or the slope of the deflection of the considered component, which is recommended by some design guidelines when we deal with mechanical failure due to excessive deformation.

Component condition index L_C is a variable that reflects the component conditions under rated loading. It can be different variables for different modes of failure. For mechanical failure due to static stress, it can be maximum normal stress, maximum shear stress, or maximum Von Mises stress at the possible weakest locations. For mechanical failure due to fatigue damage, it can be the cyclic stress amplitude or the number of cycles of cyclic stress. For mechanical failure due to excessive deformation, it can be the maximum deflection or the maximum slope of deflection at the possible weakest locations.

For the calculation of a factor of safety n, the component allowable strength index S_S and component condition index L_C must be the same type of variable. For example, for a simple straight bar made of a ductile material under axial loads, the component allowable strength index S_S is materials yield strength, and the component condition index L_C is the maximum normal stress.

The factor of safety calculated by Equation (1.33) is the actual factor of safety that a designed component possesses. The actual factor of safety of a designed component must be larger than 1. Otherwise, it will be considered a mechanical failure.

If material properties and loading conditions can be fully described by deterministic values, and the calculation models for stress/strain can provide accurate solutions, we just need to make the actual factor of safety slightly larger than 1. However, there are always some uncertainties in material mechanical properties, design loadings & working environments, and strain/stress calculation models with certain errors [2, 7], the factor of safety of a component will be typically much bigger than 1. This factor of safety can be called the rated factor of safety.

The rated factor of safety is a recommended or required value for the factor of safety according to some design codes or after the considerations of some uncertainties of design parameters during the mechanical design such as material mechanical properties, design loading, type of materials, and consequence of failures.

The rated factor of safety is one of the design specifications. When the actual factor of safety is larger than or equal to the rated factor of safety, the designed component is classified as a qualified design or a qualified component. When the actual factor of safety is less than the rated factor of safety, the component is classified as an unqualified design, and, in such a case, it needs to be redesigned.

The rated factor of safety is used to determine component minimum dimensions. The bigger rated safety factor of a component means that the component will have a bigger dimension. Therefore, the selection of the properly rated factor of safety of a component is typically a compromise process. We want the designed component to be able to safely perform its specified functions, but we don't want it to be too strong.

For mechanical failure due to excessive deformation, the design governing equation is

$$n = \frac{S_S}{L_C} > 1 \quad \text{or} \quad S_S > L_C \tag{1.34}$$

In this case, there is no recommendation for the rated factor of safety. We just typically make the actual factor of safety slightly larger than 1, that is, the component maximum deformation is less than the allowable maximum deformation, which can be determined by some design standards and typically is specified in the design specifications of the component.

For mechanical failures due to fatigue damage and static loads, the design governing equation is

$$n_R = \frac{S_S}{L_C} \tag{1.35}$$

n_R is the rated factor of safety.

For mechanical failure due to fatigue damage, the rated factor of safety is typically not very big. The typical value is

$$n_R = 1.0 \sim 1.5 \tag{1.36}$$

The rated factor of safety for a mechanical component failure due to static stress in some specific applications, for example, pressure vessels and boilers, is often mandated by law, policy, or industry standards. When the rated factor of safety of a component needs to be selected by designers, the following six aspects should be considered.

(1). If there is a corresponding design code for this component design, the rated factor of safety should be consistent with this design code.

(2). The consequence of component failure is an important aspect of selecting the rated factor of safety. If the failure might cause significant financial loss, serious injury, or death to operators, the rated factor of safety could be 4 or higher. However, the rated factor of safety of non-critical components might be 1.5.

(3). For the same applications, the rated factor of safety of a component of ductile or brittle material will have a different value. Since the component allowable strength index S_S for a brittle material component is ultimate strength, the failure due to static load might be a sudden fracture. While the component allowable strength index S_S for ductile material under static loads is yield strength, the failure due to static stress means yielding and necking and then eventually fractures. Therefore, for the same applications, the rated factor of safety of brittle material components might be twice the value that would be used for a ductile material.

(4). Component allowable strength index S_S is typically determined by the material mechanical properties. There are always some uncertainties related to material mechanical properties. For the same applications, a higher level of confidence level of the material mechanical properties could result in a smaller value of the rated factor of safety.

(5). Component design is based on design loadings, which are provided by customers as part of design specifications. If the accuracy of predictions on the loadings is provided, the rated factor of safety could be smaller for the same applications.

(6). Design engineers need to obtain the component condition index L_C such as maximum normal stress, maximum shear stress, or the maximum Von Mises stress based on the provided loading through calculation models or systems. For the same applications,

Table 1.1 Some general guidelines about the rated factor of safety for a ductile material [19]

The rated factor of safety	Applications
1.25~2.0	Components under static loading with a high level of confidence in loadings, material mechanical properties, or stress analysis models
2.0~2.5	Components under dynamic loading with an average confidence level in loadings, material mechanical properties, or stress analysis models
2.5~4	Components under static or dynamic loading with uncertainty about one of the loadings, material mechanical properties, or stress analysis models
4.0 or higher	Components under static loading or dynamic loading with uncertainty about some combinations of loads, material mechanical properties, and stress analysis models Critical components whose failure might result in significant financial loss, serious injury, or death to its operators

Table 1.2 Some general guidelines about the rated factor of safety for a brittle material [19]

The rated factor of safety	Applications
3.0~4.0	Components under static loads with a high level of confidence in all design data
4.0~8.0	Components under static or dynamic loading with uncertainty about loads, materials mechanical properties, stress analysis models

the rated factor of safety with a more accurate stress calculation model could be smaller.

Some general guidelines for the rated factor of safety of a component against failure due to static stress for ductile material and brittle materials are listed in Table 1.1 and Table 1.2 [19].

1.9 Principal Stresses and Maximum Shear Stresses

The general state of stress at a point shown in Fig. 1.3 includes six stress components. Each of these will make some contributions to mechanical failures. This section will briefly discuss the principal stresses and maximum/minimum shear stresses, which are the variables for failure theory.

A real component is typically subjected to several different loads, so component stresses induced due to several loads will be combined stresses. The superposition principles can be used to calculate the combined stresses.

The superposition principle: Typical approaches for calculating component combined stress is through the application of the superposition principle, which states that if the deformation is small, that is, in the material elastic region, the combined stresses of a component under the combination of several loads are equal to the sum of stresses caused by each load [2, 9, 19].

After the general state of stress at a point has been calculated, we can calculate the principal stresses and the maximum/minimum shear stresses of the point.

Principal stress is the maximum or minimum normal stress on specific orientated surfaces of a point, where the shear stress is zero.

For a general state of stress of a point, there are three principal stresses, which will be on three mutually perpendicular planes. These principal stresses are usually denoted as σ_1, σ_2 and σ_3. They are arranged in such a way $\sigma_1 \geq \sigma_2 \geq \sigma_3$ that σ_1 is the largest principal stress and σ_3 is the smallest principal stress.

There are several approaches for determining three principal stresses such as Mohr's circle diagram, the eigenvalue method, and the cubic formula. The cubic formula for calculating principal stress is listed here.

$$I_1 = \sigma_x + \sigma_y + \sigma_z$$
$$I_2 = \sigma_x\sigma_y + \sigma_y\sigma_z + \sigma_z\sigma_x - \tau_{xy}^2 - \tau_{yz}^2 - \tau_{zx}^2$$
$$I_3 = \sigma_x\sigma_y\sigma_z + 2\tau_{xy}\tau_{yz}\tau_{zx} - \sigma_x\tau_{yz}^2 - \sigma_y\tau_{zx}^2 - \sigma_z\tau_{xy}^2$$
$$R = \frac{1}{3}I_1^2 - I_2; Q = \frac{1}{3}I_1 I_2 - I_3 - \frac{2}{27}I_1^3; T = \left(\frac{1}{27}R^3\right)^{\frac{1}{2}}; S = \left(\frac{1}{3}R\right)^{\frac{1}{2}}; \alpha = cos^{-1}\left(-\frac{Q}{2T}\right)$$

$$\sigma_a = 2S\left[cos\left(\frac{\alpha}{3}\right)\right] + \frac{1}{3}I_1 \tag{1.37}$$

$$\sigma_b = 2S\left[cos\left(\frac{\alpha}{3} + \frac{2\pi}{3}\right)\right] + \frac{1}{3}I_1 \tag{1.38}$$

$$\sigma_c = 2S\left[cos\left(\frac{\alpha}{3} + \frac{4\pi}{3}\right)\right] + \frac{1}{3}I_1 \tag{1.39}$$

Where σ_x, σ_y, σ_z, τ_{xy}, τ_{yz}, and τ_{zx} are six cartesian stress components of the general state of stress of a point as shown in Fig. 1.3.

The quantities I_1, I_2, and I_3 are the stress invariants, which are completely determined by the six Cartesian stress components of a point. R, Q, S, T, and α are all just several intermediate variables for calculating principal stresses. These intermediate variables are all completely determined by three stress invariants. Equations (1.37)~(1.39) are used to calculate three principal stresses. The largest and smallest values of σ_a, σ_b, and σ_c are σ_1 and σ_3, respectively. The rest is σ_2.

There are three maximum or minimum shear stresses at a general state of stress of a point, which will be on three mutually perpendicular planes, which are different from those for principal stresses. After three principal stresses σ_1, σ_2, and σ_3 have been determined, the three maximum or minimum shear stresses can be calculated by following equations.

$$\tau_{1/2} = \frac{\sigma_1 - \sigma_2}{2} \tag{1.40}$$

$$\tau_{2/3} = \frac{\sigma_2 - \sigma_3}{2} \tag{1.41}$$

$$\tau_{1/3} = \frac{\sigma_1 - \sigma_3}{2} \tag{1.42}$$

Where $\tau_{1/2}$, $\tau_{2/3}$ and $\tau_{1/3}$ are three maximum or minimum shear stresses. For mechanical component design, the largest value of $\tau_{1/2}$, $\tau_{2/3}$ and $\tau_{1/3}$ is the main concern, which will be needed in Chapter 1.10: Failure theories due to static loads. Because of the arrangement of three principal stress $\sigma_1 \geq \sigma_2 \geq \sigma_3$, the largest shear stress will be certainly determined by Equation (1.42). Therefore, we have:

$$\tau_{max} = \frac{\sigma_1 - \sigma_3}{2} \tag{1.43}$$

1.10 Failure Theories Due to Static Loads

For a general state of stress at a point, each of the six stress components might have some contributions to component failures. For a component under multi-axial stresses, failure theories are needed. This section will briefly describe and explain the failure theories due to static loads.

When a component is under uniaxial stress such as a bar under axial load, we can directly compare the stress conditions with the stress conditions of a tensile test specimen. However, for a component under multi-axial stresses, failure theories are needed because direct experimental verifications could be very difficult.

Failure theories due to static loads are some experimentally verified hypotheses that describe the criteria of a component failure due to multi-axial stresses.

For ductile materials, two typical failure theories are the maximum shear stress theory and the distortion energy theory. For brittle materials, two typical failure theories are the Maximum normal stress theory and the Coulomb-Mohr theory.

Maximum shear stress (MSS) theory for ductile materials predicts that the yielding begins whenever the maximum shear stress in any element equals or exceeds the maximum shear stress in a simple tension-test specimen of the same material when that specimen begins to yield.

Per Equation (1.43), the maximum shear stress of a point is equal to $\frac{\sigma_1 - \sigma_3}{2}$ under multi-axial stresses, where σ_1 and σ_3 are the largest and smallest of the principal stresses of the point. The maximum shear stress in a tension-test specimen when yielding is equal to $\frac{S_y}{2}$, where S_y is material yield strength. Therefore, the yielding condition, that is, the failure criteria for ductile materials is:

$$\frac{\sigma_1 - \sigma_3}{2} \geq \frac{S_y}{2} \tag{1.44}$$

With the consideration of a factor of safety, the design equation based on the MSS theory is

$$n = \frac{S_y}{\sigma_1 - \sigma_3} \tag{1.45}$$

Equation (1.45) is used to determine the minimum component dimension with a given n. It can also be used to calculate the actual component factor of safety.

Distortion-energy (DE) theory for ductile materials predicts that yielding occurs when the distortion strain energy per unit volume under multi-axial stresses reaches or exceeds the distortion strain energy per unit volume for yielding in a simple tension-test specimen of the same material.

When the state of stress of a point is described by three principal stress σ_1, σ_2 and σ_3, the distortion strain energy per unit volume is $\frac{1+v}{3E}\left[\frac{(\sigma_1-\sigma_2)^2+(\sigma_2-\sigma_3)^2+(\sigma_3-\sigma_1)^2}{2}\right]$, where v and E are material Poisson's ratio and Young's modulus. For a simple tensile test at yield, the distortion strain energy per unit volume is $\frac{1+v}{3E}S_y^2$. Therefore, the yielding condition, that is, the failure criteria for ductile materials is:

$$\frac{1+v}{3E}\left[\frac{(\sigma_1 - \sigma_2)^2 + (\sigma_2 - \sigma_3)^2 + (\sigma_3 - \sigma_1)^2}{2}\right] \geq \frac{1+v}{3E}S_y^2 \tag{1.46}$$

The von Mises stress σ_{von} is defined as:

$$\sigma_{von} = \sqrt{\frac{(\sigma_1 - \sigma_2)^2 + (\sigma_2 - \sigma_3)^2 + (\sigma_3 - \sigma_1)^2}{2}} \tag{1.47}$$

The yielding condition, that is, the failure criteria for ductile materials per DE theory is:

$$\sigma_{von} = \sqrt{\frac{(\sigma_1 - \sigma_2)^2 + (\sigma_2 - \sigma_3)^2 + (\sigma_3 - \sigma_1)^2}{2}} \geq S_y \tag{1.48}$$

From Equation (1.48), the physical meaning of the von Mises stress σ_{von} is a value used to determine whether a material will yield. It is also typically treated as equivalent stress of a point under multi-axial stresses based on the DE theory. The von Mises stress can be directly calculated by using the six Cartesian stress components.

$$\sigma_{von} = \sqrt{\frac{\left(\sigma_x - \sigma_y\right)^2 + \left(\sigma_y - \sigma_z\right)^2 + (\sigma_z - \sigma_x)^2 + 6\left(\tau_{xy}^2 + \tau_{yz}^2 + \tau_{zx}^2\right)}{2}} \tag{1.49}$$

With the consideration of a factor of safety, the design equation based on the DE theory is

$$n = \frac{S_y}{\sigma_{von}} \tag{1.50}$$

Equation (1.50) is used to determine the minimum component dimension with a given n. It can also be used to calculate the actual component factor of safety.

Both MSS theory and DE theory have been experimentally verified and widely used in industries for mechanical component design. There is no preference between these two failure theories. But, typically, it can be noticed that the MSS theory is more conservative and the DE theory is more accurate.

Maximum normal stress (MNS) theory for brittle materials predicts that failure or fracture occurs whenever one of three principal stresses equals or exceeds the component's material strength.

Three principal stresses of a point can be tensile or compressive. The MNS theory can be expressed as follows in three different cases.

Case one: When all principal stresses are tensile stresses, that is, $\sigma_1 \geq \sigma_2 \geq \sigma_3 \geq 0$, the component failure criterion is:

$$\sigma_1 \geq S_{ut} \tag{1.51}$$

where S_{ut} is material ultimate tensile strength.

With a factor of safety, the design equation based on the MNS theory in Case One is:

$$n = \frac{S_{ut}}{\sigma_1} \tag{1.52}$$

Case Two: When all principal stresses are compressive stresses, that is, $0 \geq \sigma_1 \geq \sigma_2 \geq \sigma_3$, the component failure criterion is:

$$-\sigma_3 \geq S_{uc} \tag{1.53}$$

Where S_{uc} is the material compressive ultimate strength.

With a factor of safety, the design equation based on the MNS theory in Case Two is:

$$n = \frac{S_{uc}}{-\sigma_3} \tag{1.54}$$

Equations (1.52) and (1.54) are used to determine the minimum component dimension with a given n. They can also be used to calculate the actual factor of safety.

Case Three: When three principal stresses contain tensile and compressive stresses, that is, $\sigma_1 \geq 0$ and $\sigma_3 \leq 0$, the component failure criteria are:

$$\begin{aligned} \sigma_1 &\geq S_{ut} \\ -\sigma_3 &\geq S_{uc} \end{aligned} \qquad (1.55\text{a,b})$$

The experiments have verified that the MNS theory can correctly predict component failures in Case One and Case Two. However, experiments also indicate that the MNS theory cannot correctly predict component failures when three principal stresses contain both tensile and compressive stresses. The main issue with the MNS theory is that it does not consider the contribution of both tensile stress and compressive stress together to the failure. Therefore, the implementation of the MNS theory in mechanical design should be used with caution.

Brittle Coulomb -Mohr (BCM) theory for brittle materials is an experimentally verified hypothesis that describes component failure criteria for brittle materials. This hypothesis considers the contributions of both tensile and compressive stresses together. It can be mathematically expressed by following three failure criteria.

Case one: When all principal stresses are tensile stresses, that is, $\sigma_1 \geq \sigma_2 \geq \sigma_3 \geq 0$, the component failure criterion is:

$$\sigma_1 \geq S_{ut} \qquad (1.56)$$

Case Two: When all principal stresses are compressive stresses, that is, $0 \geq \sigma_1 \geq \sigma_2 \geq \sigma_3$, the component failure criterion is:

$$-\sigma_3 \geq S_{uc} \qquad (1.57)$$

Case Three: When three principal stresses contain both tensile and compressive stresses, that is, $\sigma_1 \geq 0$ and $\sigma_3 \leq 0$, the component failure criterion is:

$$\frac{\sigma_1}{S_{ut}} + \frac{-\sigma_3}{S_{uc}} \geq 1 \qquad (1.58)$$

It is noticed that the failure criteria of BCM theory for brittle materials in Case One and Case Two are the same as those of MNS theory, but in Case Three, the BCM theory considers the contribution of both tensile and compressive principal stresses together.

With a factor of safety, the corresponding design equations based on the BCM theory are:

Case one: When all principal stresses are tensile stresses, that is, $\sigma_1 \geq \sigma_2 \geq \sigma_3 \geq 0$, the component failure criterion is:

$$n = \frac{S_{ut}}{\sigma_1} \qquad (1.59)$$

Case Two: When all principal stresses are compressive stresses, that is, $0 \geq \sigma_1 \geq \sigma_2 \geq \sigma_3$, the component failure criterion is:

$$n = \frac{S_{uc}}{-\sigma_3} \tag{1.60}$$

Case Three: When three principal stresses contain tensile and compressive stresses, that is, $\sigma_1 \geq 0$ and $\sigma_3 \leq 0$, the component failure criteria are:

$$\frac{1}{n} = \frac{\sigma_1}{S_{ut}} + \frac{-\sigma_3}{S_{ut}} \tag{1.61}$$

Equations (1.59) to (1.61) can be used to determine the component minimum dimension with a given n. They can also be used to calculate the actual component factor of safety.

1.11 Concepts of Simulation-Based Mechanical Design (SBMD)

This section will briefly explain what simulation-based mechanical design is and brief descriptions of each chapter of this book.

Mechanical design has two main tasks: design check and dimension design. Design check is to check whether designed components or systems satisfy a given factor of safety. Dimension design is to determine the minimum geometric dimensions according to a given factor of safety. The design methodology and theory for mechanical design have been concisely described in Section 1.2 to Section 1.10 of this chapter. The key techniques for both tasks are to calculate the general state of stress and the factor of safety at the possible weakest points or locations of components. Since mechanical components in real products have complicated geometrical shapes and are typically under complicated loads, it is very difficult or sometimes impossible for theoretical calculations of the general state of stress at the possible weakest points or locations.

After powerful personal computers became available to all, Computer-Aided Design (CAD) and Finite Element Analysis (FEA) have become the two most useful and powerful tools for mechanical engineers. CAD is a technique to create a virtual component or virtual assembly (products) that are stored in a computer but have the same geometric shapes, dimensions, and types of materials as those of the real components and products. FEA is a technique to simulate the stress/ strain/ deformation and the factor of safety on virtual components and virtual products under service loads. By using CAD and FEA, we can obtain strain/ stress/ deformation data, factors of safety of components, as well as the possible weakest points of the product.

Simulation-Based Mechanical Design (SBMD) is a new approach to mechanical design that designs mechanical components, devices, or systems to satisfy design specifications by using current mechanical design theory, but the FEA (Finite Element Analysis) simulation is the main tool for determining stress/strain/factor of safety of components and products under specified service loadings. Five features of Simulation-Based Mechanical Design are listed here.

- SBMD is just an alternative and electronic version of mechanical design. The design theory and methodology of SBMD are the same as those in traditional mechanical design.
- CAD technology is used to create virtual mechanical components or assemblies which are presented by and saved as electronic files.
- FEA (Finite Element Analysis) simulation is the main tool for determining strain/ stress/deformation and the factor of safety of components and products. For FEA simulation, if the virtual models are meshable, there are no differences between simple components and complicated components, or between simple loading and complicated loading because FEA simulations always break components and products into thousands of tiny elements. After completing an FEA simulation, strain/ stress/deformation and factors of safety at any points of components or products can be obtained and viewed through post-processing tools. This is a big advantage in the simulation-based mechanical design when compared with the theoretical calculation of strain/stress/ deformation in traditional mechanical design.
- A design check task in Simulation-Based Mechanical design is just a comparison of the minimum factor of safety of components or products obtained from FEA simulation with the rated factor of safety.
- For the dimension design of components, Simulation-Based Mechanical Design will be an interactive process. After geometric shapes and initial dimensions of a virtual component have been created, it will be used in FEA simulation to obtain its weakest points or locations and factors of safety. Per the simulation results, we can modify the shapes and dimensions of the virtual components and rerun the FEA simulation until the components or products have satisfied the rated factor of safety.

Simulation-Based Mechanical Design is the main topic of this book. After this first chapter, the book will explain and present simulation-based mechanical design in detail. The following are concise descriptions of each chapter of this book.

Chapter 2 Fundamentals of finite element analysis (FEA): The FEA commercial software is a black box with user interfaces. It is a tool to produce outputs based on inputs. This chapter does not try to fully explain the FEA theory but offers a concise introduction to the Finite Element Analysis (FEA) theory including fundamental concepts of the finite element analysis, approximate deformation functions of an element and shape functions, element properties of a 1D element, and an example of uniaxial stepped bar under axial loads manually solved by FEA theory. The understanding of fundamental concepts of FEA will significantly facilitate us in using FEA simulation and interpreting simulation results.

Chapter 3 FEA simulation on components: This chapter will explain and demonstrate how to use FEA simulation as a tool to determine stress/strain distributions, and factors of safety of components. This chapter demonstrates each step of the general procedure for FEA simulation with lots of step-by-step guided examples. This chapter also explains how to make the best use of visualization tools of FEA simulation for redesigning or modifying components. The skills learned in this chapter are implemented to conduct some real design cases and some open-ended minor design projects.

Chapter 4 FEA simulation on assemblies: This chapter will explain how to define contacts on an interface in an assembly and then conduct the FEA simulation to evaluate the stress/strain distributions and factors of safety of the assembly. When a component's strength is not the main concern in FEA simulation, it can be simplified as a mathematic model, which is called a connector. This chapter will cover several typical connectors for mechanical design. After completing Chapter 3 and Chapter 4, the stress/strain distributions of any component /assembly under any loading can be determined by FEA simulation. When the stress/strain distributions of components or assemblies are known, failure theory can be used to complete mechanical design. The skills learned in this chapter will be implemented to conduct open-ended major design projects which are typically real practical design projects.

Chapter 5 Natural frequency analysis: Prevention of resonance is one of the main tasks for design engineers. This chapter will explain how to determine the first five natural frequencies of a component.

Chapter 6 Thermal analysis and thermal stress analysis: Thermal stress is one of the common stresses in mechanical design. This chapter will discuss how to run a thermal analysis to obtain temperature distributions and then run a thermal stress analysis to evaluate the stress/strain caused by temperature loadings.

Chapter 7 Fatigue analysis by FEA simulation: Mechanical devices are always subjected to some cyclic loading due to vibrations or repeated activities. Cyclic stress will cause fatigue damage and fatigue failure to metal components. This chapter will first briefly discuss cyclic loading, fatigue damage, S-N curves, modification factors, and Miner's rule. Then, it will explain and demonstrate how to use the fatigue analysis module in SolidWorks Simulation to simulate and evaluate the fatigue damage of a component.

Exercises

1-1 What is the five-phase engineering design process? What are the main tasks of each phase?

1-2 Do literature research to find another description of an engineering design process. Compare it with the five-phase design process.

1-3 Use one of your design projects to check whether you used all of the five phases. If not, why or what happened?

1-4 List a possible need for engineering design and then run the needs assessment.

1-5 Use one of your design examples to describe how you conduct a conceptual design.

1-6 Search the design handbook or other sources to list yield strength, ultimate strength, Young's modulus, shear Young's modulus, and passion ratio of three ductile materials.

1-7 Search the design handbook or other sources to list the ultimate strength, Young's modulus, shear Young's modulus, and passion ratio of three brittle materials.

1-8 Use one design example to explain free dimensions and mating (assembly) dimensions.

1-9 Explain the definition of a clearance fit with one example.

1-10 Explain the definition of interference fit with one example.

1-11 Explain the definition of transition fit with one example.

1-12 Use one of your design projects to explain how you select materials for components.

1-13 Use one of your design projects to explain how you select manufacturing processes for components.

1-14 Explain the general procedure for selecting a material.

1-15 What are the three main factors for selecting the manufacturing process for components?

1-16 Describe three manufacturing processes and explain possible applications.

1-17 Explain with one example what is mechanical failure.

1-18 Explain one example of mechanical failure due to static stress.

1-19 Explain one example of mechanical failure due to excessive deformation.

1-20 What is a rated factor of safety? Explain the difference between a rated factor of safety and the component factor of safety.

1-21 How to select a rated factor of safety?

1-22 List and explain the general guidelines for selecting rated factors of safety for ductile materials.

1-23 List and explain the general guidelines for selecting rated factors of safety for brittle materials.

1-24 List and explain one failure theory for ductile materials.

1-25 List and explain one failure theory for brittle materials.

References

1. Gerard Voland (2004) Engineering By Design, Pearson Prentice Hall, Upper Saddle River, N.J.
2. Richard G. Budynas, J. Keith Nisbett (2015) Shrigley's mechanical engineering design, McGraw-Hill Education, New York
3. Robert L. Norton (2006) Machine design: an integrated approach, Pearson Prentice Hall, Upper Saddle River, N.J.

4. Clive L. Dym, Patrick Little (2000) Engineering design: a project-based introduction, John Wiley, New York
5. David G. Uiiman (2010) The Mechanical design process", McGraw-Hill Higher Education, Boston
6. James G. Skakoon (2000) Detailed mechanical design: a practical guide, ASME Press, New York
7. Xiaobin Le (2019) Reliability-based mechanical design, Volume 1, component under static load, Morgan & Claypool, San Rafael, California
8. Xiaobin Le, Anthony Duva, Richard Roberts, Ali Moazed (2011) Instructional Methodology for Capstone Senior Mechanical Design, Paper presented at the 2011 ASEE international conference, Vancouver, BC, Canada, June 26 - 29, 2011
9. Xiaobin Le, Anthony Duva, John Voccio, Richard Roberts, Ali Moazed (2017) The proposed approach for determining combined stresses of a component Paper presented at 2017 ASEE Annual Conference & Exposition, Columbus, Ohio. https://doi.org/10.18260/1-2-28994
10. ASTM E8/E8M-22 Standard test methods of tension testing of metallic materials
11. The American Society of Mechanical Engineers (1994) Dimensioning and Tolerancing Standard ASME Y14.5M -1994
12. Gary K. Griffith (2002) Geometric dimensioning and tolerancing applications and inspection, Second edition, Prentice Hall, Upper Saddle River, New Jersey.
13. Xiaobin Le, Anthony Duva, and Richard L. Roberts (2021) Teaching GD&T fundamentals in the course "Design of Machine Elements" The 2021 ASEE annual conference, July 26–29, Virtual online conference.
14. Erik Oberg, Franklin D. Jones, Holbrook L. Horton, and Henry H. Ryffel (2016) Machinery's handbook, Industrial Press, Incorporated, South Norwalk.
15. William D. Callister (2014) Materials Science and engineering: an introduction, John Wiley and Sons, Inc., Hoboken, NJ
16. Michael F. Ashby (2011) Materials selection in mechanical design, Butterworth-Heinemann, Burlington, MA
17. Corrado Poli (2001) Design for manufacturing: a structured approach, Butterworth-Heinemann, Boston, MA
18. K.G Swift, J.D. Booker (2013) Manufacturing process selection handbook, Butterworth-Heinemann, Oxford
19. M.F. Spotts, T.E. Shoup and L.E. Hornberger (2003) Design of Machine elements, Prentice Hall, Upper Saddle River, New Jersey

Fundamentals of Finite Element Analysis (FEA)

2

Abstract

Finite Element Analysis (FEA) is most associated with FEA software, which is generally treated as a black box that generates useful outputs when the necessary inputs are provided. However, a basic understanding of FEA theory can make this black box much easier to use. This chapter is a concise summary and description of the Finite Element Analysis theory. It will introduce and discuss the fundamental concepts of FEA theory to help in understanding the theory behind the software driving this black box. Finite Element Analysis is a numerical approximation technique for solving deformations of objects. Every object is divided into many small blocks, called elements. Each element has several nodes at the corners, edges, and outer surfaces of the element. Approximation deformation functions inside a finite element can be guessed and expressed in terms of nodal deformation. Scientific principles, such as the minimum potential energy principle, are used to minimize the total error in constructing element properties, which converts the set of partial differential equations into a set of linear algebraic equations. The shared nodes of elements are pinned together so that an assembly of many elements can be used to represent the original continuous objects. This chapter provides several examples of how to guess a deformation function inside an element by using the concept of a shape function. One example of a uniaxial stepped bar under axial loads is fully demonstrated and solved manually by using the FEA method. This example will significantly facilitate us to have a better understanding of the fundamentals of FEA theory.

Supplementary Information The online version contains supplementary material available at https://doi.org/10.1007/978-3-031-64132-9_2.

X. Le, *Simulation-Based Mechanical Design*, Synthesis Lectures on Mechanical Engineering, https://doi.org/10.1007/978-3-031-64132-9_2

2.1   Introduction

Finite Element Analysis (FEA) software to most design engineers, is a black box that can be used to generate some outputs when necessary inputs are provided. Knowing some basic concepts about FEA theory will make FEA software significantly easier to use. This chapter will explore and explain the fundamental concepts of FEA theory which will help set up a proper simulation, and properly interpret the results. Concise descriptions of each section of this chapter are listed below.

Chapter 2.2 History of Finite Element Analysis (FEA) will provide a brief history of FEA theory.

Chapter 2.3 Fundamental concepts of FEA will explain the fundamental concepts of FEA theory. This will help us to understand how and why FEA simulation can provide approximate solutions to complex physical models.

Chapter 2.4 Deformation functions and shape functions will explain how to guess the deformation functions of a tiny element based on nodal deformation and how to use shape functions to construct deformation functions.

Chapter 2.5 Types of Typical Elements will list and explain several typical types of elements and their applications.

Chapter 2.6 Element stiffness matrix (Element properties) will use a one-dimensional example to show how to minimize the error in the FEA method so that the corresponding relationships between element geometry, material properties, loads, and deformations can be established. This relationship is called element stiffness matrix or element properties.

Chapter 2.7 An example of a uniaxial steppe bar under axial loading will demonstrate how to manually solve a one-dimensional example by using the FEA method. This example will significantly help us to have a better understanding of FEA theory.

2.2   History of Finite Element Analysis (FEA)

For mechanical design, it is almost impossible to obtain theoretical analysis solutions of stress/strain/factor of safety on real-world components because they are typically subjected to muti-axial stresses due to complex geometry and boundary conditions. However, Finite Element Analysis (FEA) has become a powerful alternative for obtaining stress/strain and factors of safety of components.

Finite Element Analysis (FEA) is a numerical tool for solving physical models governed by partial differential equations under boundary conditions. The terms FEA (Finite Element Analysis) and FEM (Finite Element Method) are typically exchangeable. FEM typically refers to the mathematical method and FEA typically refers to the program in which FEM is used.

It is difficult to get an exact date of the invention of the Finite Element Method. However, it is generally accepted that its development can be traced back to the work of A. Hrennikoff and R. Courant in the early 1900s [1, 2]. The first paper about FEM was published by Turner, M.J., Clough, R.W., and Martin H.C and Topp, L.J. in 1956 [3]. The first book "Finite element method in Structural & continuum mechanics" about FEM was published by Drs. O.C. Zienkiewicz and Y.K. Cheung in 1967 [4]. The first FEA software was developed by Computer Sciences Corporation (CSC) and released to NASA (The National Aeronautics and Space Administration) as a FEA program called NASTRAN (NASA Structural Analysis).

With the emergence of powerful personal computers available to general engineers, FEA has quickly become an extremely important and practical tool for millions of engineers and scientists and is widely used to simulate behaviors of structural, mechanical, thermal, electrical, and chemical systems for both design and performance analysis. There is a wide range of commercially available FEA software such as ABAQUS, ANSYS, COMSOL Multiphysics, LS-DYNA, SolidWorks Simulation, NASTRAN, and so on. There are also large collections of technical papers and books about FEA or FEM at different levels such as the references [5–9].

2.3   Fundamental Concepts of FEA

For general mechanical design, components or products are required to work properly throughout a specified service life when these components operate in their material's elastic regions. This is a typical small-deformation assumption. Per the definitions of strain, we have the following six strains.

$$\varepsilon_x = \frac{\partial u(x, y, z)}{\partial x} \tag{2.1}$$

$$\varepsilon_y = \frac{\partial v(x, y, z)}{\partial y} \tag{2.2}$$

$$\varepsilon_z = \frac{\partial w(x, y, z)}{\partial z} \tag{2.3}$$

$$\gamma_{xy} = \frac{1}{2}\left(\frac{\partial u(x, y, z)}{\partial y} + \frac{\partial v(x, y, z)}{\partial x} \right) \tag{2.4}$$

$$\gamma_{yz} = \frac{1}{2}\left(\frac{\partial v(x, y, z)}{\partial z} + \frac{\partial w(x, y, z)}{\partial y} \right) \tag{2.5}$$

$$\gamma_{zx} = \frac{1}{2}\left(\frac{\partial w(x, y, z)}{\partial x} + \frac{\partial u(x, y, z)}{\partial z}\right) \tag{2.6}$$

where $u(x, y, z)$, $v(x, y, z)$ and $w(x, y, z)$ are the deformation functions of a general point of a component along x-, y- and z-axes. ε_x, ε_y, and ε_z are three normal strains along x-, y- and z-axes. γ_{xy}, γ_{yz}, and γ_{zx} are three shear strains in the x–y plane, y–z plane, and z–x plane of a Cartesian Coordinate system.

In the elastic region, the generalized Hooke's law exists and is listed as the following.

$$\sigma_x = \frac{E}{(1 + v)(1 - 2v)}\left[(1 - v)\varepsilon_x + v\left(\varepsilon_y + \varepsilon_z\right)\right] \tag{2.7}$$

$$\sigma_y = \frac{E}{(1 + v)(1 - 2v)}\left[(1 - v)\varepsilon_y + v\left(\varepsilon_z + \varepsilon_x\right)\right] \tag{2.8}$$

$$\sigma_z = \frac{E}{(1 + v)(1 - 2v)}\left[(1 - v)\varepsilon_z + v\left(\varepsilon_y + \varepsilon_x\right)\right] \tag{2.9}$$

$$\tau_{xy} = 2G\gamma_{xy} \tag{2.10}$$

$$\tau_{yz} = 2G\gamma_{yz} \tag{2.11}$$

$$\tau_{zx} = 2G\gamma_{zx} \tag{2.12}$$

where σ_x, σ_y, σ_z, τ_{xy}, τ_{yz}, τ_{zx} are six Cartesian stress components of a general state of stress. E and G are material Young's and shear Young's modulus. v is Poisson's ratio.

Equations (2.1) to (2.12) are the governing equations for stresses and strains of a general point of a component and are represented as twelve partial differential equations. If the deformation functions of a component $u(x, y, z)$, $v(x, y, z)$ and $w(x, y, z)$ are known, the strain and stress of any point of a component can be determined by these twelve partial differential equations. However, it is typically impossible to describe the deformation functions of a real-world component. For example, the component in Fig. 2.1a is fixed on the bottom surface and subjected to an external force normal to the front-end surface of a cylinder. The deformation of a point inside the top cylinder will be quite different from the deformation of a point at the bottom plate. It will be almost impossible to find explicit functions to describe the deformation functions of this component. However, when the component is broken into lots of small blocks or elements as shown in Fig. 2.1b, which consists of 19,633 tiny elements, it becomes possible to guess the deformation functions for a small element. This is because the deformation functions of a small element are known to be continuous and the deformation variations inside a tiny element will not be very big. For example, we could use a constant function, a linear function, or a second-order function of coordinates as the deformation functions of this small element. After we know the deformation functions of this small element, we can

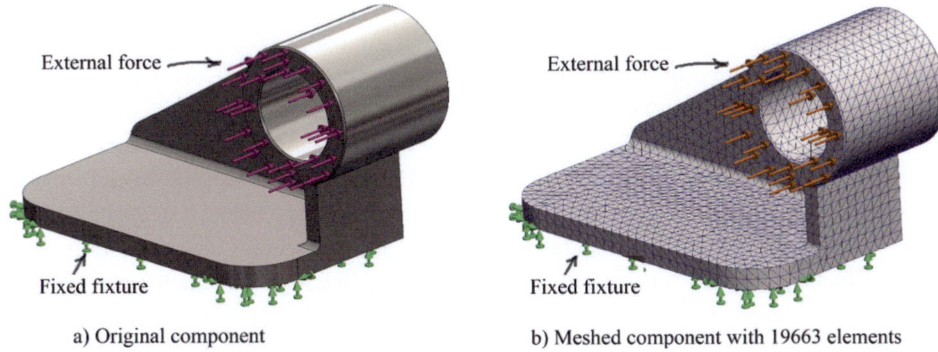

Fig. 2.1 Schematics of an original component and its meshed component

use Eqs. (2.1) to (2.12) to determine the strain and stress of any point inside this element. So, while it is almost impossible to find explicit functions for the deformation functions of a whole component, it is quite possible to find explicit functions for the deformation functions of a tiny element. This is one of the core concepts of FEA theory.

However, two new issues arise when a component is broken into thousands of smaller elements. There is a huge amount of calculations required with this new representation, due to the thousands of equations related to these elements. Without a powerful computer, FEA theory is not a practical approach for solving stress/strain problems. But this is not as big a problem anymore because powerful personal computers are increasingly available to general engineers and engineering students. The second issue is that guessed deformation functions of a small element are only approximate functions, and not the real deformation functions, which will cause some error. If the error is too big, using this approach will be meaningless. FEA theory can effectively control the errors from this approximation and provide simulation results within an acceptable level of error.

The purpose of this book is not to explore FEA theory, but to explore the fundamental concepts of FEA theory in preparation for using FEA software for setting up the simulation and properly interpreting the FEA simulation results. Fundamental concepts of FEA theory for solving strain/ stress problems are listed and concisely explained here.

(1) The finite element analysis (FEA) or finite element method (FEM) is a numerical approximation technique for solving deformations of the object (component or assembly) because the deformation functions of each element are approximations of the real deformation functions. The assembly of thousands of elements is used to represent the original component or assembly of components. Calculation of stress /strain / deformation is only a by-product of FEA theory because they can be directly calculated after the deformation functions of each element have been fully determined.

Therefore, when FEA simulation is used, we expect some error. But we want results within an acceptable level of error.

(2) An object is divided into a lot of small blocks (elements). Each element has several nodes, which will be at corners, edges, and outer surfaces. Figure 2.2 shows several examples of 1D elements, 2D elements, and 3D elements, where black dots represent nodes. The number of nodes in an element will determine the type of deformation functions. For example, a triangle element with 3 nodes will have a linear function of coordinates. With 6 nodes, the element will have a second-order function of the coordinates, which will be discussed and explained in Chap. 2.4. Each node will have corresponding nodal deformations. For example, the node i in a 3D element will have three nodal deformations u_i, v_i and w_i which are deformation along the x-, y-, and z-axis, respectively.

(3) Approximation deformation functions inside a finite element are expressed in terms of nodal deformation. For example, Fig. 2.3 shows a 2D triangle element with 3 nodes. u_i and v_i are the nodal deformation along the x- and y-axis of the i node. The deformation functions inside this element can be expressed as:

$$u(x, y) = f_1(u_1, u_2, u_3, x, y) \tag{2.13}$$

$$v(x, y) = f_2(v_1, v_2, v_3, x, y) \tag{2.14}$$

where $u(x, y)$ is the deformation function of any point inside this element along the x-axis and is a function of $u_1, u_2, u_3, x,$ and y. $v(x, y)$ is the deformation function of any point inside this ement along the y-axis and is a function of $v_1, v_2, v_3, x,$ and y. If the nodal deformations are known, the deformation functions of this element will be fully specified and can be used to calculate deformations of any point inside this element. One logical conclusion is that the smaller element will result in more accurate

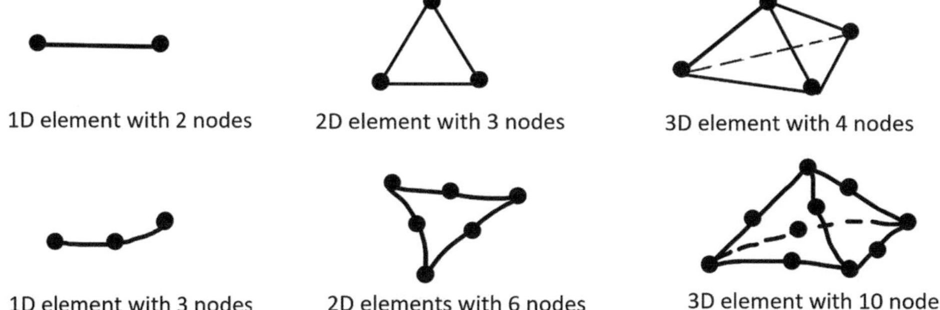

Fig. 2.2 Several elements with different nodes

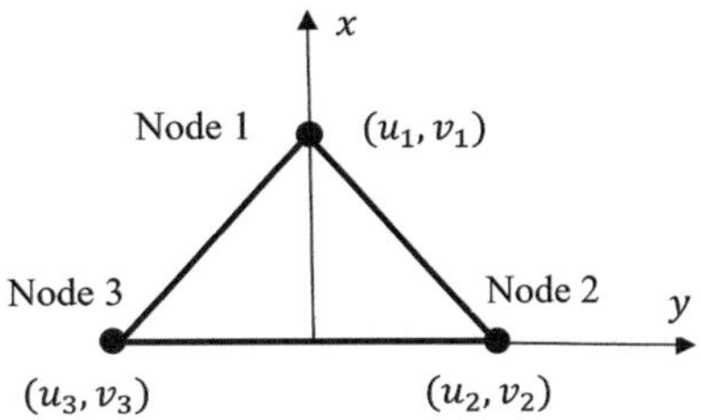

Fig. 2.3 A schematic of a 2D isosceles triangle element with 3 nodes

simulation results because the deformation functions are closer to the real deformation functions of the elements. Chapter 2.4 will discuss and explain how to determine the deformation functions.

(4) The adjacent finite elements are connected by the shared nodes as pivot points. Shared nodes refer to nodes that are at the same coordinate points but belong to different elements. For example, Fig. 2.4a is a continuous plate. Figure 2.4b shows that the plate is broken into 4 triangle elements. Each element has 3 nodes. Figure 2.4c is the assembly of 4 elements in which the shared nodes are connected. In Fig. 2.4c, AB is the interface edge between element 3 and element 4, but this edge is only connected at the shared nodes on both ends. So, when the assembly of elements is under external loads and is deformed, the deformation of the AB edge of element 3 and element 4 might be different except for the same deformations on the shared nodes on both ends. A logical conclusion is that having more nodes in elements will result in a better representation of the original continuous objects as shown in Fig. 2.5. The interface edge AB in Fig. 2.4c is only connected at two shared nodes on two ends. However, the interface edge AB in Fig. 2.5c is connected at three shared nodes.

(5) FEA simulation results will always be approximate results, but FEA theory can minimize the total error with proper changes in the FEA simulation setting such as using smaller elements or elements with more nodes. Another core technique for FEA theory is that many different scientific principles or mathematical tools are used to make the total error inside each element minimal. Therefore, the total error of the whole

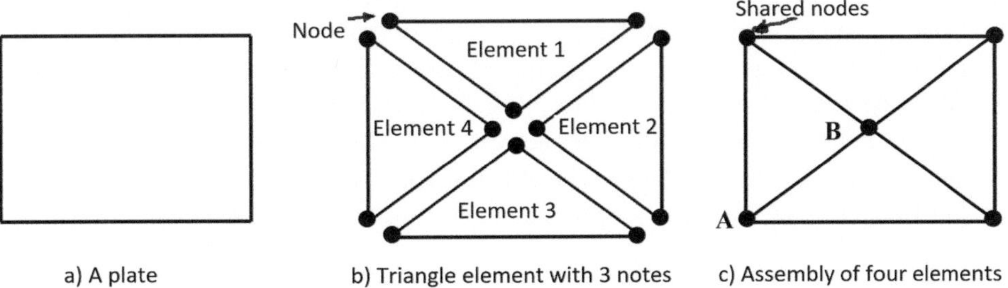

a) A plate   b) Triangle element with 3 notes   c) Assembly of four elements

Fig. 2.4 Schematics of a plate and the assembly of elements with 3 nodes

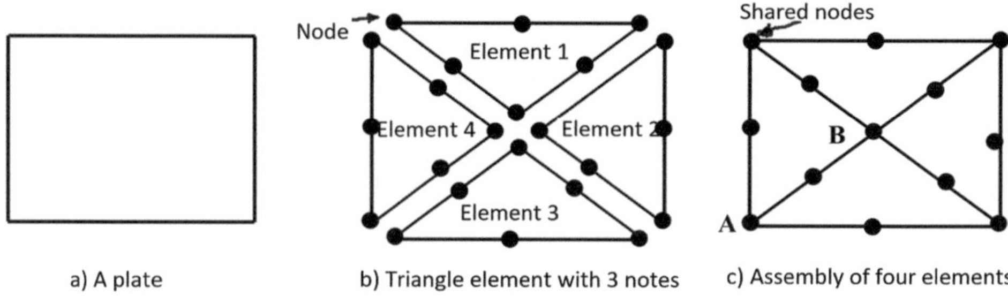

Fig. 2.5 Schematics of a plate and the assembly of elements with 6 nodes

representation of the original object by the assembly of all elements will be minimal. Some commonly used mathematical tools are the Galerkin method, the Rayleigh–Ritz method, the least squares method, and the principle of minimum potential energy. Through the implementation of one of these mathematical tools, element properties can be obtained. These element properties are the relationships between element geometric variables, material properties, loads on each node, and nodal deformations. Element properties are a set of linear algebraic equations of nodal deformations. This will be discussed and demonstrated in detail in Chap. 2.6.

(6) The global linear algebraic equations of all nodal deformation will be obtained by building the equations of equilibrium at each node of the assembly of all elements, which is the model for FEA simulation and is the representation of the original continuous objects. The original stress/strain problems governed by partial differential equations have been converted into a set of linear algebraic equations of all nodal deformation. This conversion allows us to use a powerful computer to solve thousands of linear algebraic equations. Boundary conditions will be applied to the global linear algebraic equations, which represent support and load conditions for stress/strain analysis. The nodal deformation values of nodes on the support area will be known per the support conditions. The resultant forces of nodes on the area where external loads are applied will also be determined by the loading conditions. Resultant forces on the nodes that are not on the external loading area will be zero. Nodal deformation values of nodes that are not on the support area, will be solved through the global linear algebraic equations. This will be discussed and demonstrated in Chap. 2.7.

(7) After the nodal deformation values of each node have been solved, there are now explicit deformation functions for each element. Then we can use strain definitions and Hooke's law to calculate stress and strain at any point per governing Eqs. (2.1) to (2.12). This will be discussed and demonstrated through an example in Chap. 2.7.

2.4 **Deformation Functions and Shape Functions**

Deformation functions of an element are approximate functions of the real deformation and are expressed by nodal deformation values of all nodes in this element. Since elements are typically tiny blocks and deformation functions are continuous functions, we can guess or select some types of functions as deformation functions. A polynomial function is one choice. Let's use several examples to demonstrate how to guess or determine deformation functions.

Example 1

An element has a constant cross-section bar with a length of L_p as shown in Fig. 2.6. This element has two nodes. Nodal deformation values of node 1 and node 2 are u_1 and u_2, respectively. Build a deformation function for this 1D element.

Solution

We could assume a polynomial deformation function $u_p(x)$, which is valid for this element only, that is, for $0 \leq x \leq L_p$. At node 1 and node 2 in this example, we will have the following conditions.

$$u_p(x = 0) = u_1, \quad \text{at the node 1 with } x = 0 \tag{E1-2.1}$$

$$u_p(x = L_p) = u_2, \quad \text{at the node 2 with } x = L_p \tag{E1-2.2}$$

We could guess the deformation function as a constant, a first-order, second-order, or higher-order polynomial function. However, we have two conditions (E1-2.1) and (E1-2.2) on nodes 1 and 2 and can be used to determine two unknowns. Therefore, the appropriate deformation function for this example will be a first-order polynomial function.

$$u_p(x) = c_1 + c_2 x, \quad 0 \leq x \leq L_p \tag{E1-2.3}$$

where c_1 and c_2 are two unknowns, which will be determined by two conditions (E1-2.1) and (E1-2.2).

Per two conditions (E1-2.1) and (E1-2.2), we have:

Fig. 2.6 Schematics of a 1D element with two nodes

$$u_p(x = 0) = u_1 = c_1 + c_2 \times 0 = c_1 \tag{E1-2.4}$$

$$u_p(x = L_p) = u_2 = c_1 + c_2 \times L_p \tag{E1-2.5}$$

From Eqs. (E1-2.4) and (E1-2.5), we have:

$$c_1 = u_1, \quad c_2 = \frac{u_2 - u_1}{L_p} \tag{E1-2.6}$$

Thus, the deformation function for this element will be:

$$u_p(x) = u_1 + \frac{u_2 - u_1}{L_p}x, \quad 0 \leq x \leq L_p \tag{E1-2.7}$$

The deformation value at any point x can be calculated per this deformation function. Since the deformation of this element as described in Eq. (E1-2.7) is a linear function, the strain and stress inside this element will be constant.

For this example, the second-order polynomial function $u_p(x) = c_1 + c_2 x + c_3 x^2$ cannot be selected because two conditions at two nodes cannot determine three unknowns c_1, c_2 and c_3. If we want to have a more accurate description of the real deformation of an element, we could use more nodes and thus could use higher-order polynomial functions. The next example will be a 1D element with three nodes.◄

Example 2

The schematic of a 1D element with 3 nodes is shown in Fig. 2.7. The nodal deformation values for node 1, node 2, and node 3 are u_1, u_2 and u_3, respectively. The coordinates of node 1, node 2, and node 3 are -1, 0, and 1 (mm), respectively. Build the deformation function of this element.

Solution

In this example, we have three nodes with specified nodal deformation values. Use $u_p(x)$ to be the deformation function of this element, we have:

$$u_p(x = -1) = u_1, \quad \text{at the node 1 with } x = -1 \tag{E2-2.1}$$

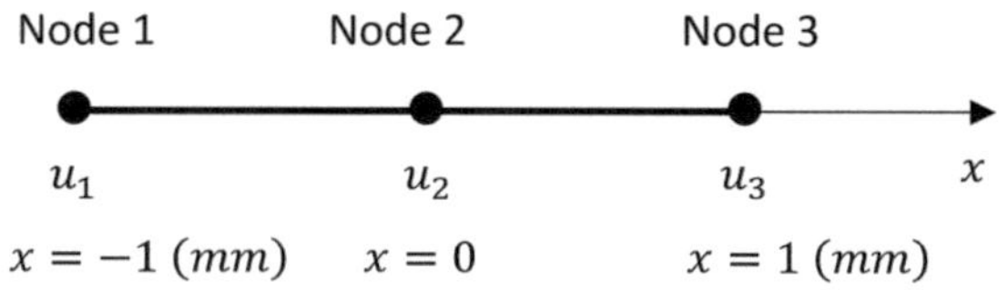

Fig. 2.7 Schematic of a 1D element with 3 nodes

$$u_p(x = 0) = u_2, \quad \text{at the node 2 with } x = 0 \tag{E2-2.2}$$

$$u_p(x = 1) = u_3, \quad \text{at the node 3 with } x = 1 \tag{E2-2.3}$$

Since there are 3 conditions, we can solve 3 unknowns. Therefore, the appropriate deformation function can be a second-order polynomial function.

$$u_p(x) = c_1 + c_2 x + c_3 x^2, \quad -1(\text{mm}) \le x \le 1(\text{mm}) \tag{E2-2.4}$$

where c_1, c_2 and c_3 are three unknowns, which will be determined by three conditions (E2-2.1), (E2-2.2), and (E2-2.3).

Per three conditions, we have:

$$u_1 = c_1 + c_2 \times (-1) + c_3(-1)^2 = c_1 - c_2 + c_3 \tag{E2-2.5}$$

$$u_2 = c_1 + c_2 \times (0) + c_3(0)^2 = c_1 \tag{E2-2.6}$$

$$u_3 = c_1 + c_2 \times (1) + c_3(1)^2 = c_1 + c_2 + c_3 \tag{E2-2.7}$$

Solving Eqs. (E2-2.5), (E2-2.6) and (E2-2.7), we have:

$$c_1 = u_2, \quad c_2 = \frac{u_3 - u_1}{2}, \quad c_3 = \frac{u_1 + u_3 - 2u_2}{2} \tag{E2-2.8}$$

Thus, the deformation function for this element will be:

$$u_p(x) = u_2 + \left(\frac{u_3 - u_1}{2}\right)x + \left(\frac{u_1 + u_3 - 2u_2}{2}\right)x^2, \quad -1(\text{mm}) \le x \le 1(\text{mm}) \tag{E2-2.9}$$

The deformation value at any point x inside this element can be described by a quadratic function. The strain and stress inside this element will be changed as a linear function of the coordinate. So, it will be a more accurate approximation of real deformation compared to the deformation function with two nodes in Example 1.◄

For 2D or 3D elements, the same techniques can be used to build approximate deformation functions. One example of a 2D triangle element with 3 nodes is shown in the following.

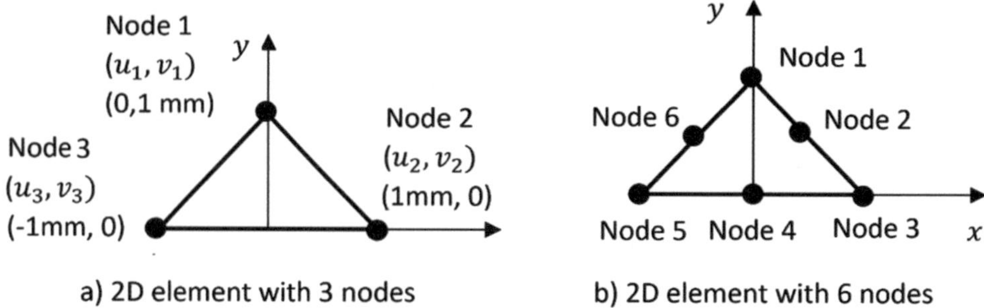

Fig. 2.8   Schematics of a 2D triangle element with 3 nodes or 6 nodes

Example 3

A 2D triangle element with 3 nodes is shown in Fig. 2.8a. (u_i, v_i, $i = 1,2, 3$) are nodal deformation of the node i along the x- and y-axis. Build deformation functions of this element.

Solution

For a 2D element, there are two deformation functions. Let's use $u_p(x, y)$ and $v_p(x, y)$ to represent x- and y-deformation functions along x- and y-axis, respectively.

Let's build the deformation $u_p(x, y)$ first. Three conditions at nodes can solve three unknowns. So, we can assume $u_p(x, y)$ to be a linear function.

$$u_p(x, y) = c_1 + c_2 x + c_3 y \tag{E3-2.1}$$

$$\text{At node 1} \quad u_1 = c_1 + c_2(0) + c_3(1) = c_1 + c_3 \tag{E3-2.2}$$

$$\text{At node 2} \quad u_2 = c_1 + c_2(1) + c_3(0) = c_1 + c_2 \tag{E3-2.3}$$

$$\text{At node 3} \quad u_3 = c_1 + c_2(-1) + c_3(0) = c_1 - c_2 \tag{E3-2.4}$$

Solving Eqs. (E3-2.2), (E3-2.3) and (E3-2.4), we have:

$$c_1 = \frac{u_2 + u_3}{2}, \quad c_2 = \frac{u_2 - u_3}{2}, \quad c_3 = \frac{2u_1 - u_2 - u_3}{2} \tag{E3-2.5}$$

Thus, the x-deformation function of this 2D element with 3 nodes is

$$u_p(x, y) = \frac{u_2 + u_3}{2} + (\frac{u_2 - u_3}{2})x + (\frac{2u_1 - u_2 - u_3}{2})y \tag{E3-2.6}$$

We can repeat the same process for the deformation function along the y-axis. The deformation function $v_p(x, y)$ is

$$v_p(x, y) = \frac{v_2 + v_3}{2} + (\frac{v_2 - v_3}{2})x + (\frac{2v_1 - v_2 - v_3}{2})y \qquad \text{(E3-2.7)}$$

The deformation functions along the x- and y-axis of a 2D triangle element with 3 nodes described in (E3-2.6) and (E3-2.7) are linear functions. Therefore, the strain and stress of this element at any point will be constant. If we want to get more accurate deformation functions, we can use an element with more nodes. For example, a 2D triangle element with 6 nodes is shown in Fig. 2.8b. The deformation functions can be represented by second-order polynomial functions as follows.

$$u_p(x, y) = c_1 + c_2x + c_3y + c_4xy + c_5x^2 + c_6y^2 \qquad \text{(E3-2.8)}$$

$$v_p(x, y) = b_1 + b_2x + b_3y + b_4xy + b_5x^2 + b_6y^2 \qquad \text{(E3-2.9)}$$

where, $c_1\sim c_6$ and $b_1\sim b_6$ are unknowns and will be determined by nodal deformations at 6 nodes.◄

FEA theory provides a better way to express approximate deformation functions through the concept of shape functions.

Shape functions are interpolation functions that are used to interpolate a solution between discrete nodal deformation values of all nodes of an element. The number of shape functions in a deformation function is equal to the number of nodes in an element.

The shape function has a value between 0 to 1. The physical meaning of a shape function N_i is used to calculate the percentage of a contribution of a nodal deformation value of node i to the deformation value at a point where the deformation value is calculated.

The deformation function of a 1D element can be expressed as:

$$u_p(x) = \sum_{i=1}^{n} N_i(x)u_i \qquad \text{(2.15)}$$

where $u_p(x)$ is deformation function. u_i is the x-nodal deformation value of node i. n is the number of nodes of this element. $N_i(x)$ is the shape function for node i and is a polynomial function of the coordinate x.

The deformation function of a 2D element can be expressed as:

$$u_p(x, y) = \sum_{i=1}^{n} N_i(x, y)u_i$$

$$v_p(x, y) = \sum_{i=1}^{n} N_i(x, y)v_i \qquad \text{(2.16)}$$

where $u_p(x, y)$ and $v_p(x, y)$ are deformation functions along the x- and y-axis. u_i and v_i are the x- and y-nodal deformation values of node i. n is the number of nodes of this element. $N_i(x, y)$ is the shape function for node i and is a polynomial function of the coordinates x and y.

The deformation function of a 3D element can be expressed as:

$$u_p(x, y, z) = \sum_{i=1}^{n} N_i(x, y, z)u_i$$
$$v_p(x, y, z) = \sum_{i=1}^{n} N_i(x, y, z)v_i \qquad (2.17)$$
$$w_p(x, y, z) = \sum_{i=1}^{n} N_i(x, y, z)w_i$$

where $u_p(x, y, z)$, $v_p(x, y, z)$ and $w_p(x, y, z)$ are deformation functions along x-, y- and z-axis, respectively. u_i, v_i and w_i are the x-, y-, and z-nodal deformation values of node i. n is the number of nodes of this element. $N_i(x, y, z)$ is the shape function for node i and a polynomial function of the coordinates x, y, and z.

Shape functions have the following two properties.

(1) For the shape function N_i at node j, its value can be only 1 or 0.

$$N_i = \begin{cases} 1 & \text{when } i = j \\ 0 & \text{when } i \neq j \end{cases} \qquad (2.18)$$

(2) The sum of all shape functions of an element must be equal to 1.

$$\sum_{i=1}^{n} N_i = 1 \qquad (2.19)$$

Shape functions of typical elements can be directly found in FEA theory books. Here, we will use the results of Example 1 and Example 3 to demonstrate how to get shape functions and verify these two properties of shape functions.

Example 4

(1) Use the deformation function result of Example 1 to determine the shape functions of the 1D element shown in Fig. 2.6. (2) Verify their two properties. (3) if $u_1 = 0.0005(\text{mm})$, $u_2 = -0.0003(\text{mm})$, $L_p = 0.20(\text{mm})$, calculate the deformation at $x = 0.05(\text{mm})$

Solution

(1) Shape functions

From (E1-2.7), the deformation function of the 1D element with 2 nodes is:

$$u_p(x) = u_1 + \frac{u_2 - u_1}{L_p}x, \quad 0 \le x \le L_p \tag{E4-2.1}$$

According to Eq. (2.15), the deformation function can be expressed by shape functions.

$$u_p(x) = \sum_{i=1}^{2} N_i(x)u_i = N_1(x)u_1 + N_2(x)u_2 \tag{E4-2.2}$$

We can rearrange Eq. (E4-2.1) in the format of Eq. (E4-2.2) and have:

$$u_p(x) = u_1 + \frac{u_2 - u_1}{L_p}x = u_1 + \frac{u_2}{L_p}x - \frac{u_1}{L_p}x = \left(1 - \frac{x}{L_p}\right)u_1 + \left(\frac{x}{L_p}\right)u_2 \tag{E4-2.3}$$

By comparison of Eq. (E4-2.3) with Eq. (E4-2.2), the shape functions $N_1(x)$ and $N_2(x)$ are

$$N_1(x) = \left(1 - \frac{x}{L_p}\right) \quad 0 \le x \le L_p \tag{E4-2.4}$$

$$N_2(x) = \frac{x}{L_p} \quad 0 \le x \le L_p \tag{E4-2.5}$$

(2) Verification of shape function properties

For shape function $N_1(x) = \left(1 - \frac{x}{L_p}\right)$, we have:

$$N_1(x) = \left(1 - \frac{x}{L_p}\right) = \begin{cases} \left(1 - \frac{0}{L_p}\right) = 1 & x = 0 \text{ at node 1} \\ \left(1 - \frac{L_p}{L_p}\right) = 0 & x = L_p \text{ at node 2} \end{cases} \tag{E4-2.6}$$

For shape function $N_2(x) = \frac{x}{L_p}$, we have:

$$N_2(x) = \frac{x}{L_p} = \begin{cases} \frac{0}{L_p} = 0 & x = 0 \text{ at node 1} \\ \frac{L_p}{L_p} = 1 & x = L_p \text{ at node 2} \end{cases} \tag{E4-2.7}$$

The sum of two shape functions is:

$$\sum_{i=1}^{2} N_i(x) = N_1(x) + N_2(x) = \left(1 - \frac{x}{L_p}\right) + \frac{x}{L_p} = 1 \qquad \text{(E4-2.8)}$$

(3) Calculate the deformation at $x = 0.05\,(\text{mm})$

Per Eq. (E4-2.3) and $u_1 = 0.0005\,(\text{mm})$, $u_2 = -0.0003\,(\text{mm})$, $L_p = 0.20\,(\text{mm})$, we have the deformation function.

$$u_p(x) = \left(1 - \frac{x}{0.20}\right)0.005 + \left(\frac{x}{0.20}\right)(-0.003) \qquad \text{(E4-2.9)}$$

At $x = 0.05\,(\text{mm})$, we use Eq. (E4-2.9), we have:

$$u_p(0.05) = \left(1 - \frac{0.05}{0.20}\right)0.005 + \left(\frac{0.05}{0.20}\right)(-0.003)$$

$$= (0.75)0.005 + (0.25)(-0.003) = 0.003\,(\text{mm}) \qquad \text{(E4-2.10)}$$

The Eq. (E4-2.10) can be used to explain the meaning of the shape functions. The deformation value at the point $x = 0.05\,(\text{mm})$ is the sum of contributions from two nodes: 75% of nodal deformation is from $u_1 = 0.0005\,(\text{mm})$. and the remaining 25% of nodal deformation is from $u_2 = -0.0003\,(\text{mm})$.

Now, let's use one 2D triangle element to show how to calculate strain and stress. ◄

Example 5

Nodes, nodal deformations, and their coordinates of a 2D triangle element with 3 notes are shown in Fig. 2.9. (1) Build the deformation functions, (2) Determine the shape functions, (3) The Young's modulus and Poisson ratio of this element material are $E = 2\times10^5\,(Mpa)$ and $\upsilon = 0.29$. If $(u_1 = 0,\ v_1 = 0)$; $(u_2 = 0.001\text{ mm},\ v_2 = -0.001\text{ mm})$; $(u_3 = -0.002\text{ mm},\ v_3 = 0.003\text{ mm})$, calculate normal strains and normal stresses at the point $(x = 0.5\text{ mm},\ y = 0.5\text{ mm})$.

Solution

(1) Deformation functions

    With a total number of 3 nodes, the appropriate deformation functions along the x- and y-axis will be linear.

$$u_p(x, y) = c_1 + c_2 x + c_3 y \qquad \text{(E5-2.1)}$$

Fig. 2.9 A schematic of a 2D right-angle triangle element with 3 nodes

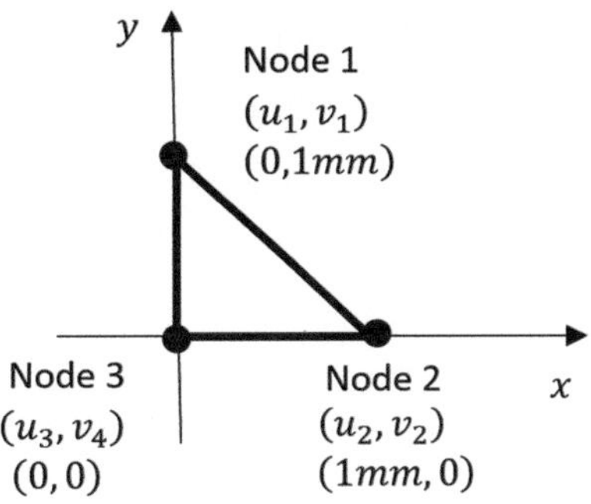

$$v_p(x, y) = b_1 + b_2 x + b_3 y \qquad \text{(E5-2.2)}$$

Three conditions at nodes for the deformation $u_p(x, y)$ are:

$$\text{At node 1} \quad u_1 = c_1 + c_2(0) + c_3(1) = c_1 + c_3 \qquad \text{(E5-2.3)}$$

$$\text{At node 2} \quad u_2 = c_1 + c_2(1) + c_3(0) = c_1 + c_2 \qquad \text{(E5-2.4)}$$

$$\text{At node 3} \quad u_3 = c_1 + c_2(0) + c_3(0) = c_1 \qquad \text{(E5-2.5)}$$

Solving Eqs. (E5-2.3), (E5-2.4) and (E5-2.5), we have:

$$c_1 = u_3, \quad c_2 = u_2 - u_3, \quad c_3 = u_1 - u_3 \qquad \text{(E5-2.6)}$$

Thus, the x-deformation function of this 2D element with 3 nodes is

$$u_p(x, y) = u_3 + (u_2 - u_3)x + (u_1 - u_3)y \qquad \text{(E5-2.7)}$$

We can repeat the same process for the deformation function along the y-axis. The deformation function $v_p(x, y)$ is:

$$v_p(x, y) = v_3 + (v_2 - v_3)x + (v_1 - v_3)y \qquad \text{(E5-2.8)}$$

(2) Shape functions

We can use either one of Eqs. (E5-2.7) or (E5-2.8) to determine shape functions. Let's rearrange Eq. (E5-2.7) as the following:

$$u_p(x, y) = u_3 + u_2 x - u_3 x + u_1 y - u_3 y = (y)u_1 + (x)u_2 + (1 - x - y)u_3 \qquad \text{(E5-2.9)}$$

By comparison of Eq. (E5-2.9) with the definition (2.16), the three shape functions are:

$$N_1(x, y) = y \tag{E5-2.10}$$

$$N_2(x, y) = x \tag{E5-2.11}$$

$$N_3(x, y) = 1 - x - y \tag{E5-2.12}$$

Now, the deformation functions of this element can be expressed as:

$$u_p(x, y) = N_1(x, y)u_1 + N_2(x, y)u_2 + N_3(x, y)u_3 \tag{E5-2.13}$$

$$v_p(x, y) = N_1(x, y)v_1 + N_2(x, y)v_2 + N_3(x, y)v_3 \tag{E5-2.14}$$

(3) Normal strains and normal stresses at the point ($x = 0.5\,\text{mm}$, $y = 0.5\,\text{mm}$)
Per definitions of normal strains as shown in Eqs. (2.1) and (2.2), we have:

$$\varepsilon_x(x, y) = \frac{\partial u_p(x, y)}{\partial x} = (u_2 - u_3) = 0.001 - (-0.002) = 0.003 \tag{E5-2.15}$$

$$\varepsilon_y(x, y) = \frac{\partial v_p(x, y)}{\partial y} = (v_1 - v_3) = 0 - 0.003 = -0.003 \tag{E5-2.16}$$

Since the normal strain's functions are constant, the normal strains at the point ($x = 0.5\,\text{mm}$, $y = 0.5\,\text{mm}$) will be:

$$\varepsilon_x(0.5\,\text{mm}, 0.5\,\text{mm}) = 0.003 \tag{E5-2.17}$$

$$\varepsilon_y(0.5\,\text{mm}, 0.5\,\text{mm}) = -0.003 \tag{E5-2.18}$$

Per definitions of normal stresses as shown in Eqs. (2.7) and (2.8), we have:

$$\sigma_x(x, y) = \frac{E}{(1 + v)(1 - 2v)}\left[(1 - v)\varepsilon_x(x, y) + v\varepsilon_y(x, y)\right] = 82.05\,\text{(Mpa)} \tag{E5-2.19}$$

$$\sigma_y(x, y) = \frac{E}{(1 + v)(1 - 2v)}\left[(1 - v)\varepsilon_y(x, y) + v\varepsilon_x(x, y)\right] = -82.05\,\text{(Mpa)} \tag{E5-2.20}$$

Since the normal stress functions are constant, the normal stresses at the point ($x = 0.5\,\text{mm}$, $y = 0.5\,\text{mm}$) will be:

$$\sigma_x(0.5\,\text{mm},\ 0.5\,\text{mm}) = 82.05(Mpa) \qquad (\text{E5-2.21})$$

$$\sigma_y(0.5\,\text{mm},\ 0.5\,\text{mm}) = -82.05(Mpa) \qquad (\text{E5-2.22})$$

◀

2.5 Types of Typical Elements

In FEA simulation, all components will be meshed into lots of small elements. Every commercial FEA simulation software such as ANSYS, ABAQUS, NASTRAN, LS-DYNA, COMSOL, and SOLIDWORKS Simulation has a rich library of types of elements, which serve as the core of the FEA simulation software. This section will concisely describe typical types of elements and show some examples of them.

All mechanical components are 3D objects. Naturally, they can always be meshed by 3D elements. However, the purpose of FEA simulation is to obtain the stress and strain of mechanical components for mechanical design. For some situations, meshing as simplified 1D elements or 2D elements can still obtain results with an acceptable accuracy while significantly reducing the computation time. Therefore, the selection of a type of element in FEA simulation will be dependent on the types of analysis problems. Generally, elements in FEA simulation can be grouped into 1D elements, 2D planar elements, 2D surface (shell) elements, and 3D elements.

1D elements as shown in Fig. 2.10 are line elements in which physical variables of each element such as stress, strain, and deformation are the function of the coordinate along the line. When one of the dimensions of a mechanical component is very large compared to the other two dimensions such as a pipe or a rod, it can be meshed as 1D elements. A truss member can also be meshed as 1D elements. A beam can also be meshed as 1D elements as well because the stress/strain along the cross-section of the beam will be calculated by beam theory.

2D planar elements are area elements in a two-dimensional plane where physical variables of this element such as stress, strain, and deformation are the functions of 2D coordinates in the plane. The simplest 2D element is a 2D triangle element with 3 nodes

Fig. 2.10 Examples of 1D elements

Fig. 2.11 Examples of 2D planar elements

Fig. 2.12 Example of 2D surface elements

Fig. 2.13 Example of 3D elements

as shown in Fig. 2.11. In a triangular 2D planar element with 6 nodes, the edges of the triangle can be a curved line. Typically, plane stress, plane strain, and an axisymmetric problem can be meshed as 2D planar elements.

2D surface (shell) elements are area elements on a surface that is typically a mid-surface of a component. The simplest 2D surface element is a 2D triangle element with 3 nodes as shown in Fig. 2.12. This element uses a flat surface to represent a curved surface in a meshing process. In a triangular 2D surface element with 6 nodes, the edges can be a curved line, and this 2D element can be a curved surface. Typically, a sheet metal part, a thin vessel, or a membrane can be meshed as 2D surface elements.

3D elements are volume elements in which physical variables of this element such as stress/ strain and deformation are the functions of 3-D coordinates. The simplest 3D element is a tetrahedral element with 4 nodes as shown in Fig. 2.13. The 3D tetrahedral element with 6 nodes can have curved edges. There are a variety of different 3D elements. Any mechanical components can be meshed as 3D elements.

2.6   Element Stiffness Matrix (Element Properties)

Element stiffness matrix, also known as element properties, is used to construct relationships between nodal displacements and nodal forces of the element including element geometric parameters and the element's material properties.

After a component is divided into lots of elements and the deformation functions of each element are specified, we must construct the relationships between nodal displacements and nodal forces of this element to complete the numerical simulation. A lot of approaches such as the Galerkin method, the Rayleigh–Ritz method, the least squares method, and the principle of minimum potential energy can be used to construct an element stiffness matrix. Here, we will use the principle of minimum potential energy to demonstrate how to construct an element stiffness matrix.

The principle of minimum potential energy states that deformations at the equilibrium position occur such that the total potential energy of a stable system is the minimum value.

The schematics shown in Fig. 2.14 can be used to explain this principle. Both (a) and (b) in Fig. 2.14 represent the equilibrium positions of a ball. For Fig. 2.14a, if a tiny disturbance is applied to the ball, the ball will roll down along the surface and will never return to its equilibrium position. So, the original position is not a stable equilibrium position. The potential energy in this case is the maximum potential energy. For Fig. 2.14b, after a tiny disturbance is applied to the ball, the ball will roll back to its equilibrium position. So, the original position is a stable equilibrium position. The potential energy in this case is the minimum potential energy. Based on this principle, a stable deformed element will always have a minimum potential energy.

The total potential energy of a deformed element, that is the element strain energy minus the mechanical works done by nodal forces, is a function of nodal deformations. Let's assume that the total potential energy of a general element with n nodes is $\Pi(u_1, v_1, w_1, \ldots, u_n, v_n, w_n)$. Per the principle of minimum potential energy, we can have the following equations:

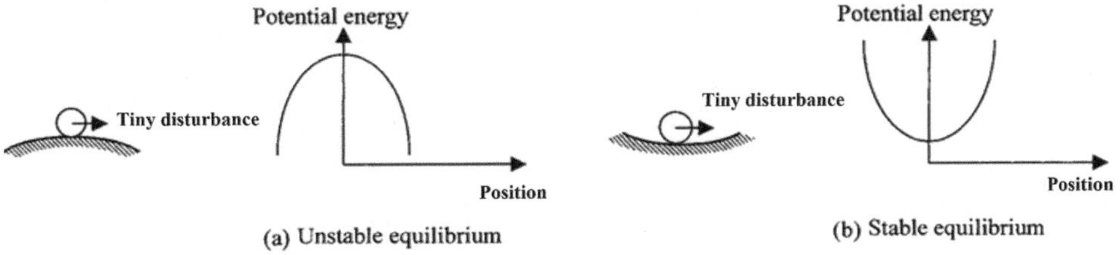

Fig. 2.14   Static equilibrium states

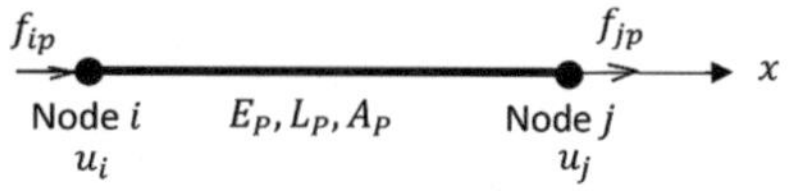

Fig. 2.15 A schematic of a 1D element with two nodes

$$\frac{\partial \Pi(u_1,v_1,w_1,\ldots,u_n,v_n,w_n)}{\partial u_i} = 0, \quad i = 1,\ldots,n$$
$$\frac{\partial \Pi(u_1,v_1,w_1,\ldots,u_n,v_n,w_n)}{\partial v_i} = 0, \quad i = 1,\ldots,n \qquad (2.20)$$
$$\frac{\partial \Pi(u_1,v_1,w_1,\ldots,u_n,v_n,w_n)}{\partial w_i} = 0, \quad i = 1,\ldots,n$$

where u_i, v_i, and w_i are the x-, y- and z-direction deformation of the ith node. n is the number of nodes in this element.

By using Eq. (2.20), the relationship between nodal deformations and nodal forces, also known as the element stiffness matrix, can be constructed.

Now let's use a 1D element with 2 nodes as an example to demonstrate how to construct an element stiffness matrix.

Example 6

A 1D element with 2 nodes is shown in Fig. 2.15. Use the principle of minimum potential energy to construct the element stiffness matrix (element properties). In Fig. 2.15, the subscripts i and j represent the nodes i and j. The subscript P represents the element P. u_i and u_j are nodadeformations of the nodes i and j. In f_{iP}, f represents an internal nodal force; the first subscript represents the node number, and the second subscript represents the element number. f_{iP} is the nodal force at the node i of the element P. f_{jP} is the nodal force at the node j of the element P. E_P, L_P and A_P are the element's material Young's modulus, the element length, and the element's cross-section area, respectively.

Solution

Use the solution in Example 1, the deformation function $u_p(x)$ of this element is,

$$u_p(x) = u_i + \frac{u_j - u_i}{L_p}x, \quad 0 \le x \le L_p \qquad (E6\text{-}2.1)$$

Per definition, the normal strain function of this element is:

$$\varepsilon_p(x) = \frac{\partial u_p(x)}{\partial x} = \frac{u_j - u_i}{L_p} \quad 0 \le x \le L_p \qquad (E6\text{-}2.2)$$

Using Hooke's law, the normal stress function of this element is:

$$\sigma_p(x) = E_P \times \varepsilon_p(x) = E_P \frac{u_j - u_i}{L_p} \quad 0 \le x \le L_p \qquad (E6\text{-}2.3)$$

The potential energy of this element is:

$$\Pi\left(u_i, u_j\right) = \int_{V_P} \frac{1}{2} \varepsilon_p(x)\sigma_p(x)dv - \left(f_{iP}u_i + f_{jP}u_j\right)$$

$$= \int_0^{L_P} \frac{E_P A_P}{2} \left(\frac{u_j - u_i}{L_p}\right)^2 dx - \left(f_{iP}u_i + f_{jP}u_j\right)$$

$$= \frac{E_P A_P}{2L_P}\left(u_j - u_i\right)^2 - \left(f_{iP}u_i + f_{jP}u_j\right) \qquad \text{(E6-2.4)}$$

Implementing the principle of minimum potential energy, we have:

$$\frac{\partial \Pi\left(u_i, u_j\right)}{\partial u_i} = 0, \quad \frac{E_P A_P}{L_P}\left(u_i - u_j\right) = f_{iP} \qquad \text{(E6-2.5)}$$

$$\frac{\partial \Pi\left(u_i, u_j\right)}{\partial u_j} = 0, \quad \frac{E_P A_P}{L_P}\left(u_j - u_i\right) = f_{jP} \qquad \text{(E6-2.6)}$$

Equations (E6-2.5) and (E6-2.6) are the relationships between nodal deformations and nodal forces. These two equations can be called the element properties. We can also use a matrix form to express Eqs. (E6-2.5) and (E6-2.6) and get the element stiffness matrix $[k]$ as shown in Eq. (E6-2.7).

$$\begin{bmatrix} \frac{E_P A_P}{L_P} & -\frac{E_P A_P}{L_P} \\ -\frac{E_P A_P}{L_P} & \frac{E_P A_P}{L_P} \end{bmatrix} \begin{Bmatrix} u_i \\ u_j \end{Bmatrix} = [k]\begin{Bmatrix} u_i \\ u_j \end{Bmatrix} = \begin{Bmatrix} f_{iP} \\ f_{jP} \end{Bmatrix} \qquad \text{(E6-2.7)}$$

The element stiffness matrix contains a variable E_P (a material property) and variables A_P (cross-section area) and L_P (length) of element geometry. It links the nodal deformation with the nodal forces of this element.◄

For 2D or 3D elements, we can follow the same process to get their element stiffness matrix. These can be found in most FEA books and as a result, they will not be discussed in this book.

2.7   An Example of a Uniaxial Stepped Bar Under Axial Loading

This section will manually solve an example of a stepped bar under axial load by using FEA theory, which will allow us to gain a better understanding of the fundamental concepts of FEA theory.

Finite element analysis or finite element method is an approach for solving a component's stress/strain under external loading. Even though there are many different procedures for this process, the general procedure for finite element analysis can be described as the following.

Step 1: Components (models). We need to specify the geometrical shapes, dimensions, and type of material for a component or an assembly.

Step 2: Discretize the object (meshing). We select a proper type of element for the analysis project, then mesh the component or assembly into many small elements. The shared nodes will be pinned together by imaginary infinite-strength pivot pins. Now, the original continuous components or assembly is represented by an assembly of small elements through shared nodes.

Step 3: Construct element deformation functions by using shape functions. The element deformation functions are approximate functions of the element's true deformation and are described and determined by nodal deformations.

Step 4: Construct an element stiffness matrix (element properties). Through the element stiffness matrix, we can establish a set of linear algebraic equations between nodal forces and nodal deformations. This set of equations also incorporates the element material properties and element geometric dimensions into the calculations.

Step 5: Build the global stiffness matrix or build the global set of linear algebraic equations that establish the relationships between all nodal deformation and external forces at each node of the object. At the shared node, the sum of all nodal forces at the shared nodes will be equal to the total external forces at this shared node. Now, the original stress/strain problem of the project becomes a set of global linear algebraic equations between nodal deformation and external forces at each node. There are no difficulties in solving this set of linear algebraic equations except for the large number of linear algebraic equations that need to be solved.

Step 6: Specify boundary conditions. Boundary conditions are the supporting conditions and loading conditions. The supporting conditions will specify given values such as the zero of nodal deformation of the nodes on the supporting area. The reaction forces on these nodes can be calculated when the global set of linear algebraic equations of the project are solved. The loading conditions will specify the external forces applied to the nodes. The nodal deformation of the nodes on which external forces are applied will be solved through the global set of linear algebraic equations. The majority of nodes will not belong to the supporting area or loading area. For these nodes, external forces on the nodes will be zero and the nodal deformation will be solved through the global set of linear algebraic equations.

Step 7: Apply the boundary conditions. After the boundary conditions are applied in the global set of linear algebraic equations, we can solve them to get the nodal deformations of every node of the object. Since the nodal deformations of every node are known, we have fully determined a set of explicit deformation functions for every element.

Step 8: Compute additional results. After the deformation functions of every element are known, we can use the definitions of strain to calculate the strain and then use Hooke's law to calculate the stresses.

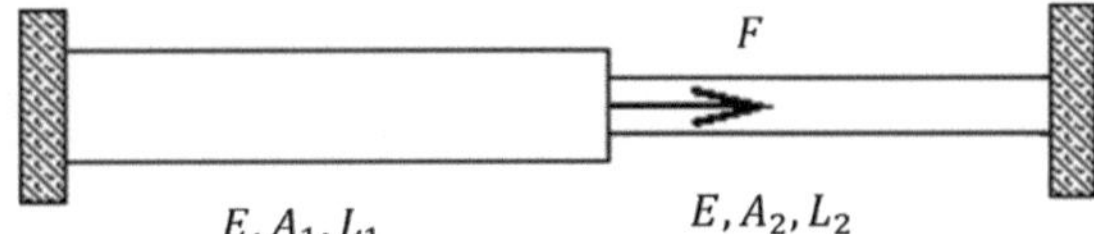

Fig. 2.16 Schematic of a stepped bar with fixed ends

It is not recommended to complete FEA calculations by hand calculation because of the number of algebraic equations. It is typically completed by FEA software. Here, we will use the concepts and procedures of FEA to complete a hand calculation for a simple problem. The purpose of this is to gain a better understanding of the fundamental concepts of finite element analysis.

Example 7 [10]: A stepped bar is fixed on both ends and subjected to an external load $F = 1000(N)$ at the middle of the stepped cross-section as shown in Fig. 2.16. E is the bar's material Young's modulus. A_1 and L_1 are the cross-section area and the length of the first bar. A_2 and L_2 are the cross-section area and the length of the second smaller bar. The information for this bar is: $L_1 = L_2 = 0.1(m)$; $E = 2 \times 10^{11}(Pa)$. $A_1 = 1 \times 10^{-4}(m^2)$; and $A_2 = 5 \times 10^{-5}(m^2)$. Use the finite element analysis to manually calculate deformation, strain, and stress at the middle of each step bar and reaction force on the left and right ends.

Solution

Step 1: The component (model)

The component is displayed in Fig. 2.16. Its related information is described in the problem.

Step 2: Discretize the object (meshing)

This is a straight stepped bar under axial load. For a hand calculation, we can use a 1D element with two nodes as the type of element. And we can discretize the bar into two elements as shown in Fig. 2.17. So, there are a total of 3 nodes. Here, u_1, u_2 and u_3 are the nodal deformations of nodes 1, 2, and 3. F_1, F_2 and F_3 are external nodal forces at nodes 1, 2, and 3. f_{ij} is the internal nodal force at the node i in the element j. For example, f_{21} is the internal nodal force at node 2 in element 1 and f_{22} is the internal nodal force at node 2 in element 2.

Step 3: Construct element deformation functions

Per Eq. (E1-2.7) in Example 1, the deformation function $u_{p1}(x)$ for element 1 is:

$$u_{p1}(x) = u_1 + \frac{u_2 - u_1}{L_1}x, \quad 0 \leq x \leq L_1 \tag{E7-2.1}$$

The deformation function $u_{p2}(x)$ for element 2 is:

$$u_{p2}(x) = u_2 + \frac{u_3 - u_2}{L_2}x, \quad 0 \leq x \leq L_2 \tag{E7-2.2}$$

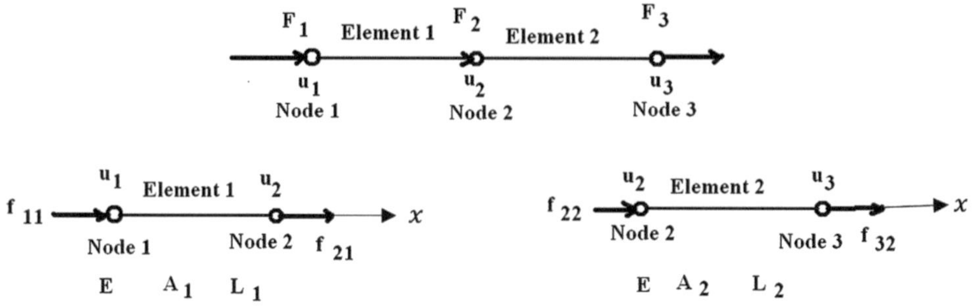

Fig. 2.17 The meshing

Step 4: Construct an element stiffness matrix (element properties).

Per Eqs. (E6-2.5) and (E6-2.6) in Example 6, we have two linear algebraic equations for element 1.

$$\frac{EA_1}{L_1}(u_1 - u_2) = f_{11}, \quad or, \quad 2 \times 10^8(u_1 - u_2) = f_{11} \tag{E7-2.3}$$

$$\frac{EA_1}{L_1}(u_2 - u_1) = f_{21}, \quad or, \quad 2 \times 10^8(u_2 - u_1) = f_{21} \tag{E7-2.4}$$

We can also have two linear algebraic equations for element 2.

$$\frac{EA_2}{L_2}(u_2 - u_3) = f_{22}, \quad or, \quad 1 \times 10^8(u_2 - u_3) = f_{22} \tag{E7-2.5}$$

$$\frac{EA_2}{L_2}(u_3 - u_2) = f_{32}, \quad or, \quad 1 \times 10^8(u_3 - u_2) = f_{32} \tag{E7-2.6}$$

We could also use an element stiffness matrix to express the equations between nodal deformation and nodal forces for each element. However, the above expressions will be much more convenient for this simple example.

Step 5: Build the global set of linear algebraic equations

We can draw the free-body diagram for each node. The free-body diagrams for this example are shown in Fig. 2.18.

Node 1 is not shared with any other element and belongs to element 1 only. Per the free body diagram in Fig. 2.18 for node 1, we have:

$$F_1 = f_{11} \tag{E7-2.7}$$

Node 2 is a shared node that connects Element 1 with Element 2. Per the free-body diagram in Fig. 2.18 for node 2, we have:

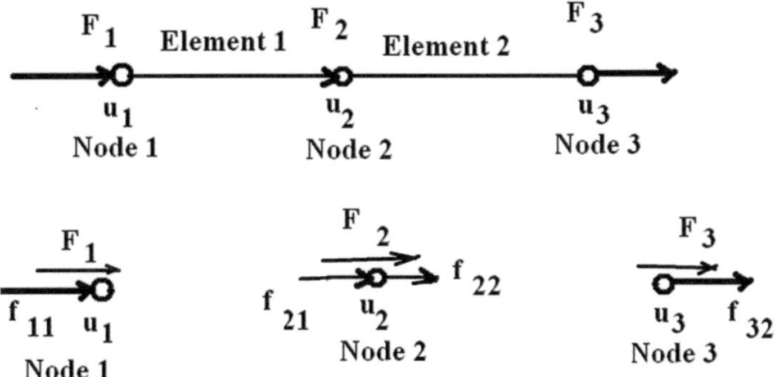

Fig. 2.18 Free-body diagrams of 3 nodes

$$F_2 = f_{21} + f_{22} \qquad \text{(E7-2.8)}$$

Node 3 is not shared with other elements and belongs to element 2 only. Per the free-body diagram in Fig. 2.18 for node 3, we have:

$$F_3 = f_{32} \qquad \text{(E7-2.9)}$$

Per Eqs. (E7-2.7) to (E7-2.9) with Eqs. (E7-2.3) to (E7-2.6), we can get the global set of linear algebraic equations.

$$2 \times 10^8 u_1 - 2 \times 10^8 u_2 = F_1 \qquad \text{(E7-2.10)}$$

$$-2 \times 10^8 u_1 + 3 \times 10^8 u_2 - 1 \times 10^8 u_3 = F_2 \qquad \text{(E7-2.11)}$$

$$-1 \times 10^8 u_2 + 1 \times 10^8 u_3 = F_3 \qquad \text{(E7-2.12)}$$

Step 6: Specify boundary conditions

For the supporting conditions, we have:

$$u_1 = 0, \quad u_3 = 0 \qquad \text{(E7-2.13)}$$

F the loading conditions, we have:

$$F_2 = 1000(N) \qquad \text{(E7-2.14)}$$

Step 7: Apply the boundary conditions and solve the global set of equations

By applying the boundary conditions to the global set of Eqs. (E7-2.10) to (E7-2.12), we have

$$-2 \times 10^8 \times u_2 = F_1 \tag{E7-2.15}$$

$$3 \times 10^8 \times u_2 = 1000 \tag{E7-2.16}$$

$$-1 \times 10^8 \times u_2 = F_3 \tag{E7-2.17}$$

Solving this global set of equations, we have:

$$u_2 = 3.333 \times 10^{-6}(m), \ F_1 = -667(N), \ F_3 = -333(N) \tag{E7-2.18}$$

where $F_1 = -667(N)$ is the reaction force at the left end. $F_3 = -333(N)$ is the reaction force at the right end.

Now, the nodal deformations of all 3 nodes are known. They are:

$$u_1 = 0, \quad u_2 = 3.333 \times 10^{-6}(m), \quad u_3 = 0 \tag{E7-2.19}$$

Step 8: Compute additional results

For FEA, additional calculations will be solved based on each element.

Per Eqs. (E7-2.21) and (E7-2.19), the deformation function $u_{p1}(x)$ of element 1 is:

$$= u_1 + \frac{u_2 - u_1}{L_1}x = 3.333 \times 10^{-5}x \ \ 0 \le x \le 0.1(m) \tag{E7-2.20}$$

Per the definition of normal strain, the strain function $\varepsilon_{p1}(x)$ of element 1 is:

$$\varepsilon_{p1}(x) = \frac{du_{p1}(x)}{dx} = 3.333 \times 10^{-5} \ \ 0 \le x \le 0.1(m) \tag{E7-2.21}$$

By using Hooke's law, the normal stress function $\sigma_{p1}(x)$ of element 1 is:

$$\sigma_{p1}(x) = E\varepsilon_{p1}(x) = 6.666 \times 10^6(Pa) \ \ 0 \le x \le 0.1(m) \tag{E7-2.22}$$

The deformation at the middle of the first bar (element 1) at $x = 0.05(m)$ per Eq. (E7-2.20) is

$$u_{p1}(x = 0.05) = 1.667 \times 10^{-6}(m)$$

From Eqs. (E7-2.21) and (E7-2.22), the strain and stress are constant inside this element. So, strain and stress in the middle of the first bar are:

$$\varepsilon_{p1} = 3.333 \times 10^{-5}, \quad \sigma_{p1} = 6.666 \times 10^{6} (Pa)$$

Per Eqs. (E7-2.2) and (E7-2.19), the deformation function $u_{p2}(x)$ for element 2 is:

$$u_{p2}(x) = 3.333 \times 10^{-6} - 3.333 \times 10^{-5}x \quad 0 \leq x \leq 0.1 \ (m) \tag{E7-2.23}$$

We can follow the above procedure to obtain the deformation, strain, and stress at the middle of the second bar. They are:

$$u_{p2}(x = 0.05) = 3.166 \times 10^{-6}(m), \quad \varepsilon_{p2} = -3.333 \times 10^{-5}, \quad \sigma_{p2} = -6.666 \times 10^{6}(Pa)$$

When FEA is used to calculate the deformation, strain, and stress of real-world parts, it is almost impossible to use this method with hand calculations for acceptable results for anything but the simplest problems. It must be done using FEA software on a computer. Nowadays, there are lots of commercial FEA software available, which serve as wonderful tools for mechanical engineers to conduct mechanical design. In the rest of this book, we will discuss and demonstrate how to use these tools to conduct FEA simulations on components or assemblies for mechanical design.

Exercises

2-1 List and explain the history of FEA.

2-2 List and explain the fundamental concepts of FEA.

2-3 Explain the difference between the original components and an assembly of tiny elements that are connected at the shared nodes.

2-4 Use one example to explain a shape function.

2-5 What are the two properties of shape functions? Use an example to explain it.

2-6 A 2D planar triangle element with 3 nodes is shown in Fig. 2.19. The coordinates of nodes 1, 2, and 3 are (0, 1) mm, (1, 0) mm and (−1, 0) mm, respectively. The deformations along the x and y axis of Nodes 1, 2, and 3 are (u_1, v_1), (u_2, v_2) and (u_3, v_3), respectively.

   (1) Use the nodal deformations to build this element's deformation functions $u(x, y)$ and $v(x, y)$.

   (2) Express the deformation functions $u(x, y)$ and $v(x, y)$ by using shape functions.

   (3) Verify the two properties of the shape functions obtained in the above step.

   (4) If $(u_1 = 0, v_1 = 0.01mm)$, $(u_2 = 0, v_2 = 0)$ and $(u_3 = 0.02 \, mm, v_3 = 0)$, calculate the deformation, the normal strain ε_x and ε_y at the point (0.25, 0.25).

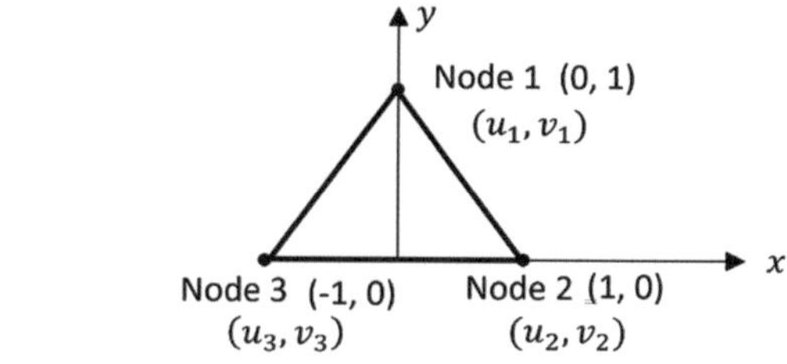

Fig. 2.19 A 2D planar triangle element with 3 nodes

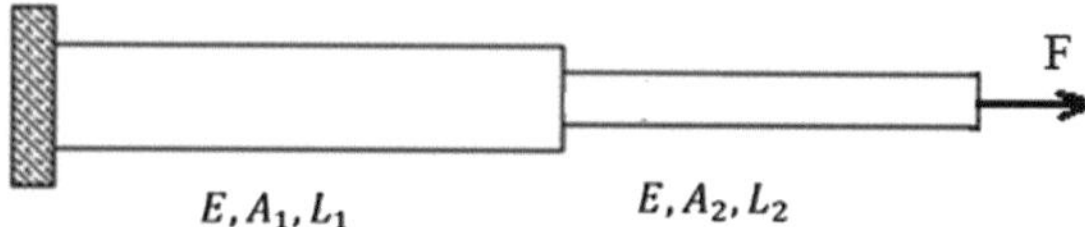

Fig. 2.20 Schematic of a stepped bar with one fixed end

2-7 List and briefly describe 5 commercial FEA software and their available types of elements.

2-8 What is a 1D element? Provide one example of using 1D elements.

2-9 What is a 2D planar element? Provide one example of using 2D planar elements.

2-10 What is a 2D shell element? Provide one example of using 2D shell elements.

2-11 What is a 3D element? Explain one example of using 3D elements.

2-12 A stepped bar is fixed on the left end and subjected to an external load $F = 1000(N)$ at the right end as shown in Fig. 2.20. In the Figure, E is the bar's material Young's modulus. A_1 and L_1 are the cross-section area and the length of the first bar. A_2 and L_2 are the cross-section area and the length of the second smaller bar. The information for this steppe bar is: $L_1 = L_2 = 0.1(m)$; $E = 2 \times 10^{11}(Pa)$. $A_1 = 1 \times 10^{-4}(m^2)$; and $A_2 = 5 \times 10^{-5}(m^2)$. Use the FEA method to calculate deformation, strain, and stress at the middle of each step bar and reaction force on the left end.

References

1. Hrennikoff, Alexander (1941) Solution of problems of elasticity by the framework method, Journal of Applied Mechanics. **8** (4): 169–175.
2. Courant, R. (1943) Variational methods for the solution of problems of equilibrium and vibrations, Bulletin of the American Mathematical Society. **49**: 1–23.
3. Turner, M.J., Clough, R.W., Martin H.C., and Topp, L.J. (1956) Stiffness and deflection analysis f complex structures, Journal of the Aeronautical sciences, Vol. 23 No. 9, 1956 pp. 805–823.
4. O.C. Zienkiewicz and Y.K. Cheung (1967) Finite element method in structural & continuum mechanics, McGraw Hill, 1967
5. John O. Dow (2015) A concise overview of the Finite Element Method, Momentum Press, LLC, New York
6. Jacob Fish and Ted Belytschko (2007) A first course in finite elements, John Wiley & Sons, Ltd, West Sussex PO19 8SQ, England

7. M. Asghar Bhatti (2005) Fundamental finite element analysis and application with MATHE-MATICA and MATLAB computations, John Wiley & Sons, INC, Hoboken, New Jersey.
8. Richard H. Macneal (1994) Finite Elements: their design and performance, Marcel Dekker, Inc, New York.
9. M. Moatamedi and H. Khawaja (2018) Finite element analysis, CRC Press, New York
10. Xiaobin Le, Richard Roberts, and Anthony Duva, A. W. (2019) Teaching Finite Element Analysis for Mechanical Undergraduate Students Paper presented at 2019 ASEE Annual Conference & Exposition, Tampa, Florida. https://doi.org/10.18260/1-2--33348

FEA Simulation on Components

3

Abstract

This chapter discusses and demonstrates how to use the "Static" module in an FEA simulation tool: SolidWorks Simulation, to simulate and display the stress/strain and factors of safety of a component for mechanical design. First, there is a brief description of SolidWorks Simulation and its four types of elements: truss elements, beam elements, shell elements, and tetrahedral solid elements. Then, the general procedure for FEA simulation is explained in detail, which includes: Step 1—Pre-processing; Step 2—Set up a project; Step 3—Select the type of element and assign a material to each component; Step 4—Connections; Step 5—Fixtures; Step 6—Loadings; Step 7 – Meshing; Step 8—Run Simulation; Step 9—Post-processing; and Step 10—Interpretation and verification of simulation results. Some key steps such as pre-processing, selecting types of elements, fixtures, loads, meshing, and post-processing are discussed in detail and demonstrated in several examples with step-by-step instructions. Then the concept of convergence conditions is introduced and discussed in detail. Convergence conditions can be used to check whether the FEA simulation results are acceptable. Several examples are presented to demonstrate convergence conditions. Finally, a minor design project is presented, which is a real-industrial design project. We need to implement all the skills learned in this chapter to complete this real design project.

Supplementary Information The online version contains supplementary material available at https://doi.org/10.1007/978-3-031-64132-9_3.

3.1   Introduction

There are lots of different types of design calculations for mechanical component design such as design calculation due to static loads, design calculation due to dynamic loads, design calculation due to thermal loads, design calculation for fatigue, and so on. First of all, we must start with design calculations due to static loads. We use this calculation to determine geometric shapes and dimensions. Then we will continue to run other design calculations based on the actual situations. This chapter will explain and demonstrate how to use the "Static" module in SolidWorks Simulation to conduct FEA simulation, to determine and display stress/strain and factors of safety of a component for mechanical design. We will start with a general procedure and explain and demonstrate each step by example. Two important notes about SolidWorks Simulation are (1) For the same task, there can be several different ways to complete it in SolidWorks Simulation, and (2) https://help.solidworks.com [1] is a website as reference tools with rich information about FEA simulation. After the completion of this chapter, we will be able to conduct an FEA simulation on a component under static loading for mechanical design.

The brief descriptions of each section in this chapter are as follows.

Chapter 3.2 Introduction to SolidWorks Simulation will be a brief discussion about SolidWorks Simulation, its types of elements, and its modules.

Chapter 3.3 General procedure for FEA simulation and post-processing will explain each step of a general procedure. Then we will use an example to demonstrate the general procedure and the ninth step: post-processing.

Chapter 3.4 Errors in FEA simulation and convergence conditions will explain and demonstrate how to determine the convergence of FEA simulation results.

Chapter 3.5 Pre-processing and creating a project will explain how to prepare component models and create a simulation project, which are the first and second steps of the general procedure.

Chapter 3.6 Type of Elements and Assigning Materials will explain and demonstrate how to select a type of element and assign a material to the component, which is the third step of the general procedure.

Chapter 3.7 Fixtures will explain and demonstrate how to define typical types of supports for an FEA simulation project of a component, which is the fifth step of the general procedure. The fourth step of the general procedure, connections, will be discussed in Chap. 4: FEA simulation on assemblies.

Chapter 3.8 Loading will explain and demonstrate how to apply typical types of loads on a FEA simulation project of a component, which is the sixth step of the general procedure.

Chapter 3.9 Meshing will explain and demonstrate how to determine global and local element sizes for meshing, which is the seventh step of the general procedure.

Chapter 3.10 Minor Design project will introduce some minor design projects which are practical design projects on components. Conducting such projects will help us to integrate all skills for FEA simulation into a real-world design project.

3.2 SolidWorks Simulation

A FEA simulation software is an executable computer program, which typically consists of three sections as shown in Fig. 3.1.

The first box in Fig. 3.1 is a user interface where we can specify a project, boundary conditions such as loading conditions & supporting conditions, and input some information about the type of elements and meshing of the project. The rest of this book will mainly be focused on how to properly specify the input information to obtain acceptable simulation results for a simulation project.

The middle box in Fig. 3.1 is the program codes, which are compiled using FEA theory and will carry out all necessary calculations such as dividing the components into small elements, building element stiffness matrices, constructing global stiffness matrices, constructing & applying boundary conditions, and solving global algebraic equations. This section is typically called a black box because most of us typically have no clues about it. However, some understanding of it will help us to use the FEA software. Chapter 2 of this book provides concise discussions of some fundamental concepts of FEA theory that can provide a basic understanding of this black box.

The last box in the schematic is the output, also called post-processing, which provides various tools to display the simulation results. After an FEA simulation on a project is completed, all nodal deformations are known, so all other variables such as stress, strain, deformation, and factor of safety at any point can be calculated and displayed. This book will only discuss several frequently and typically used tools for displaying simulation results.

SolidWorks is developed by the Dassault Systems SOLIDWORKS Corporation and is a parametric modeling 3D design, analysis, and product data management software. It is one of the most popular design and engineering software and can be used to do all sorts of tasks for engineering design, such as creating part and assembly models, generating mechanical drawings, simulating the stress/strain of components and assemblies, and so on. SolidWorks Simulation 2023 version, a module of the SolidWorks platform, will be

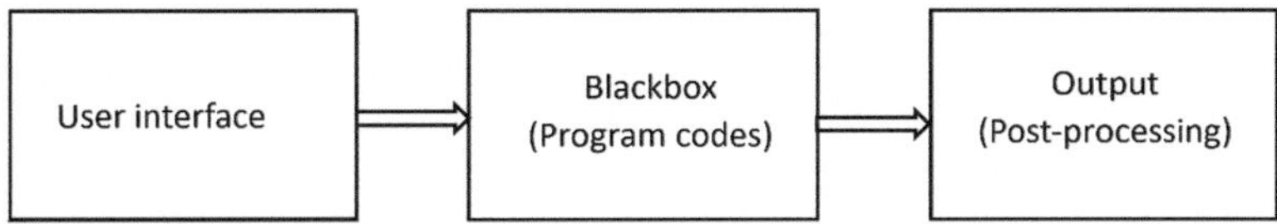

Fig. 3.1 Schematic of a FEA simulation software

Table 3.1 Symbols of four types of elements in SolidWorks Simulation

Type of element	Truss element (1D)	Beam element (1D)	Shell element (2D)	Tetrahedral element (3D)
Symbol				

used as an FEA simulation tool in this book to demonstrate simulation-based mechanical design.

SolidWorks Simulation only uses four types of elements. These are truss elements and beam elements, which are 1D elements. Shell elements represent 2D elements, and tetrahedral elements (3D), which will be discussed in later sections. The corresponding symbols for each type of element in SolidWorks Simulation are displayed in Table 3.1.

To run FEA simulation analysis in the SolidWorks platform, we must enable this module through the tool "add-ins". Open SolidWorks and follow the steps shown in Fig. 3.2 to add SolidWorks Simulation: (1) Click the "Tools" tab; (2) Select "Add-ins" in the drop-down list; (3) Check the checkboxes under "Active Add-ins" and "Start Up" for "SOLIDWORKS Simulation"; and (4) Click "Ok". After this, the "Simulation" tab will appear in the toolbar as shown in Fig. 3.3 after you open any part.

There are 12 different programs or modules for SolidWorks Simulation as shown in Fig. 3.4: (1) Click the "Simulation" tab; (2) Click "New Study"; (3) The "Study" Property Manager will appear.

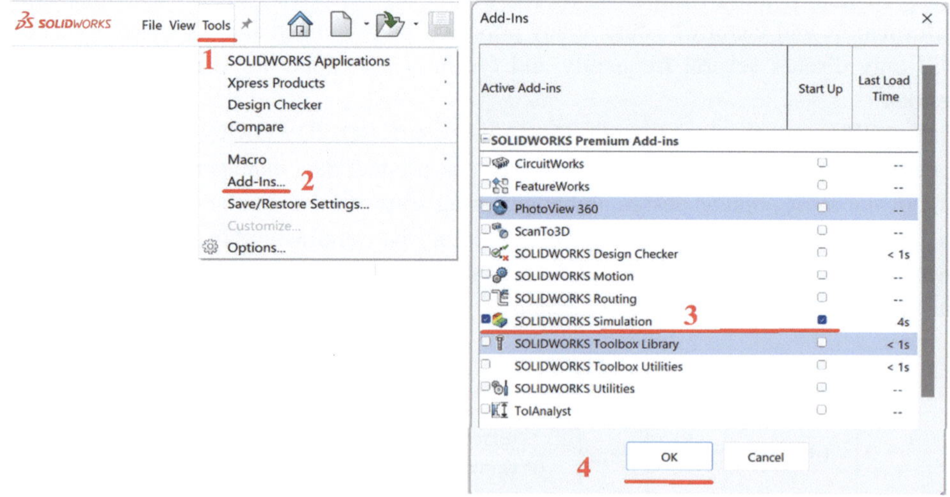

Fig. 3.2 Schematics for adding in the SOLIDWORKS Simulation

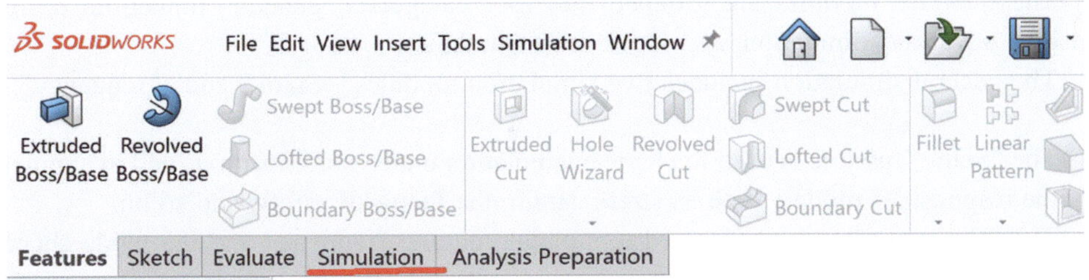

Fig. 3.3 Schematic of the toolbars

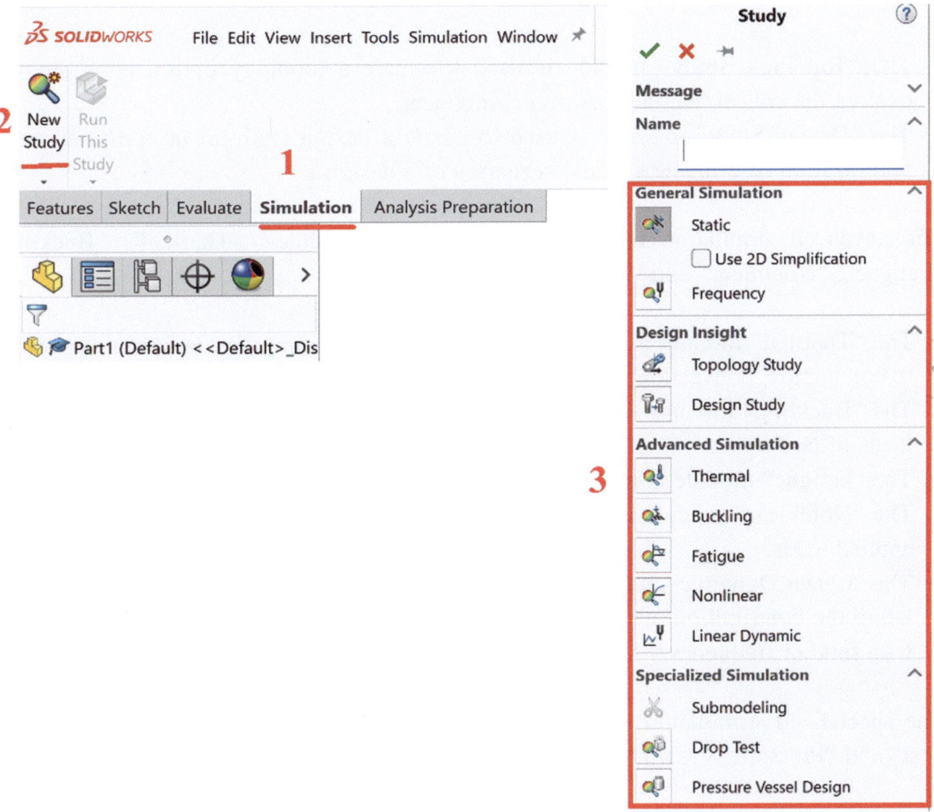

Fig. 3.4 Different simulation modules

These twelve modules are grouped into four categories: general simulation, design insight, advanced simulation, and specialized simulation.

The general simulation includes two simulation modules: "Static" and "Frequency".

- The "Static" module is used to create a simulation under static loadings and to simulate the response of models such as stress, strain, the factor of safety, and so on.
- The "Frequency" module is used to simulate natural frequency and the mode shapes of models.

The design insight includes two simulation modules: "Topology study" and "Design study".

- The "Topology Study" module is used to create a topology optimization study to explore the conceptual designs of a component.
- The "Design Study" module is used to create a design study to perform parametric optimization or evaluate specific scenarios of a design.

The advanced simulation includes five simulation modules: "Thermal", "Buckling", "Fatigue", "Nonlinear" and "Linear Dynamic".

- The "Thermal" module is used to simulate the temperature distribution in bodies due to conduction, convection, and radiation.
- The "Buckling" module is used to simulate the buckling modes and critical buckling loads of bodies.
- The "Fatigue" module is used to simulate the fatigue damage due to cyclic loading.
- The "Nonlinear" module is used to simulate the nonlinear response of models due to applied loads.
- The "Linear Dynamic" module is used to simulate the response of models by accumulating the contribution of each mode to the loading environment. These loads change with time or frequency.

The specialized stimulation includes three simulation modules: "Submodeling", "Drop Test" and "Pressure Vessel Design".

- The "Submodeling" module is used to create a submodeling study to improve the results in local regions of large models without having to rerun the analysis for the whole model.
- The "Drop Test" module is used to evaluate the impact of a part or an assembly with a rigid or flexible planar surface.
- The "Pressure Vessel Design" module is used to create a pressure vessel design study to combine the results of static studies with the specified factors.

This book will only discuss and demonstrate the "Static", "Thermal", "Fatigue" and "Frequency" modules, as these are important modules for mechanical design.

In this chapter, we will only discuss and demonstrate how to use the "Static" module, which is generally used to determine stress/strain and factors of safety of a component under static loadings.

3.3 A General Procedure for FEA Simulation and Post-Processing

This section will explore two topics: (1) a general procedure for an FEA simulation and (2) post-processing. Post-processing is the ninth step in the general procedure. But we need to use it in our first example, so we will discuss and explain it in this section.

The general procedure for a static FEA simulation typically consists of 10 steps. The concise descriptions of these steps are listed as follows.

Step 1: **Pre-processing**. FEA simulation is a virtual experiment on virtual models, such as part models or assembly models. Models in FEA simulation must be divided into lots of small elements. So the two main tasks in the Pre-processing step are (1) creating a virtual component or assembly model of a real component or assembly, and (2) preparing or modifying these components for FEA numerical simulations so that they are meshable. A more detailed discussion of step 1 will be discussed in Chap. 3.5: Pre-Processing and Creating a Project.

Step 2: **Set up a project**. The two main tasks in Step 2 are: (1) defining a project name and (2) determining the type of analysis per the objective of a design project. After the project name is specified, all related information for the simulation results will be linked to this project name and can be retrieved when the project is reopened. Different analyses can be carried out on the same component or assembly as shown in Fig. 3.4. For this book, we will only discuss analyses using the "Static", "Thermal", "Fatigue" and "Frequency" modules for mechanical design. In this chapter, the "Static" module is used to run FEA simulations on components under static loading, the main task for mechanical component design. A detailed discussion of this step can be found in this section.

Step 3: **Select the type of element and assign a material to each component**. The two main tasks in Step 3 are: (1) to select one type of element for a component for meshing, and (2) to assign a material to a component. A more detailed discussion of this step will be presented in Chap. 3.6: Type of Elements and Assigning Material.

Step 4: **Connections**. This step is where the possible interface relationships among components are specified. This step will not be used in this chapter and will instead be discussed in Chap. 4: FEA simulation on assemblies.

Step 5: **Fixtures.** This step is used to define the supports of a component so that it will not have a rigid motion. A more detailed discussion of this step will be presented in Chap. 3.7: Fixtures.

Step 6: **Loadings.** This step is used to define the external loadings applied on a component. A more detailed discussion of this step will be presented in Chap. 3.8: Loadings.

Step 7: **Meshing.** The key feature of FEA simulation is to break down components into many small elements, to obtain acceptable simulation results of component stress/strain/deformation. The two main tasks in this step are: (1) to define global element size for whole components, and/or (2) to specify local element size in possible weakest locations or areas with high stress concentrations. A more detailed discussion of this step will be presented in Chap. 3.9: Meshing.

Step 8: **Run simulation.** After all the required input information is specified, we can just click "Run" or "Run this study" to start the FEA simulation.

Step 9: **Post-processing.** After an FEA simulation is completed, all the simulation results of the project are available. Post-processing is a set of different tools that can be used for displaying simulation results such as strain, stress, displacement, and factor of safety. Since we will use several of these post-processing tools in Example 1, a detailed discussion for step 9 is present in this section.

Step 10: **Interpretation and verification of simulation results.** Since FEA simulation software is an executable computer program, it will generate simulation results as an output after all necessary input information has been entered. After each simulation, it is good practice to interpret or verify the simulation results. Theoretical calculation results, experimental results, and our own experiences can be used in this step to make reasonable conclusions about simulation results.

Now we will use Example 1 to explain and demonstrate this general procedure. We will also use this example to explain and demonstrate Step 2: Setting up a project and Step 9: Post-processing.

Example 1: A static simulation (FEA2) Download and open FEA2, which is made of AISI 1020 steel. It is fixed at one end and is subjected to a 2000 (lb) normal force on the top surface on the other end as shown in Fig. 3.5. Run FEA simulation.

(1). Get familiar with the general procedure for FEA simulation and explain Step 9: post-processing.

(2). Run the simulation results with the global standard element size 0.20″ and then generate the following distribution plots.

    a) The von Mises stress plot with maximum and minimum annotation.

    b) The von Mises stress graph along the arc ab which is marked in Fig. 3.5.

    c) The z-displacement plot with maximum and minimum annotation.

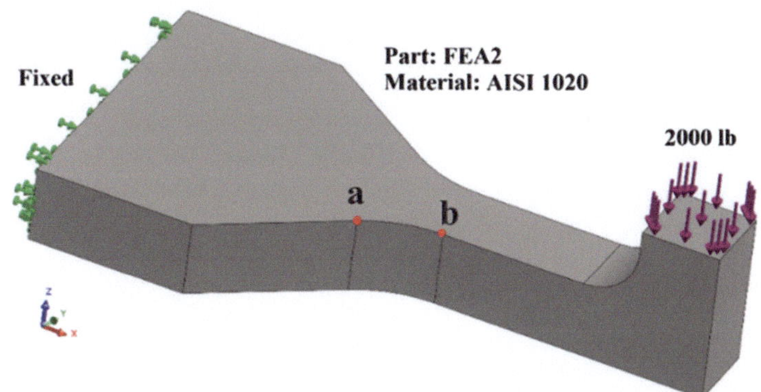

Fig. 3.5 A schematic of boundary conditions for Example one

    d) The x-normal strain plot with maximum and minimum annotation.

    e) The factor of safety plot with minimum annotation.

    f) The factor of safety plot with red areas below the factor of safety of 2.00.

(3) Show the differences in the number of nodes, the number of elements, the maximum von Mises Stress, and the maximum resultant displacement with the global standard element sizes of $1.00''$ and $0.20''$.

Solution

(1) To get familiar with the general procedure for FEA simulation and to explain Post-processing

Step 1: **Pre-processing**. For this example, the SolidWorks model (FEA2) as shown in Fig. 3.5 is already meshable and ready for FEA simulation. No further modification is needed.

Step 2: **Set up a project.** Per the problem description in Example 1, this will be a static analysis. Figure 3.6 shows how to set up a study. (1) Click the tab "Simulation"; (2) Click "New Study" which will open the "Study" Property Manager window; (3) Select the "Static" module; (4) Enter the project name "Example 1"; (5) Click the check mark to complete project creation for "Example 1".

    The simulation study tree with the project name tab "Example 1" will appear as shown in Fig. 3.7. The simulation study tree consists of 7 tabs. The first tab is the project name. The second tab shows the type of element and the name of the SolidWorks model, which will be discussed in Chap. 3.6. The third tab is "Connections" which is only used for assemblies. The fourth tab is "Fixture" and will be discussed in more detail in Chap. 3.7. The fifth tab is "External loads" and will be discussed in Chap. 3.8. The sixth tab is "Mesh", which will be discussed in

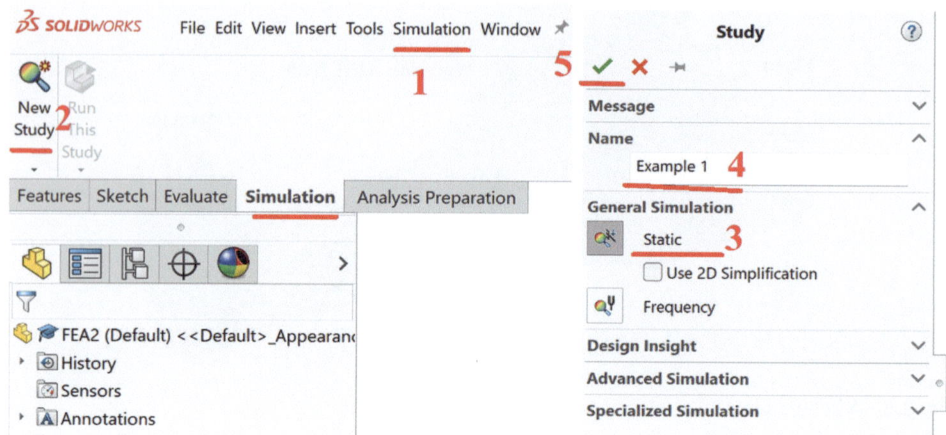

Fig. 3.6 A schematic of setting up a static analysis project

Chap. 3.9. The seventh tab is "Results options", which will be available after the simulation is completed. The second to fifth tabs are used to specify the necessary inputs for the simulation. At the bottom of the SolidWorks window, a new tab "Example 1" is added. If the simulation project is saved after its completion, all input information and simulation results about this project can be retrieved by clicking this tab.

Step 3: **Select the type of element and assign a material to each component**. For this example, tetrahedral solid elements (3D) will be used. The symbol ⬡ in the second tab in Fig. 3.8 means that the current type of element is a tetrahedral solid element, so we don't need to make any changes. The second symbol in this tab, ⬢ means a high-quality mesh which implies a non-linear polynomial deformation function. "FEA2" is the name of the model under investigation.

The material for this model is AISI 1020 steel. Follow the sequence as shown in Figs. 3.8 and 3.9 to assign material AISI 1020 to the model "FEA2". In Fig. 3.8, (1) Right-click the tab "FEA2" and a drop-down list will appear; (2) Select "Apply/

Fig. 3.7 Schematic of the simulation study tree and a project name tab

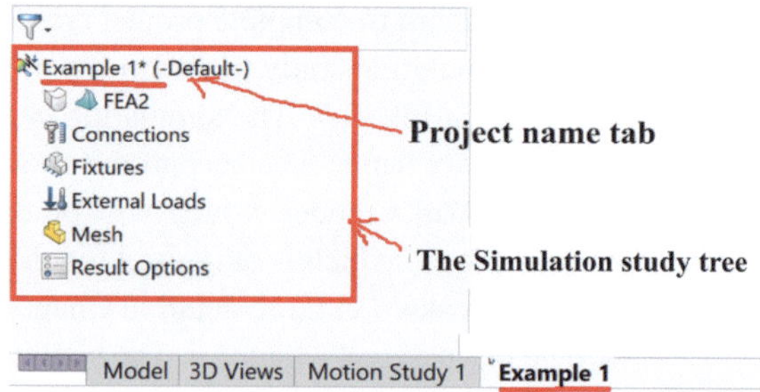

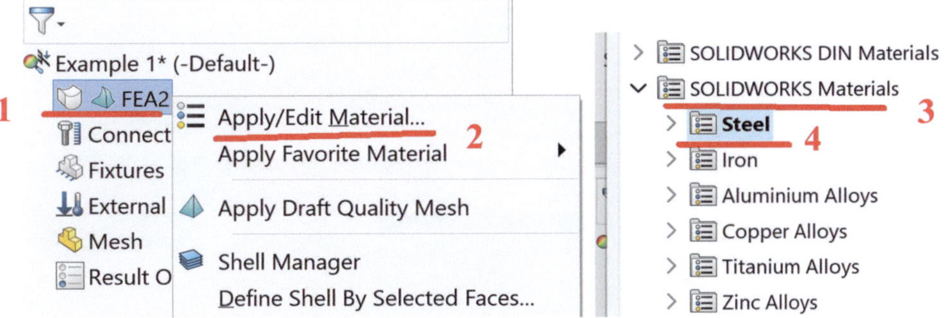

Fig. 3.8 Schematic of assigning material to a component-1

Edit material" and a window containing a list of material libraries will appear; (3) Select the proper material library, in this case, "SOLIDWORKS Materials"; (4) Select the proper category of material, which in this case is "Steel". In Fig. 3.9, (5) Select AISI 1020 as the type of material, and a pop-up window with its material properties will appear. We don't need to worry about the type of unit systems assigned in the material properties as the computer program will automatically convert them as needed; (6) Click "Apply"; (7) Click "Close". (8) After a material has been assigned to the component. A check mark appears on the element symbol and the name of the material appears after the model's name.

Step 4: **Connections**. This is about the interfaces between assemblies. So, we don't need this step for now.

Step 5: **Fixtures.** The support condition for Example 1 is shown in Fig. 3.5. A more detailed explanation of fixtures will be presented in Chap. 3.7: Fixtures. Follow the sequence

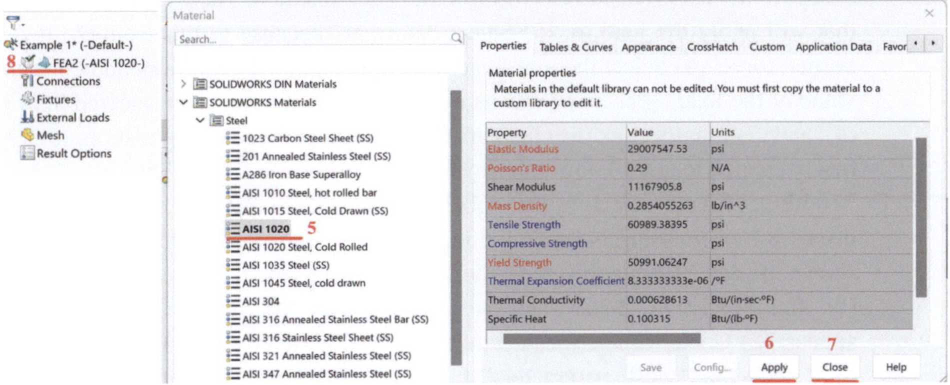

Fig. 3.9 Schematic of assigning material to a component-2

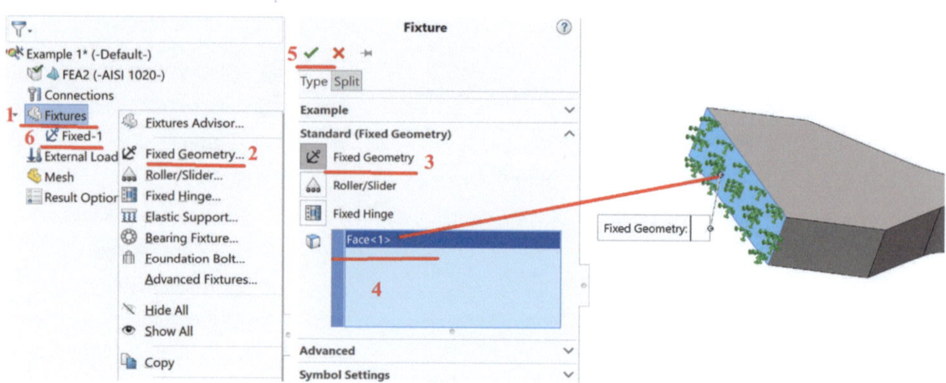

Fig. 3.10 Schematic of applying a fixed fixture

shown in Fig. 3.10 to define a fixed fixture. (1) Right-click the "Fixture" tab and a drop-down list of fixtures appears; (2) In this example, the support area is fixed, so click on "Fixed Geometry". The "Fixture" Property Manager will appear; (3) Select "Fixed Geometry" for this example; (4) Move the cursor inside the window and click inside it. Then, select the fixed surface on the model; (5) Finally click the check mark to complete the creation of a fixed surface. (6) The newly added fixture will now appear under the "Fixture" Tab.

Step 6: **Loadings.** For this example, there is a 2000 (lb) normal force applied on the top surface of the right end as shown in Fig. 3.5. A more detailed explanation of loadings will be presented in Chap. 3.6: Loads. For now, follow the sequence as shown in Fig. 3.11 to apply this load. (1) Right-click the "External Loads" tab and a drop-down list containing different types of loads will appear; (2) Click "Force" and the "Force/Torque" Property Manager appears; (3) Select "Force"; (4) Move the cursor inside the window and click inside it. Then, select the surface that will apply the load to; (5) Select "Normal" because in this example it is a normal force; (6) Select the unit, load direction, and load condition, and enter the value of the load. "Per item" means the value is applied on each selected item in (4); and (7) Finally click the check mark to complete the load application. (8) The newly created load will now appear under the "External Loads" Tab.

Step 7: **Meshing.** This step specifies the global element size for meshing. A more detailed discussion of meshing is presented in Chap. 3.9: Meshing. Follow the sequence shown in Fig. 3.12 to apply a standard mesh with a global element size of 0.20″. (1) Right-click the "Mesh" tab and a drop-down list related to meshing will appear; (2) Click "Create mesh" and the "Mesh" Property Manager will appear; (3) Check the "Mesh parameters" checkbox; (4) Select "Standard mesh"; (5) Select a unit and enter a global mesh size of 0.20″; (6) Click the check mark to

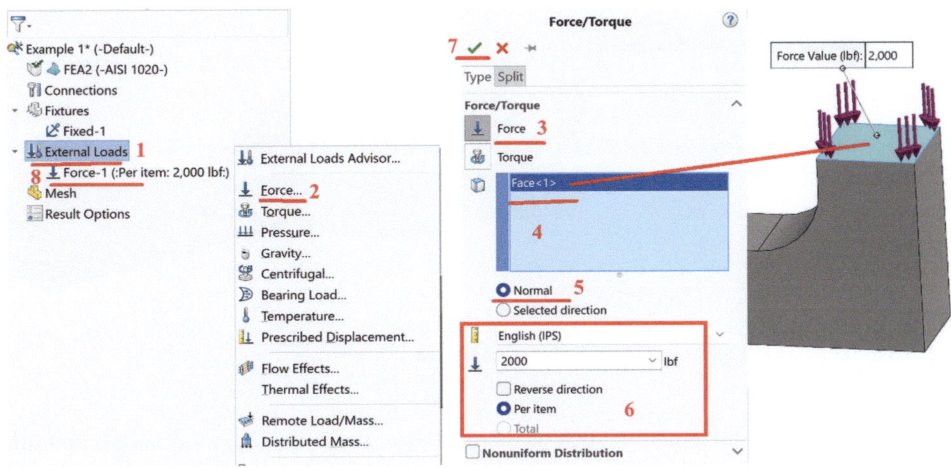

Fig. 3.11 Schematics of applying a normal force

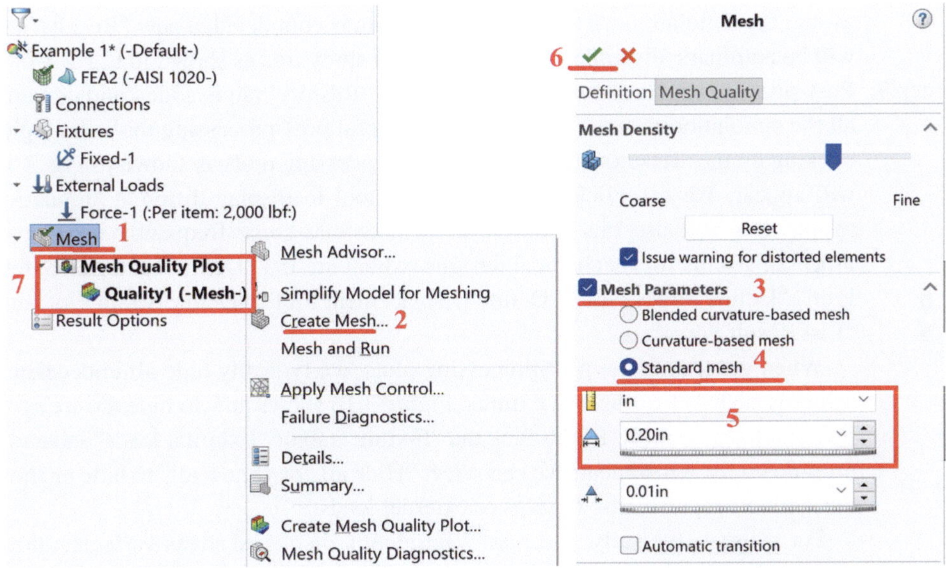

Fig. 3.12 Schematics of creating a mesh

complete this meshing. (7) The newly added entries "Mesh Quality Plot" and "Quality 1(-Mesh-)" will be automatically added under the "Mesh" tab.

By right-clicking the entry "Quality 1 (-Mesh-)", we can access the tab to hide or show the meshing. If we right-click the "Mesh" tab and select "Details", we

Fig. 3.13 The mesh details and the meshed component

can obtain information such as element size, mesh quality, total nodes, and total elements as shown in Fig. 3.13.

Step 8: **Run simulation**. Right-click on the project name tab "Example 1" in the simulation study tree and select "Run" on the drop-down list to start the FEA simulation as shown in Fig. 3.14. Alternatively, you can click "Run this study" in the toolbars to start the simulation. After the simulation run is completed, a new "Results" tab will be automatically added to the simulation study tree as shown in Fig. 3.14.

Step 9: **Post-processing**. As shown in Fig. 3.14, the "Results" tab is added and contains all the simulation results. There are many useful post-processing tools. By right-clicking on the "Results" tab, a list of post-processing tools as shown in Fig. 3.15 will appear. We can click a corresponding tool to display different simulation results. We will use this example to demonstrate some frequently used post-processing tools for mechanical designs, which are the "Define Factor of Safety Plot", "Define Stress Plot", "Define Displacement Plot", "Define Strain Plot" and "List Result Force".

When generating the post-processing plots, we typically hide all unnecessary symbols such as loading and fixtures. Figure 3.16 shows how to hide fixture symbols and load symbols. Right-click the "Fixture" tab or "External loads" tab and a drop-down list will appear. We can select "Hide all" or "Show all" to hide or show the symbols representing fixtures or external loads.

For stresses, typically, there are 7 frequently displayed stress variables: three normal stresses, three shear stresses, and the Von Mises stress. Now, let's show how to generate a Von Mises stress (psi) on an undeformed shape with maximum and minimum annotations which is shown in Fig. 3.17.

Right-click "Results" in the simulation study tree and select "Define stress plot" in the drop-down list. The "Stress Plot" Property Manager will appear. Follow the sequence as shown in Fig. 3.18 to define a stress plot with maximum and minimum annotations. (1) Specify the unit; (2) Uncheck the "Deformed shape" checkbox; (3) Left-click "VON: Von Mises Stress" and a drop-down menu will display all

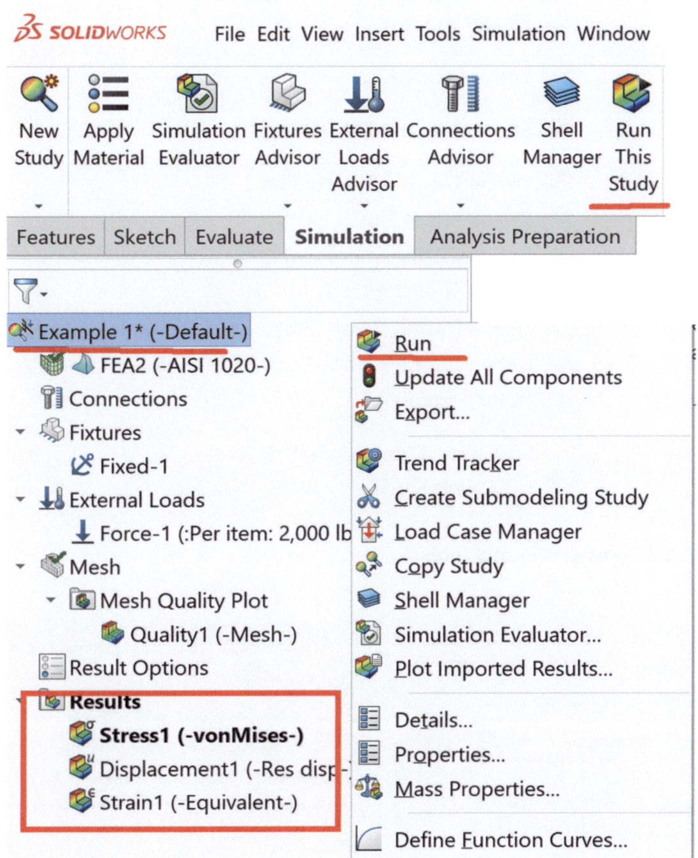

Fig. 3.14 The simulation study tree and new tab "Results"

possible stress variables such as the normal stress, shear stress, and Von Mises stress; (4) Select Von Mises stress; (5) Left click "Chart Options"; (6) Check the boxes for "Show min annotation" and "Show max annotation"; (7) Finally click the check mark to generate a stress plot. Every time a plot is generated, a new entry for this plot will be automatically added underneath the "Results" tab. Only one plot will be active at a time. Using this procedure, we can generate any type of stress plot. Double-clicking the entry for a plot will make it an actively displayed plot.

For strain, there are 7 frequently displayed strains: three normal strains, three shear strains, and an equivalent strain. Now, let's show how to generate an x-normal strain on an undeformed shape with minimum and maximum annotation as shown in Fig. 3.19.

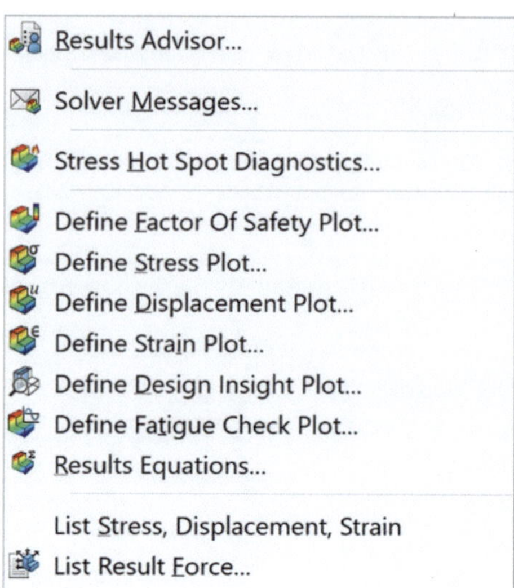

Fig. 3.15 Available post-processing tools

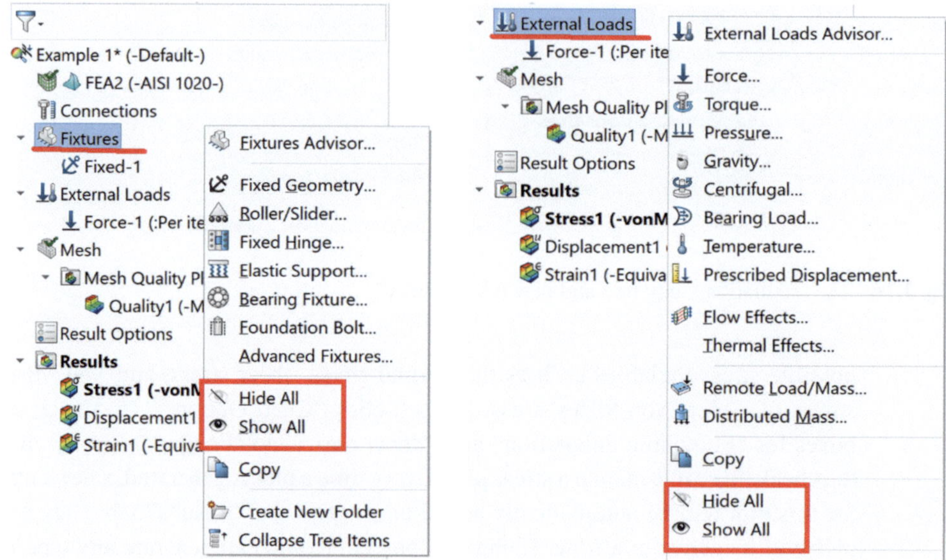

Fig. 3.16 A schematic of hiding unnecessary symbols for a plot

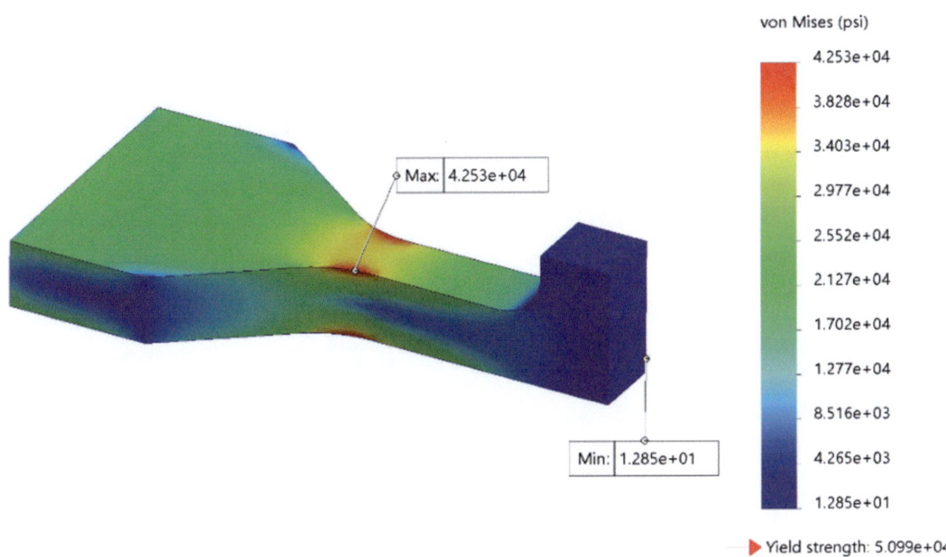

Fig. 3.17   Von Mises stress plot with the maximum and minimum annotations

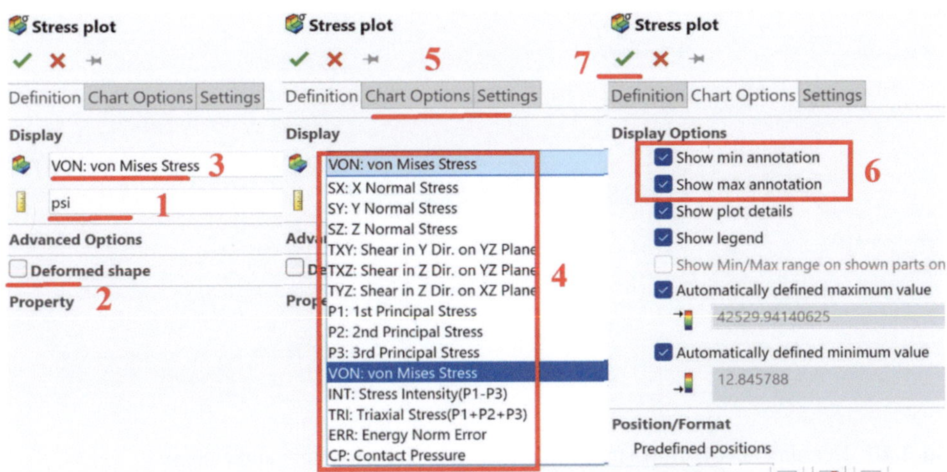

Fig. 3.18   A schematic of defining a stress plot with maximum and minimum annotations

Right-click "Results" in the simulation study tree and select "Define strain plot" in the drop-down list. The "Strain Plot" Property Manager will appear. Follow the sequence in Fig. 3.20 to define an x-normal strain plot with maximum and minimum annotations. (1) Uncheck the "Deformed shape" option; (2) Left-clicking on "ESTRN: Equivalent Strain" will display all the possible strain variables such as

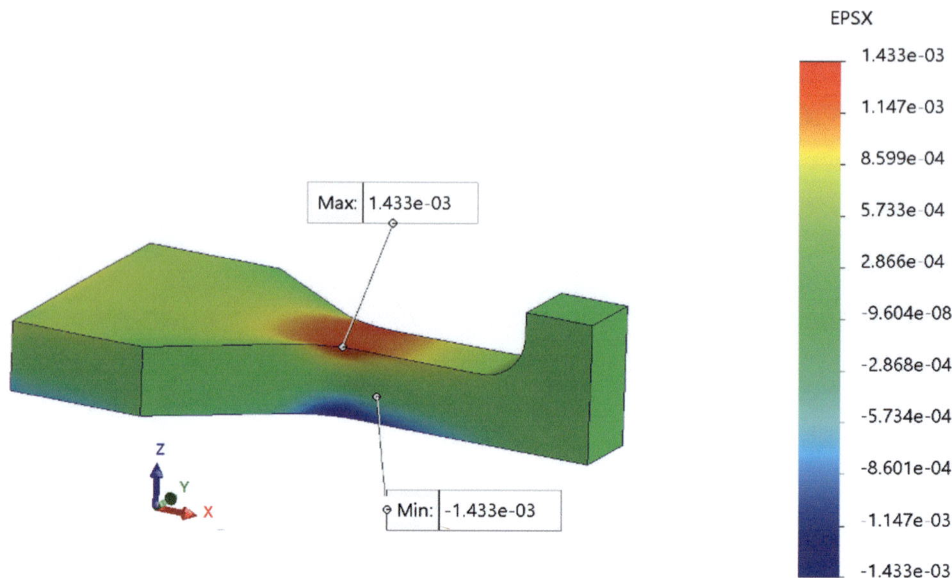

Fig. 3.19 The x-normal strain plot with minimum and maximum annotations

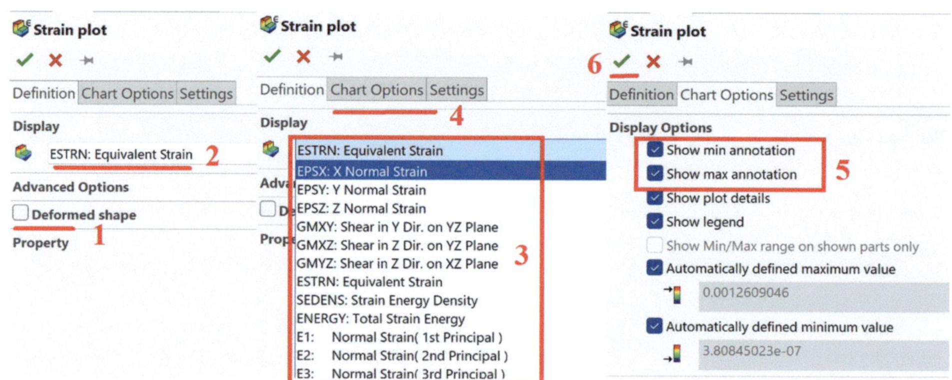

Fig. 3.20 Defining an x-normal strain with minimum and maximum annotations

normal strains, shear strains, and equivalent strain; (3) Select the x-normal; (4) Left click "Chart Options"; (5) Check the boxes for "Show min annotation" and "Show max annotation"; (6) Finally click the check mark to generate a strain distribution plot. Using this procedure, we can generate any type of strain distribution plot.

For displacements, there are 4 frequently- displayed displacements: the x-, y-. z-displacements, and the resultant displacement. Now, let's show how to generate a

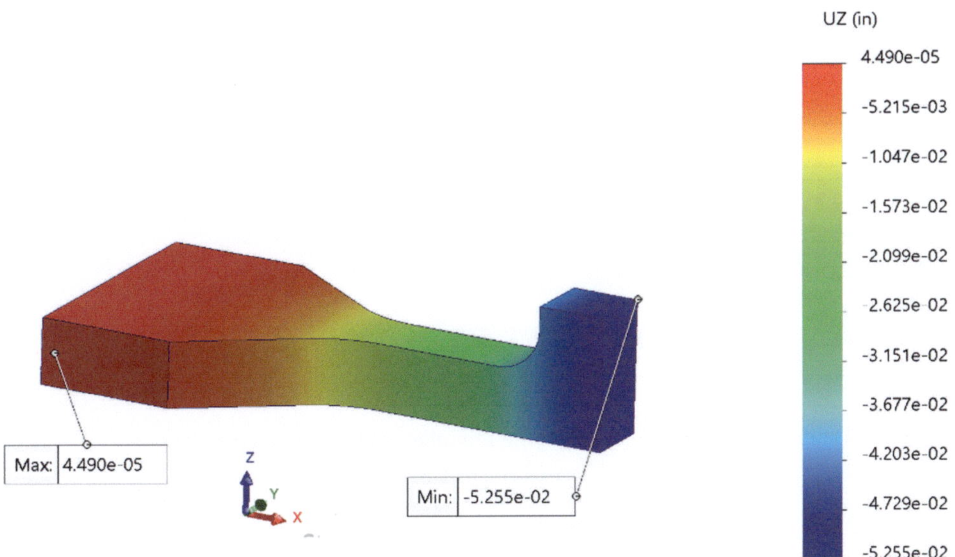

Fig. 3.21 Z-displacement plot with minimum and maximum annotations

z-displacement on an undeformed shape with minimum and maximum annotation as shown in Fig. 3.21.

Right-click "Results" in the simulation study tree and select "Define displacement plot" in the drop-down list. The "Displacement" Property Manager will appear. Follow the sequence as shown in Fig. 3.22 to define a z-displacement plot with maximum and minimum annotations. (1) Specify the unit; (2) Uncheck the "Deformed shape" option; (3) Left-click "USES: Resultant Displacement" will display all the possible displacement variables such as x-, y-, z-displacements, and resultant displacement; (4) Select z-displacement; (5) Left click "Chart Options"; (6) Check the boxes for "Show min annotation" and "Show max annotation"; (7) Finally click the check mark to generate a displacement plot. Using this procedure, we can generate any type of displacement distribution plot.

The factor of safety of a component is used to determine whether the component design satisfies the design requirements for strength issues. Two different factors of safety plots can be generated. The first one is the factor of safety distribution plot with a minimum annotation as shown in Fig. 3.23a. The second one is the factor of safety plot with red areas under a given value of the factor of safety such as 2.0 as shown in Fig. 3.23b.

For SolidWorks Simulation, the calculation of a factor of safety is based on the four failure theories, which are discussed in Chap. 1.10. But we don't need to make this selection because the option "Automatic" will automatically select a proper one as shown in Fig. 3.24.

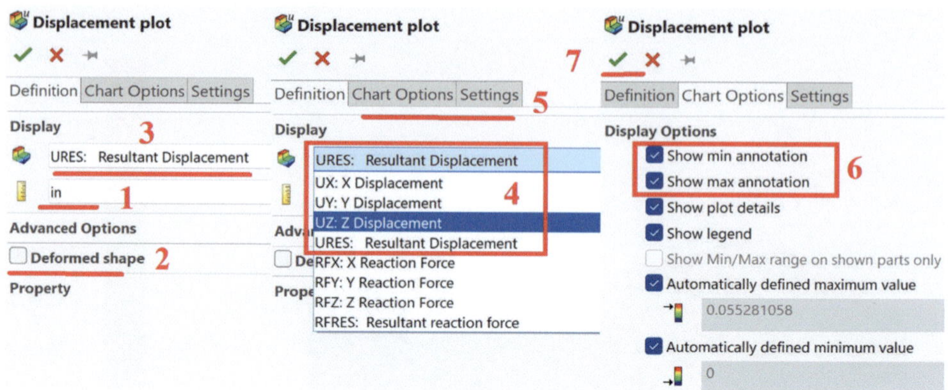

Fig. 3.22 Defining a z-displacement plot with minimum and maximum annotations

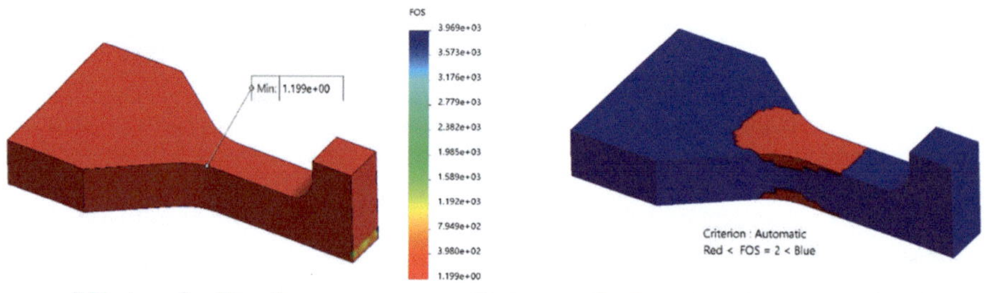

a) Factor of safety plot　　　　　　　　　　b) Factor of safety plot with areas under the value of 2

Fig. 3.23 Factors of safety plots

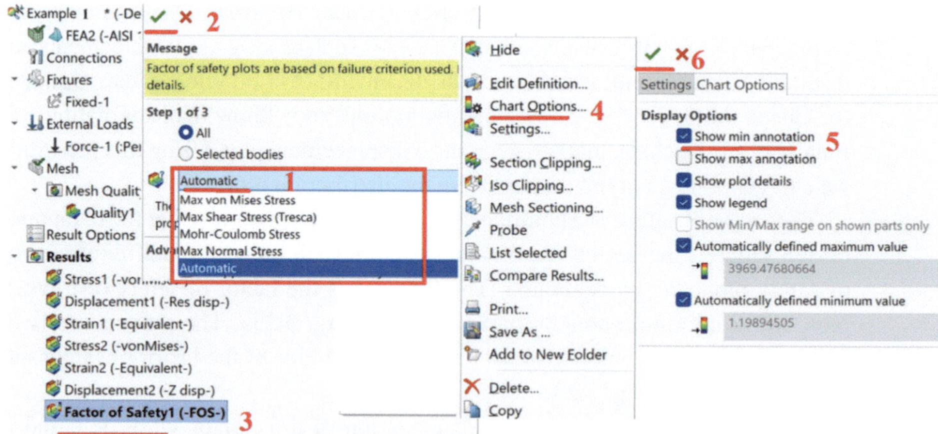

Fig. 3.24 Defining a factor of safety plot with a minimum annotation

For the factor of safety distribution plot, right-click "Results" in the simulation study tree and select "Define factor of safety plot" in the drop-down list. The "Factor of Safety" Property Manager will appear. Follow the sequence shown in Fig. 3.24 to define the factor of the safety plot with a minimum annotation. (1) Left click on the "Automatic" option to open up a list of four failure theories. Typically, we use "Automatic"; (2) Click the check mark to create the factor of safety plot. After this, the new entry "factor of safety 1″ will be automatically added underneath the "Results" tab; (3) Right-click this new entry to open a drop-down list of some related tools; (4) Select "Chart Options"; (5) Check the box for "Show min annotation"; (6) Finally click the check mark to add the annotations for the minimum factor of safety.

To create a factor of safety plot with red areas where the factor of safety is below a given value, right-click "Results" in the simulation study tree and select "Define factor of safety plot" in the drop-down list. The "Factor of Safety" Property Manager will appear. Follow the sequence shown in Fig. 3.25 to define the factor of safety plot to display red areas for areas below a value of the factor of safety. (1) Left-click the next left arrow twice; (2) Select "Areas below factor of safety"; (3) Enter the desired benchmark value for the factor of safety; (4) Finally click the check mark to generate a factor of safety plot with red areas below a value of the factor of safety. In this plot, the red areas mean that its factor of safety is less than the benchmark value. If the benchmark value is the required factor of safety, the red areas represent failures.

All the above plots are distribution plots for variables such as stress, strain, displacement, or factor of safety of a component. Another very useful post-processing tool called "List Selected" can be used to display a variable along a line or curve.

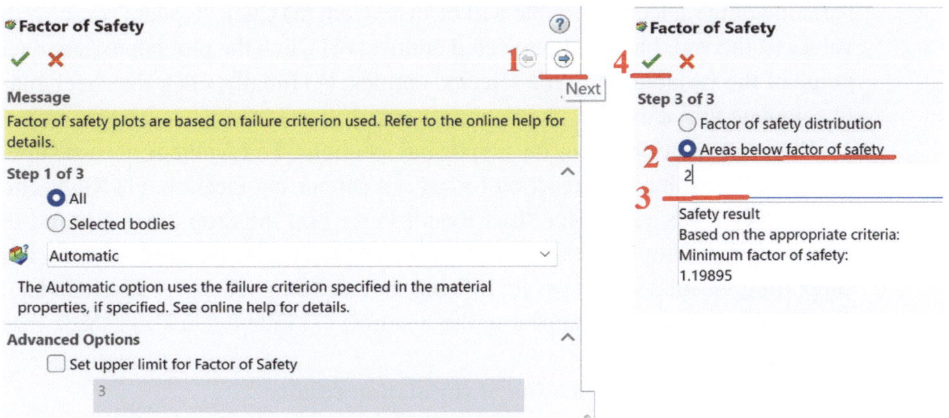

Fig. 3.25 Defining a factor of safety with areas below a given value of factor of safety

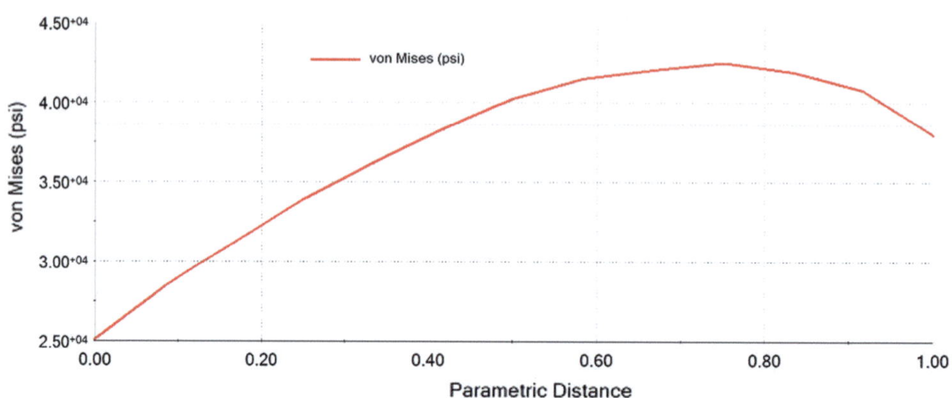

Fig. 3.26 Von Mises graph along the arc ab

Let's show how to generate a graph of Von Mises along the arc ab shown in Fig. 3.5. This graph is displayed in Fig. 3.26, where the vertical axis is the Von Mises stress, and the horizontal axis is a parametric distance along the edge. The parametric distance here is defined as the ratio of the relative distance from one end of the edge to the total length of the edge.

To generate a graph of a variable along a line or curve as shown in Fig. 3.26, we must first create the distribution plot of the variable and make that distribution plot the active plot. Follow the sequence shown in Fig. 3.27 to define this graph. (1) Right-click the tab of the corresponding distribution plot. In this case: "Stress 2 (-vonMises-)"; (2) Select "List Selected" from the new drop-down list. The "Probe Result" Property Manager will appear; (3) Select "On selected entities"; (4) Then go to the displayed model area to select the line, curve, or edge on which the variation of the selected variable will be displayed; (5) Click "Update" to show the values of the variables on the selected entities; (6) Click the plot tab to generate a graph of the variable along the selected entities; (7) Finally, click the check mark to complete this activity.

There are reaction forces on supporting locations. Follow the sequence shown in Fig. 3.28 to obtain the reaction forces at a supporting location. (1) Right-click the "Results" tab, (2) Select "List Result Force" on the drop-down list and the "Result Force" Property Manager will appear; (3) Specify the unit; (4) Select the supporting location where we are looking for reaction forces; (5) Click "Update"; (6) The result forces will appear in the window; (7) Finally click the check mark to complete this action.

Step 10: **Interpretation and verification of simulation results.**

After the FEA simulation on Example 1 has been completed, all information for mechanical design is available through the "Results" tab in the simulation study

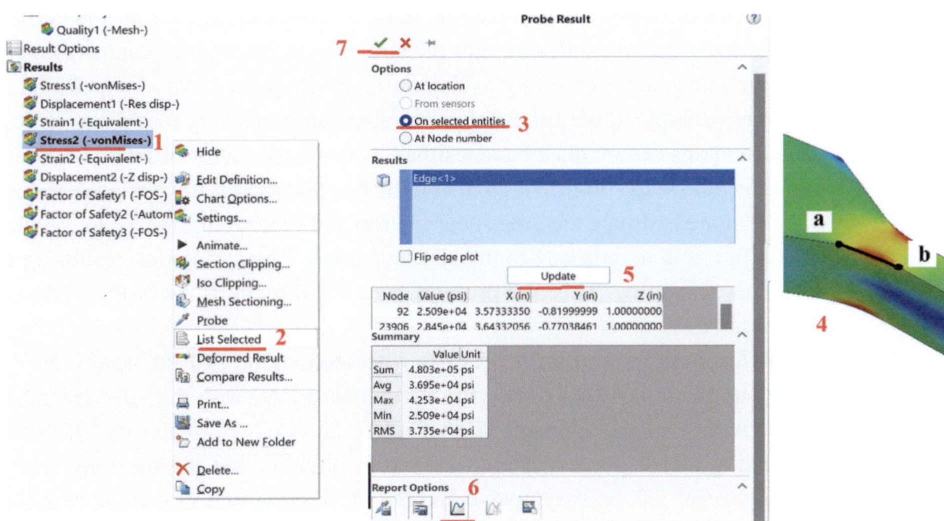

Fig. 3.27 Defining a graph of a variable along a line or curve

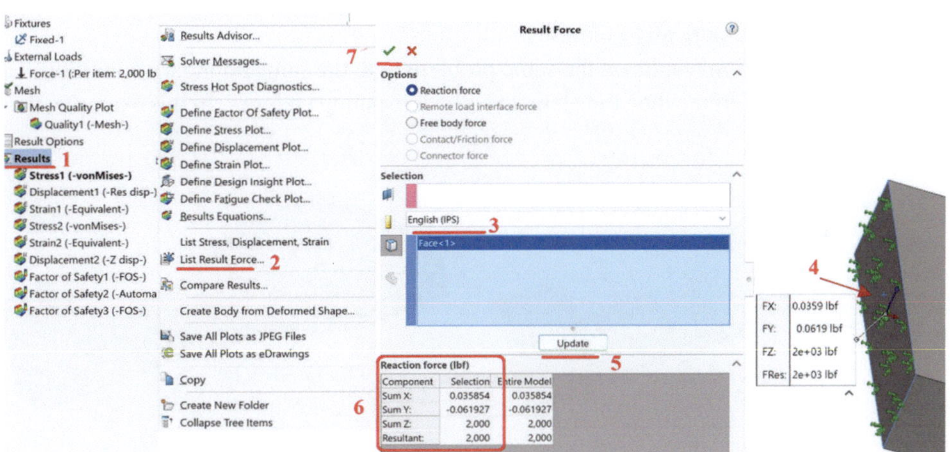

Fig. 3.28 Finding reaction forces

tree. Based on the designer's intention, we can start to interpret or verify simulation data. Here are some general comments about this example.

(a) Example 1 is mainly under bending stress due to the normal force on the top surface of the right end. Due to the fixed support on the left end, the maximum displacement should happen at the right end of the part. The Z-displacement plot as shown in Fig. 3.21 agrees with this.

(b) The factor of safety with red areas under 2 as shown in Fig. 3.23b is extremely useful for mechanical design. For the strength design of components, we can use the rated factor of safety to determine whether the design is qualified or not. For example, if we assume that the rated factor of safety for this example is 2.0., then this design must be redesigned because the factor of safety in the red areas is below 2.0. Figure 3.23 also indicates that the failure areas are around the fillet areas. Since the areas near the top and bottom of the part in the fillet areas fail, it is mainly due to the bending stress. The simulation results agree with our reasoning. This plot provides possible directions for further redesign when necessary.

(2) **Show some differences in simulation results with element sizes 1.00″ and 0.20″**
We can repeat another simulation run on the component FEA2 with the same boundary conditions but with a global element size of 1.0″. The images of the two different generated meshing are displayed in Fig. 3.29. It is obvious that the meshing with a global element size of 1.0″ is very coarse and the meshing with a global element size of 0.20″ is finer.

   After completing these two simulations, we can fill Table 3.2 with the simulation results. We can make some comments based on Table 3.2.

(a) When the element size changes from 1.00″ to 0.20″, the number of nodes and elements dramatically increases.

(b) The simulation results on the same problem with the same boundary conditions can be quite different when the element size for meshing changes. In this example, there

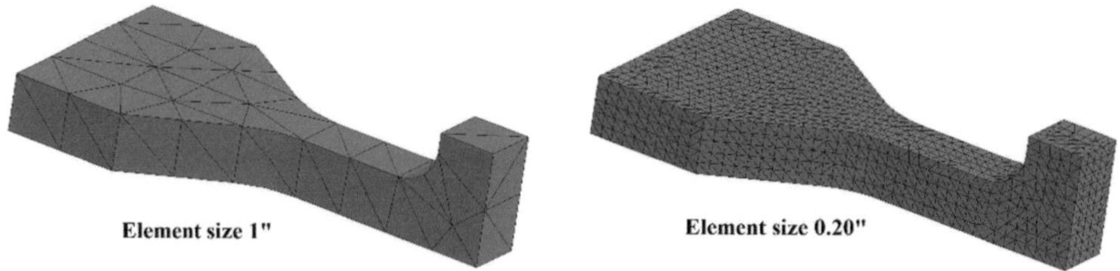

Fig. 3.29 Images of two different meshing

Table 3.2 Some simulation results for element sizes 1.0″ and 0.20″

Element size	Number of nodes	Number of elements	Maximum Von Mises (psi)	The minimum factor of safety	Maximum resultant displacement (in)
1.00″	418	193	3.763×10^4	1.355	5.415×10^{-2}
0.20″	24,495	16,012	4.253×10^4	1.199	5.528×10^{-2}

is a relative difference of 11.5% for the maximum Von Mises. The minimum factor of safety experiences a relative change of 13.0%, while the maximum displacement of the component sees a 2.04% change. These changes come from the element size change in the meshing.

For the same project, the simulation results can be quite different when the simulation settings are different. The comparison in Table 3.2 indicates that as a numerical approximation technique, FEA simulation can introduce a relative error in the simulation results compared to the true results. It will be the designers' responsibility to interpret the simulation results and to determine whether the simulation results are acceptable.

3.4 Errors in FEA Simulation and Convergence Conditions

FEA simulation will always have some degree of error because it is a numerical approximation tool. This section will discuss the sources of these errors, and how to estimate the relative error by using the convergence level.

The relative error in FEA simulation is defined as the ratio of the difference between a true value and FEA simulation results to the true value.

$$\varepsilon = \frac{abs(P_{FEA} - P_{True})}{P_{True}} \tag{3.1}$$

where P_{True} is the true value of a physical variable of a component such as stress, strain, displacement, or factor of safety. P_{FEA} is the FEA simulation result of the same type of variable. ε is the relative error.

The FEA simulation always has a relative error. The error typically comes from assumptions in four areas: geometry, material properties, loads, and fixtures [2] as shown in Fig. 3.30, where the vertical bars represent the possibility of relative errors. The two highest possibilities of errors come from assumptions about fixtures and loads. The fixtures and loads applied in FEA simulation are mathematical assumptions of the true supports and true loads, so if there is a large error in an FEA simulation, it is recommended to re-evaluate the assumptions made about fixtures and loads. Errors can also come from the assumption of material properties, and assumptions about geometries such as type of elements and element size. We will use one example to showcase the possible sources of these big errors.

Example 2: Relative errors (FEA3) Download and open FEA3, which is a bar made of Alloy steel. It is subjected to an axial load of 2000 lbs. as shown in Fig. 3.31. The line cd is a split line in the middle of the bar and is used to display the x-normal stress along this line. The global standard element size will be 0.1″.

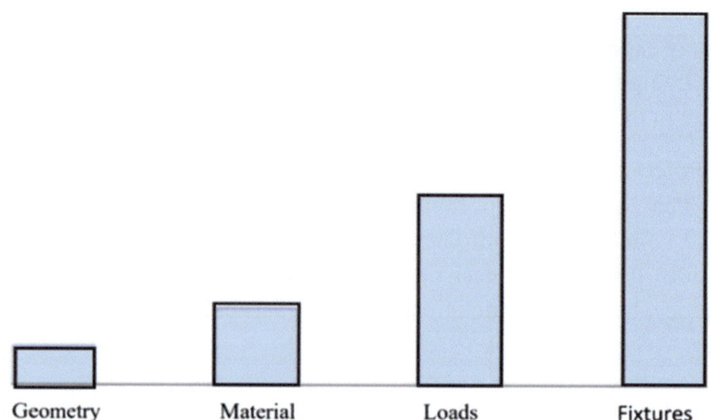

Fig. 3.30 Possible errors in FEA simulation [2]

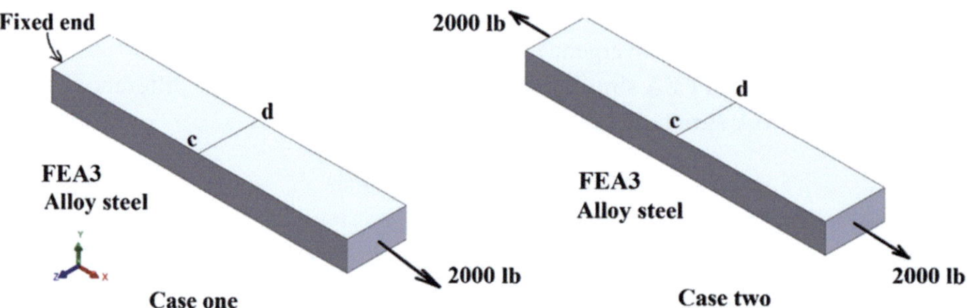

Fig. 3.31 A bar under axial loads

Case One: The left end is fixed, and a 2000 (lb) axial load is applied on the right end.

- Generate the x-normal stress distribution plot and the x-normal stress graph along the split line cd.
- Compare the FEA simulation results with the theoretical result and discuss the differences.

Case Two: Axial loads of 2000 (lb) are applied on both ends but in opposite directions.

- Generate the x-normal stress distribution plot and the x-normal stress graph along the split line cd.
- Compare the FEA simulation results with the theoretical result, and discuss the differences.

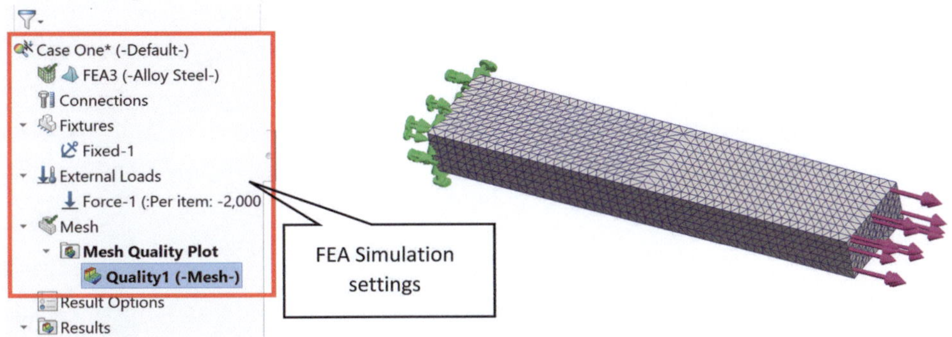

Fig. 3.32 The FEA simulation setting for Case One

Solution

Case One

Follow the descriptions in Example 1 to complete steps 1 to 7 for the FEA simulation for Case One. The expected FEA simulation settings and meshing for Case One are displayed in Fig. 3.32. From the FEA simulation settings, we can see the project name (Case One), the material for FEA3 (Alloy Steel), the fixture (Fixed-1), and loads (Force-1).

The x-normal stress distribution plot on an undeformed shape is shown in Fig. 3.33. The theoretical value of the x-normal stress will be constant with a value of $\frac{2000\ (lb)}{0.5\ (in^2)} = 4000\ (psi)$. However, the FEA simulation shows that the maximum x-normal stress is 7127 (psi). The relative error compared to the theoretical result is 78.2%. The reason for this big relative error is due to the fixture on the left end. One assumption made for the theoretical calculation equation of a bar under axial load is that the cross-section can be freely expanded. The left end of Case One is fixed, which means that any point on the fixed end will have zero displacement along the x-, y-, and z-directions. So, the assumption about the fixed end in this FEA simulation is completely different from the assumption of the freely expanded cross-section in the theoretical calculation of a bar under axial loads. Therefore, the relative error is so big, because these two results should not be compared as the conditions are not the same.

Now, let's check the x-normal stress along the split line cd. The node on the split line can be freely expanded. Use "List Selected" to generate the x-normal stress graph on the split line cd as shown in Fig. 3.34. From the graph, there are some variations of the x-normal stress along the split line cd. Theoretically, the x-normal stress along the split line cd should be a constant. However, FEA simulation is a numerical approximate technique and could cause some variations. Based on the "Summary" in the graph, the variations of the normal force are small, between 3998 and 4003 (psi). This value is very close to theoretical calculation. When the value of 4003 (psi) from the FEA simulation is used as the normal force, the relative error compared to the theoretical result is only 0.8%.

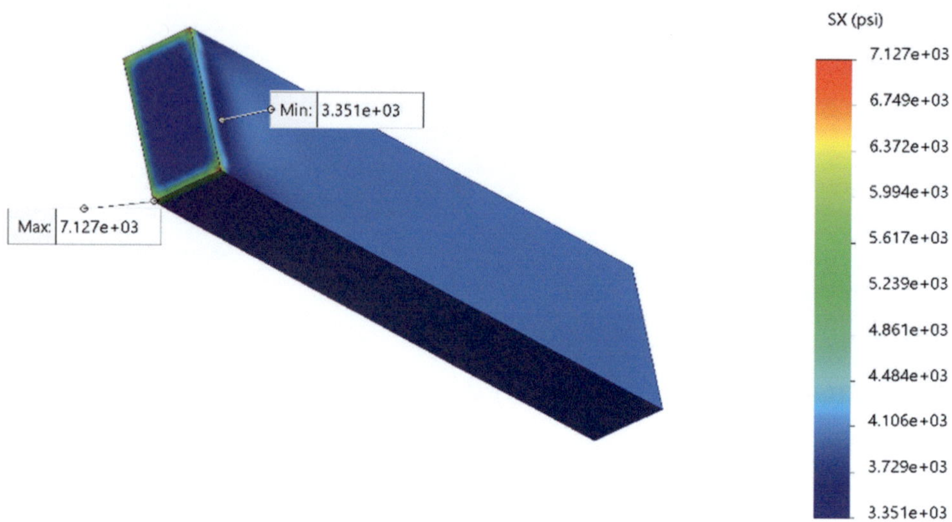

Fig. 3.33 The x-normal stress plot in Case One

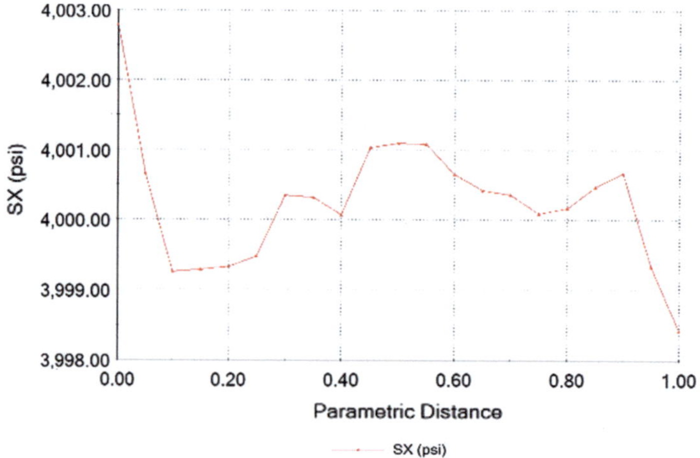

Fig. 3.34 The x-normal stresses on the split line cd in Case One

Case Two

Follow the descriptions shown in Example 1 to complete steps 1 to 7 for the FEA simulation on Case Two, except for step 5: Fixtures because in Case Two there are no fixtures.

Case Two satisfies all assumptions of the theoretical axial normal stress equation for a bar under axial load. Theoretically, the model in Case Two is in equilibrium. However, during an FEA simulation, there might be some very small, unbalanced forces, which will cause

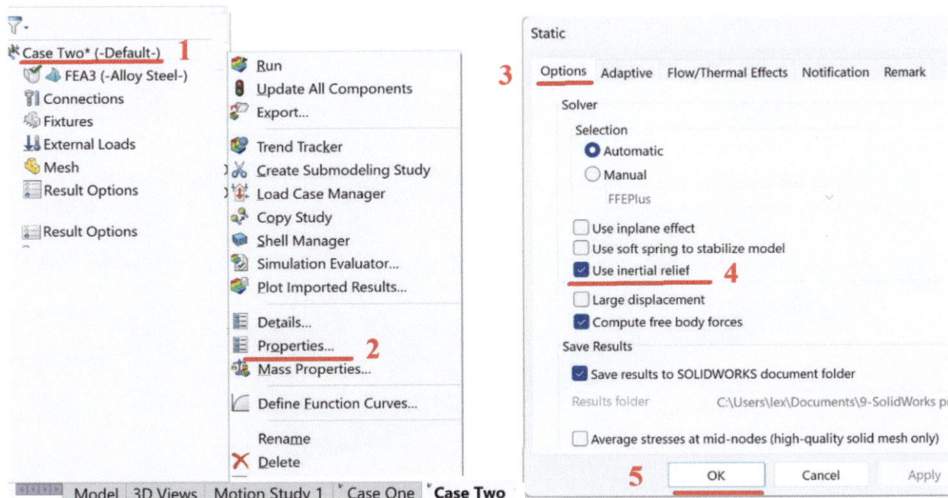

Fig. 3.35 Schematic of applying "Use inertial relief"

rigid motion in the model. This is not allowed for static analysis, and the project cannot be run. In such a case, we need to use the "Use inertial relief" option to allow the simulation to run.

"Use inertial relief" is a program setting that can automatically generate an additional force to keep models in equilibrium. This can only be used when the models are theoretically in equilibrium, and don't have enough fixtures applied.

In Step 5 for fixtures, follow the sequence shown in Fig. 3.35 to apply the "Use inertial relief" setting. (1) Right-click the tab of the project name, "Case Two"; (2) Click the "Properties" tab in the Prompted window. The properties window will open; (3) Click the "Option" tab; (4) check the box for "Use inertial relief"; (5) Click "Ok" to complete this action.

For Case Two the x-normal stress distribution plot is displayed in Fig. 3.36. The maximum and minimum values of the x-normal stress are very close to the theoretical value of 4000 psi.

The x-normal stress along the split line is shown in Fig. 3.37. The variation of the x-normal stress along the split line cd is very small. If the maximum x-normal stress (4000.02 psi) is used for the calculation of the relative error, the relative error compared to the theoretical result is 0.005%. So, the FEA results here are very close to the theoretical results.

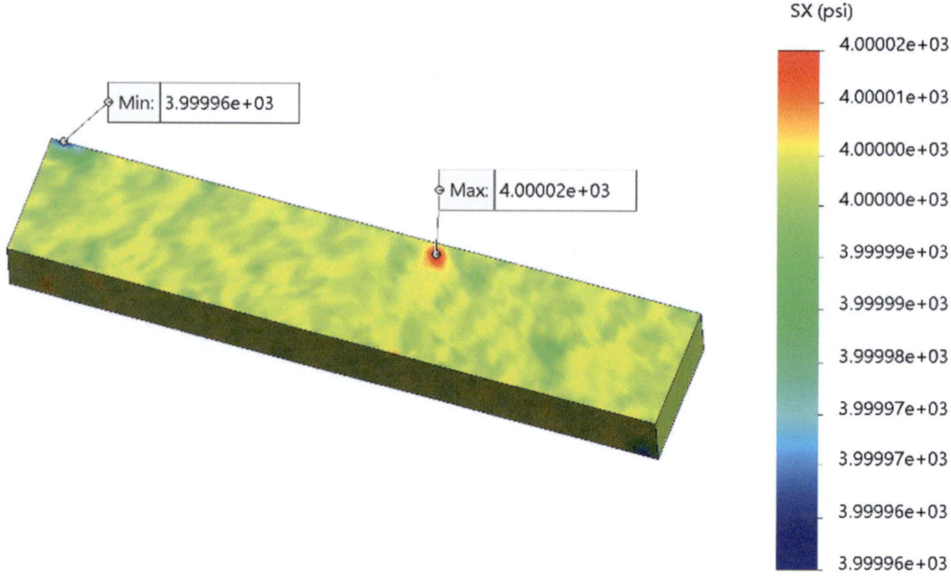

Fig. 3.36 The x-normal stress plots in Case Two

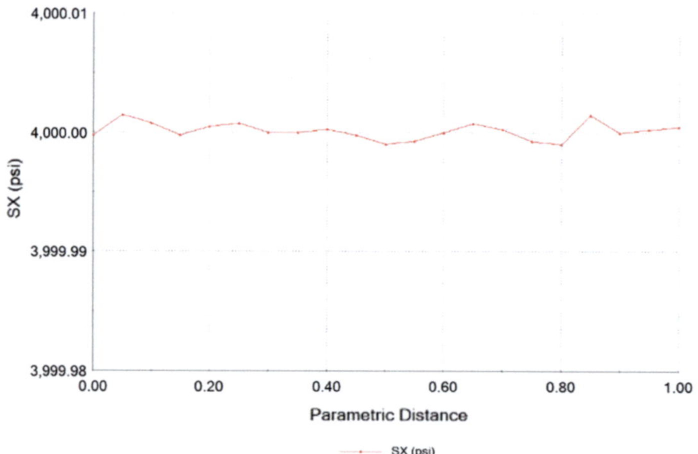

Fig. 3.37 The x-normal stresses along the split line cd

Equation (3.1) can be used to calculate the relative error of an FEA simulation result. However, the true value of a physical variable is typically unknown even though this value of the variable exists for any mechanical design project. The purpose of FEA simulation for mechanical design is not to determine the true values. We just try to obtain acceptable simulation results to complete mechanical component designs. Technically, when we use fine enough meshing or higher rank polynomial functions as the element deformation functions, the FEA simulation results will be very close to the true values and become acceptable. Since the true values are unknown, we cannot calculate the relative error, so we can use an alternative approach to check whether the simulation results are acceptable. This approach is called the FEA simulation convergence condition.

The FEA simulation convergence condition. When FEA simulation settings such as element sizes or the rank of polynomial deformation functions of elements are significantly changed or changed in a pre-described pattern, the simulation results obtained from each FEA simulation run will form a convergent series and gradually converge towards the true value. When the relative difference between two adjacent runs is less than a pre-defined convergence level, it can be said that FEA simulation results have satisfied the convergence conditions and are acceptable. The FEA simulation convergence condition can be expressed mathematically by this equation.

$$Relative\ difference = \frac{abs(P_n - P_{n-1})}{P_n} \leq \varepsilon_c \tag{3.2}$$

where the subscript n is the nth FEA simulation run; P is a selected parameter such as stress, strain, deformation, a factor of safety, or total strain energy for checking a convergence level; P_{n-1} and P_n are the values of the selected variable P in the $(n-1)$th and nth FEA simulation runs; and ε_c is a pre-defined convergence level such as 10%, 5%, 2%, or 1%.

Now, we will use an example to demonstrate how to use the FEA simulation convergence condition.

Example: Three: Convergence condition (FEA2) Download and open FEA2, which is made of AISI 1020 steel. It is fixed at the left end and is subjected to a 2000 (lb) normal force on the top surface on the right end as shown in Fig. 3.5. The maximum Von Mises stress σ_{von} is selected as a physical parameter for checking the convergence and the FEA convergence level is set to 2%. A series of FEA simulation runs will start with a global element size of $1''$ and the next global element size will be half of the previous element size. Run the FEA simulation to obtain the acceptable maximum von Mises stress of the component.

Solution

Follow the descriptions in Example 1 to complete the FEA simulation runs with standard element sizes $1''$, $0.5''$, $0.25''$, and $0.125''$. The maximum von Mises stresses of each FEA simulation run with relevant other information are shown in Table 3.3. Plots of the element size vs. the Von Mises stress the relative difference of von Mises stress are shown in Fig. 3.38. From Table 3.3, it can be seen that when element size is decreased, the number of elements is dramatically increased. Since the convergence level is specified as 2%, we cannot stop the FEA simulation in the third run because the relative difference of von Mises in this run

Table 3.3 FEA simulation results from several runs and the relative differences

Runs	Element size (inch)	Number of elements	Von Mises stress σ_{von}	Relative difference ε_c
1	1.000″	193	3.76E + 04 (psi)	/
2	0.500″	1143	4.14E + 04 (psi)	9.06%
3	0.250″	8491	4.25E + 04 (psi)	2.57%
4	0.125″	60,995	4.28E + 04 (psi)	0.817%

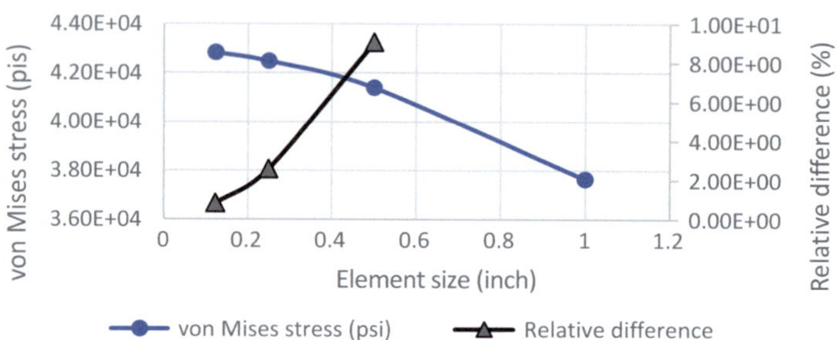

Fig. 3.38 The element size vs the von Mises stresses and its relative differences

is 2.57%. However, the von Mises stress (4.28E + 04 psi) in the fourth run can be accepted because the relative difference is only 0.817%, which is less than the specified convergence level of 2%.

3.5   Pre-Processing and Setting up a Project

This section will discuss Step 1: Pre-processing and Step 2: Set up a project of the general FEA simulation procedure.

Step 1: **Pre-processing**

Step 1 of the general FEA simulation procedure is pre-processing, which is where we prepare the models for FEA simulation. Models for creating mechanical drawings for its production are different from models for FEA simulation.

Models for mechanical drawings must include all the detailed information because the purpose of such models is to duplicate components. However, the purposes of models for FEA simulation are (1) to make the models meshable and (2) to obtain physical variables such as stress, deformation, or factor of safety through FEA simulation for mechanical design. Thus, models for FEA simulation can be slightly different from models for mechanical drawings. Two key points in pre-processing are:

- The geometries of models must be meshable because FEA simulation must break the models into lots of elements.
- Even after the geometry is meshable, some further modifications on models may be necessary to obtain simpler meshing and shorter computing times. This is because the goal of FEA simulation is to obtain acceptable simulation results for mechanical design, so that features unnecessary for stress analysis can be removed.

Four typical actions done in pre-processing on models for FEA simulation are (1) defeaturing; (2) clean-up; (3) idealization; and (4) adding some split lines.

Defeaturing is the process of removing some geometrical features that are deemed insignificant for stress/strain analysis. One defeaturing example is shown in Fig. 3.39. The original model has fillet edges, which remove the sharp edges. Based on the fixture condition and load condition shown in Fig. 3.39a, these fillet edges have an insignificant impact on the component stress. Therefore, the simplified model shown in Fig. 3.39b can be used for FEA simulation, which will result in a simple meshing and reduced computing time.

Clean-up is a process to remove un-meshable and insignificant features such as paints, logos, and nameplates such as those shown in Fig. 3.40. These features are extremely important for real productions. But they can be very thin items such as a layer of paint, or irregular & complex geometries such as engraved thin-layer logos. Features like this might not be meshable and likely have no contributions to the component strength. Therefore, these features can be cleaned up and removed from the FEA model.

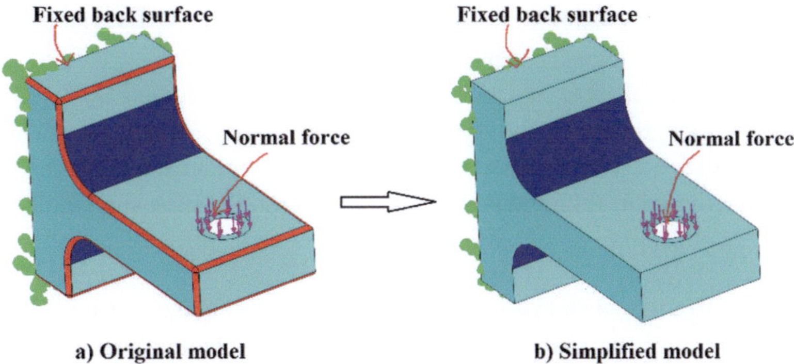

Fig. 3.39 An example of defeaturing

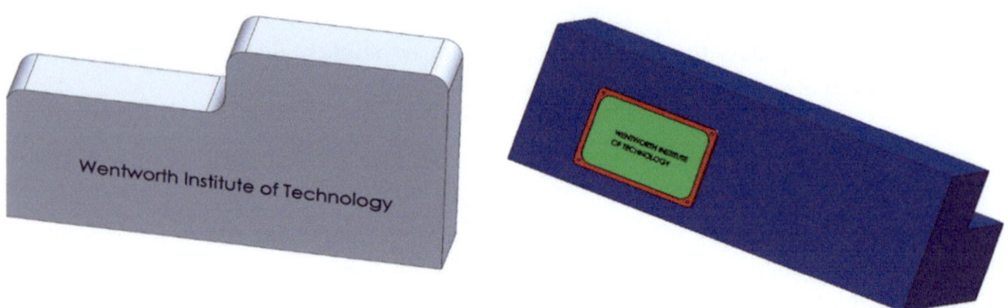

Fig. 3.40 Examples of clean-up

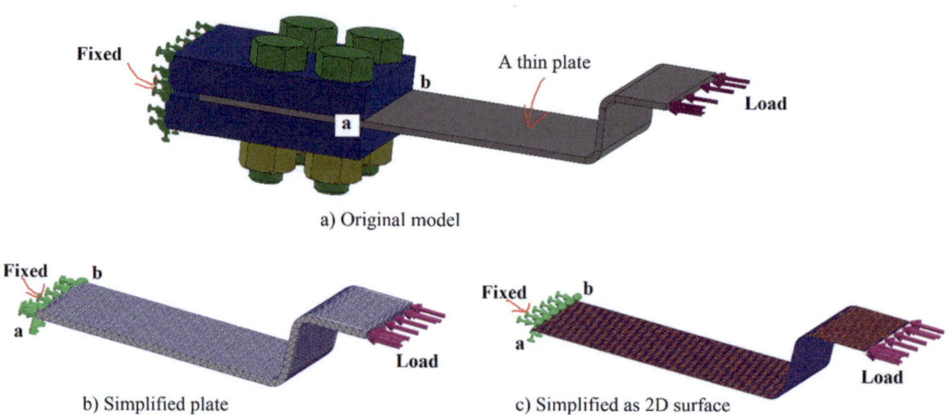

Fig. 3.41 Examples of idealizations

Idealization is a process that might significantly change the model or simplify the model into 1D elements such as beam elements, truss elements, or 2D shell elements when the main features related to stress/strain analysis are kept.

An example of idealization is shown in Fig. 3.41. The original model is shown in Fig. 3.41a. For strength issues, the thin plate will be the concerned component because two clamping plates are very thick & strong, and the bolt connection is strong. So, in this problem, we should focus on the thin plate, so we can run the FEA simulation on the thin plate alone.

- The first idealization as shown in Fig. 3.41b is that the model can be simplified as a thin plate that only starts from the clamping edge ab. The left end with the edge ab will be treated as a fixed fixture. The component will be meshed with 3D elements.
- A further idealization as shown in Fig. 3.41c is that the thin plate can be simplified as a 2D surface that will mesh with 2D shell elements.

Adding some split lines is an approach by which some auxiliary lines or curves are added to specify boundary conditions or to display some physical values along those specified lines or curves.

A split line or curve can be added by right clicking the "Insert" tab in the toolbar, selecting "Curve", selecting "Split Lines", and then following the prompt window to complete the action.

In Example 2, the split line cd as shown in Fig. 3.30 is added to display the x-normal stress along it. Another example of adding a split line for displaying stress is shown in Fig. 3.42. The original model is shown in Fig. 3.42a. Based on knowledge of the mechanics of materials, there will be a stress concentration on the hole. We can add a split line as shown in Fig. 3.42b. Now we can display the maximum concentrated stress along this split line.

A split curve for defining boundary conditions is shown in Figure 3.43. A load is applied to a circular area on the top surface as shown in Figure 3.43a. However, the whole top surface is one surface. A split curve can be added to the area where the load is applied as shown in Figure 3.43b. After the split line is added, the top surface is split into two areas: one area inside the split line and the other area outside the split line.

Step 2: **Set up a project**

After pre-processing is completed and the model for FEA simulation is ready, we can set up a simulation project, which has been explained in Chap. 3.3 General Procedure for FEA Simulation and Post-processing. It is worth knowing how to properly set up the static properties using the "Static" Property Manager as shown in Fig. 3.44. (1) Right-click the project name tab; (2) Select "Properties", which will open the "Static" Property Manager window. (3) If a project is in equilibrium, but is not fully restrained, we must select "Use initial relief" or "Use soft spring to stabilize model" for an FEA simulation to be able to run. (4) The "Adaptive" tab is used for the h-method or p-method, which is used for FEA simulation convergence and will be discussed in Chap. 3.8 Meshing.

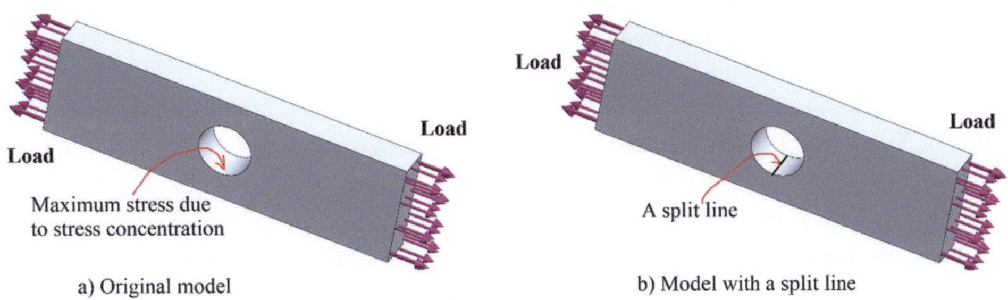

Fig. 3.42 A split line for displaying maximum concentration stress

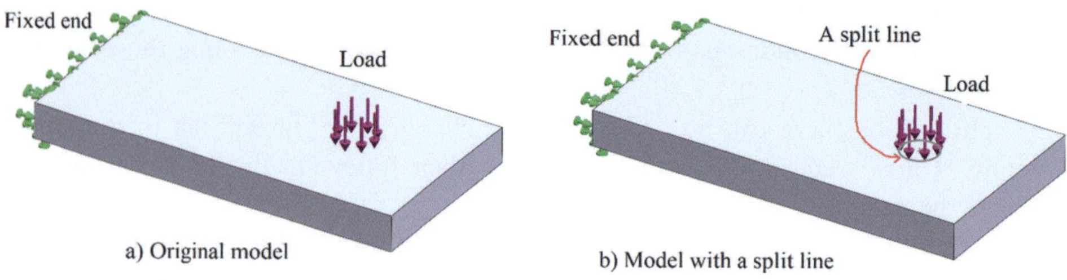

Fig. 3.43 A split line for boundary conditions

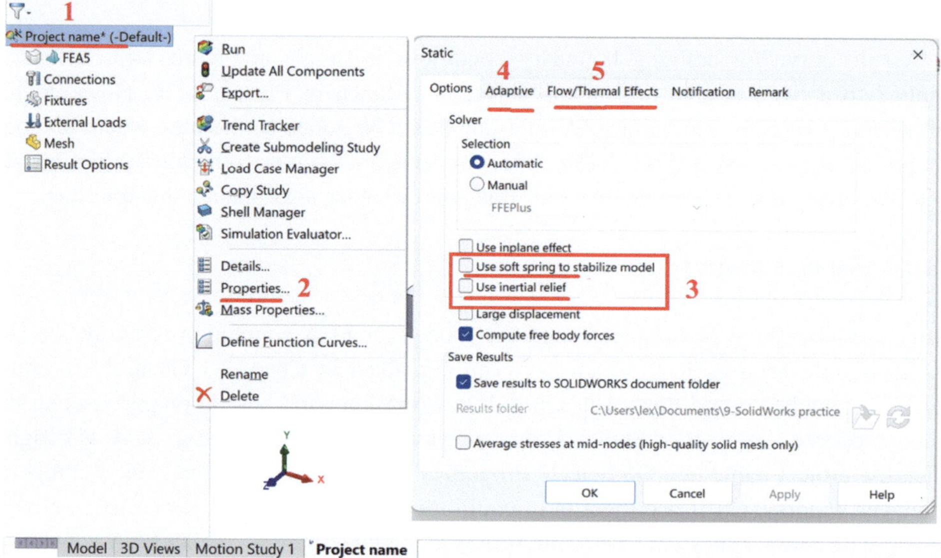

Fig. 3.44 The static properties window

(5) The "Flow/Thermal Effects" tab is used for thermal stress analysis, which will be discussed in Chap. 6: Thermal analysis and thermal stress analysis.

3.6   Type of Elements and Assigning Materials

This section will explain and discuss step 3: Type of element and assigning materials, which covers how to set a type of element for meshing, how to assign a material to a component, or how to create a custom material.

After a simulation project has been set up, the simulation FEA study tree will appear as shown in Fig. 3.7. The first tab below the project name is for Step 3, where it is to select the type of element and assign a material to each component. There are three tasks in this step: (1) Assign a material to a component, (2) select quality meshing, and (3) Select the type of elements.

(1) **Assign material to a component**: For an FEA simulation, a type of material must be assigned to each component. Without a material type assigned, the element stiffness matrix (element properties) cannot be constructed, as discussed in Chap. 2.6: Element stiffness matrix. When the type of materials is included in the SolidWorks material library, it can be directly assigned to a component as shown in Figs. 3.8 and 3.9.

Create custom material: When the desired type of material is not included in a SolidWorks material Library, we can create a custom material and then assign it to a component. For stress/strain analysis problems, at least four mechanical properties must be specified for a custom material. They are elastic Young's modulus, Poisson's ratio, mass density, yield strength, and/or tensile strength. If there is no "Custom Materials" library, we need to first create a "Custom Materials" library. Follow the sequence specified in Fig. 3.45 to create a "Custom Materials" library. (1) Right-click the first tab with the component name (FEA2); (2) Select "Apply/Edit Material" from the drop-down list. The "Material" Property Manager will appear; (3) Right-click any empty location and a pop-up menu will appear; (4) Select "New Library" and follow the prompts to specify "Custom Materials" as the name for the new material library and select "Save"; The new library will appear in the "Material" Property Manager as shown in position (5).

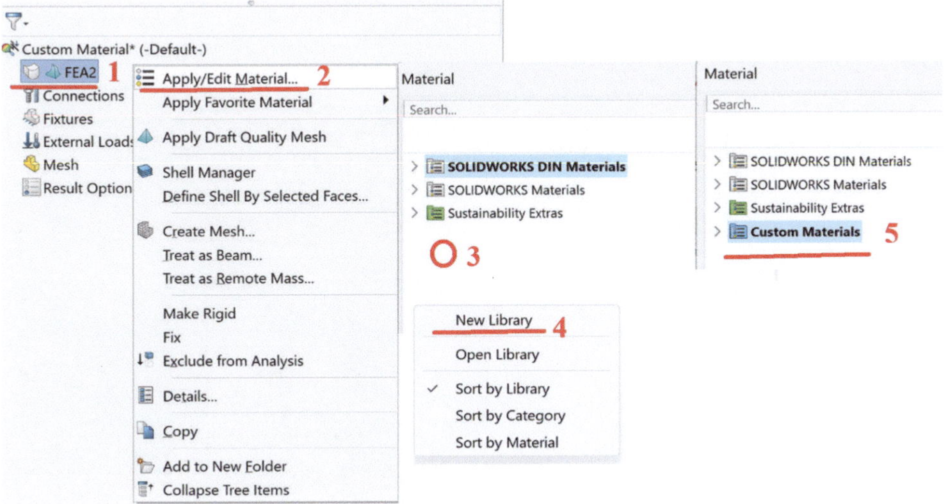

Fig. 3.45 Schematic of Creating a "Custom Materials" Library

If there is a "Custom Materials" library, we can directly add any new custom materials to this library. The structure of the material library in SolidWorks Simulation consists of two levels: different categories and different materials under each category. We will use one example to show how to specify a new material.

Example 4: Specify a new custom material (FEA2) Download and open FEA2. Create a new material "Metal #1" under the category "Solids" in the "Custom Materials" Library. The five mechanical properties of this material are the Elastic Young's modulus = 3.04E08 (psi), Poisson's ratio = 0.29, Mass Density = 0.32 (lb/in^3), Tensile strength = 8.5E4 (psi), and Yield Strength = 6.7E4 (psi).

Solution

First, right-click the component name (FEA2) tab and select "Apply/Edit material". Follow the sequence specified in Fig. 3.46 to create a new material "Metal #1" under the "Solids" category. (1) Right-click "Custom Materials" and a new pop-up menu will appear; (2) Select "New Category" and name it "Solids". The new category name will appear underneath "Custom Materials". (3) Right-click the new "Solids" category name and a new pop-up menu will appear; (4) Select "New Material" and then the material properties window will appear; (5) Select the "English (IPS)" units; (6) Enter the material name, "Metal #1"; (7) Enter the provided material mechanical properties into the corresponding rows; (8) Click "Apply". (9) After this, the new material "Metal #1" will automatically be added to the "Custom Material" Library; (10) Finally, select "Close" to exit the "Material" property window, and assign this material to a component.

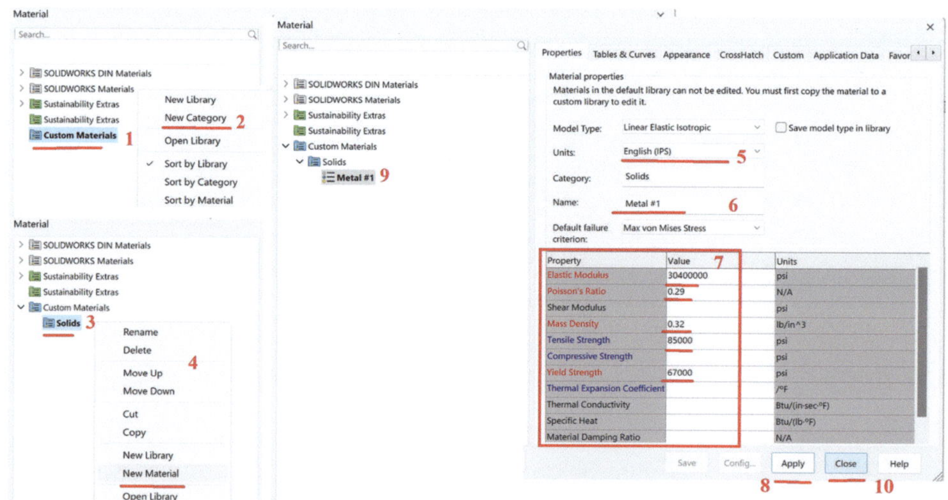

Fig. 3.46 Schematics of creating a new material

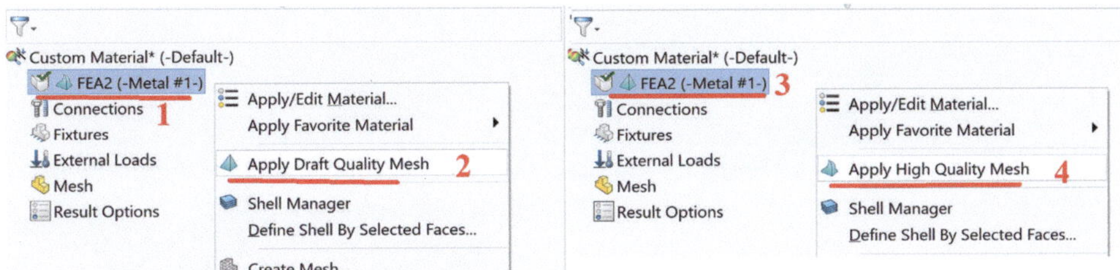

Fig. 3.47 The draft and high-quality mesh

(2) **Select quality mesh**: For SolidWorks simulation, there are two different qualities for meshes: draft quality (⬆)and high-quality mesh (⬆). The draft quality mesh (⬆) implies that the deformation function for an element will be a linear polynomial function. The strain and stress of an element with draft quality mesh will be constant. So, it will have low accuracy for the simulation results. Therefore, typically, we don't use draft-quality meshes. The high-quality mesh (⬆) means that the deformation functions will be at least second-order polynomial functions, and it is generally recommended that the high-quality mesh should be used.

If the current quality mesh is high, we can change it into "draft quality" as shown in Fig. 3.47: (1) Right-click the tab with the component name, and (2) Click "Apply Draft Quality Mesh" from the drop-down list. If the current quality mesh is a draft quality mesh, we can change it into the high-quality mesh following the method in Fig. 3.47: (3) Right-click the tab with the component name, and (4) Click "Apply High-Quality Mesh" from the drop-down list.

(3) **Select the type of elements**. As shown in Table 3.1, SolidWorks Simulation has four types of elements: 3D tetrahedral elements, 2D shell elements, 1D beam elements, and 1D truss elements. Every mechanical component is a 3D model and can always be meshed by 3D tetrahedral elements. However, some mechanical components can be meshed by 2D shell elements, 1D beam elements, or 1D truss elements based on their geometrical features and loading conditions. This allows us to reduce the simulation computing time while still yielding acceptable results.

3D tetrahedral element: SolidWorks models are typically created by two different approaches. One approach is to use typical features such as extruded base, revolved base, and extruded cut. This type of model is called a solid model. After a simulation project is set up for solid models, the default type of element will be 3D tetrahedral elements (⬡) as shown in Fig. 3.48a.

Another approach is to use sheet metal tools such as base flange, edge flange, jog, and sketched bend. These types of models can be called sheet metal models. After a

a) Solid model   b) Sheet metal model

Fig. 3.48   Schematics of normal model and sheet metal model

simulation project is set up for sheet metal models, the default type of element will be a 2D shell element ($\searrow$) as shown in Fig. 3.48b. We can convert the 2D shell elements into 3D tetrahedral elements as shown in Fig. 3.48b: (1) Right-click the tab with the component name (FEA6); (2) Select "Treat as Solid" on the drop-down list; (3) The type of element is then converted into a 3D tetrahedral element as shown in Fig. 3.48 b).

2D shell element: When the thickness of a three-dimensional solid model is very small compared to its other dimensions, and it has a mid-surface, this model can be meshed by 2D shell elements.

When a model is created by a sheet metal tool, it can be directly meshed by 2D shell elements in SolidWorks Simulation. When a sheet metal model has been converted into a 3D tetrahedral element, it can be converted back to a sheet metal model as shown in Fig. 3.49: (1) Right-click the tab with the component name (FEA6); (2) Select "Treat as sheet metal"; (3) It will become a sheet metal model again.

When a model such as tubing is created by a non-sheet metal tool and has a thin thickness, it can be converted into a model meshed by 2D shell elements. One convenient approach for converting models into a model for 2D shell elements is to first create a mid-surface in its part model. After the simulation project of a component with a mid-surface is set up, follow the sequence in Fig. 3.50 to convert the component into a model for 2D shell elements: (1) Click the expand arrow to the left of the component name FEA7 tab to show the two features of the model. One feature ($\odot$) is for 3D element meshing. The other feature ($\spadesuit$) is for 2D shell element meshing; (2) Right-click "SolidBody2"; (3) Select "Exclude from Analysis" on the drop-down list. This will exclude the SolidBody component from the simulation. (4) Now look at "SurfaceBody1" that is the mid-surface. We will use this mid-surface for FEA simulation. However, the thickness has not been defined. To define the thickness, right-click "SurfaceBody1", (5) Select "Edit Definition" in the drop-down list. The "Shell Definition" Property Manager will appear; (6) Enter the

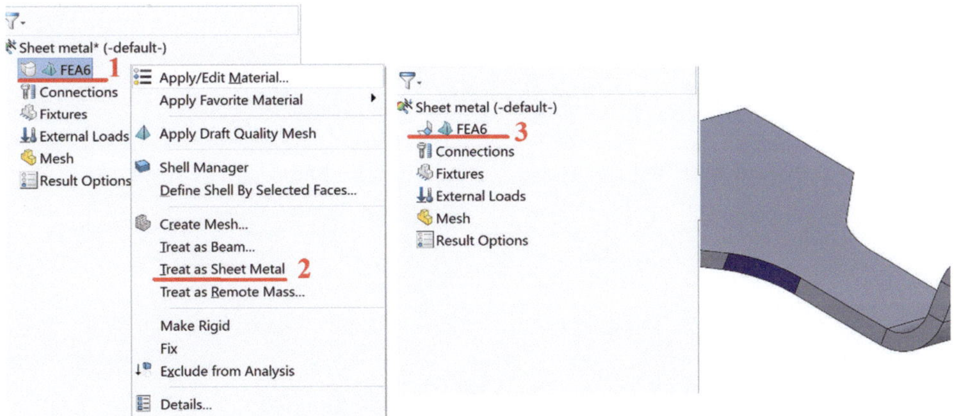

Fig. 3.49 Schematics of converting back to sheet metal model

thickness, (7) Finally, click the check mark to complete the model conversion. Now the model is ready for simulation using 2D shell elements.

By suppressing the mid-surface in the part model, the model will become a model using 3D tetrahedral elements again.

Beam Element: When a model is a slender structure where the length dimension is much larger than the other two dimensions of its cross-section, and the model is subjected to forces or/and moments it can be meshed using 1D beam elements. It can also be meshed

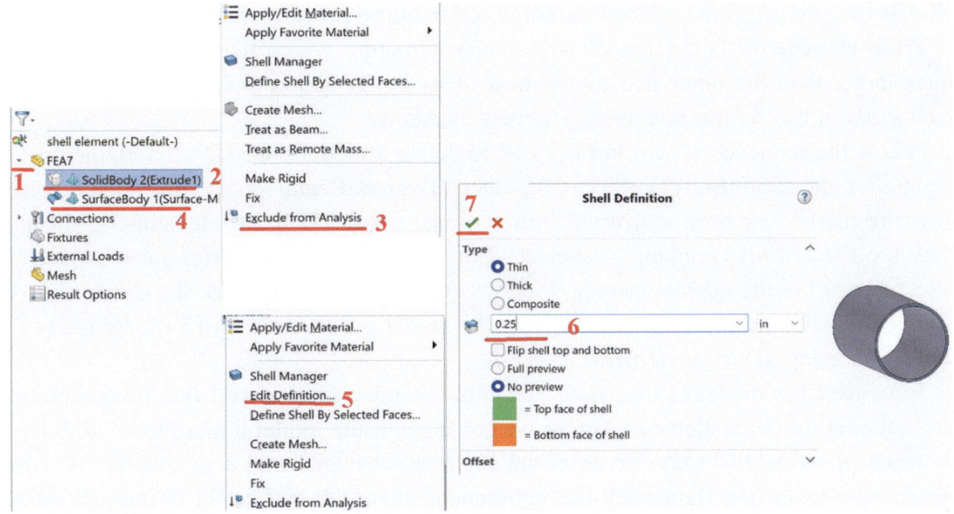

Fig. 3.50 Schematic of converting a model for 2D shell element meshing

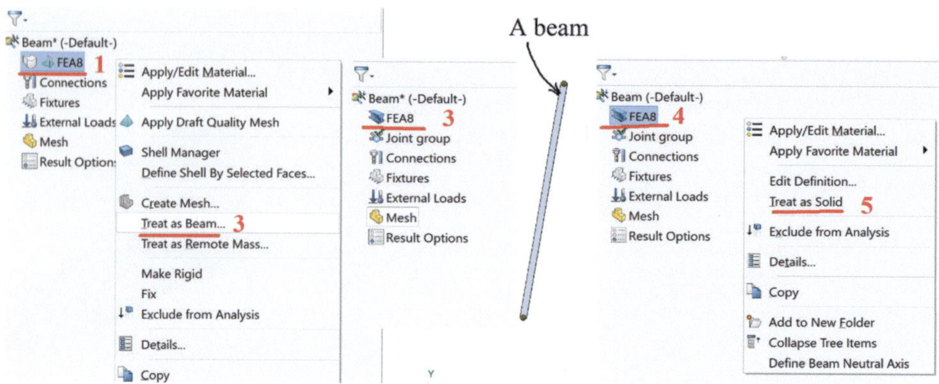

Fig. 3.51 Schematics of defining a beam element

using 3D tetrahedral elements, but with 1D beam elements, the number of elements and computing time can be significantly reduced while still providing acceptable results.

When a simulation project for a slender structure is set up, the default type of element ( ) will be the 3D tetrahedral elements. Follow the sequence shown in Fig. 3.51 to define a model for 1D beam elements: (1) Right-click the tab with the component name (FEA8); (2) Select "Treat as Beam" on the drop-down menu. (3) This will convert the 3D solid model into a model using 1D beam elements ( ). At the same time, joints will be automatically generated at both ends of the beam structure.

Following the sequence shown in Fig. 3.51, a model for beam elements can be converted back into a model using 3D solid elements: (4) Right-click the component name (FEA8) tab, and (5) Select "Treat as Solid" on the drop-down menu.

Truss element: When a model is a slender structure where the length dimension is much larger than the other two dimensions of its cross-section and is only subjected to axial loads, it can be meshed using 1D truss elements.

Follow the sequence shown in Fig. 3.52 to define a model with 1D truss elements: (1) Right-click the (SolidBody1) entry; (2) Select "Treat as Beam" on the drop-down menu. Now the model has been converted into a model using 1D beam elements; (3) Right-click the (SolidBody1) again; (4) Select "Edit Definition" in the drop-down menu; (5) Select "Truss" in the opened prompt window; (6) When finished, click the check mark to complete the definition of a truss. (7) Now the model is converted into a model ready for simulation using 1D truss elements ( ).

To convert the model to use truss elements, we must first convert the 3D component into a model for beam elements before we can convert the model into a model with truss elements. If we want to convert a model for truss elements into a model for 3D solid elements, we can just right-click the component name tab and select "Treat as Solid". This will convert a model using truss elements directly into a model using 3D elements.

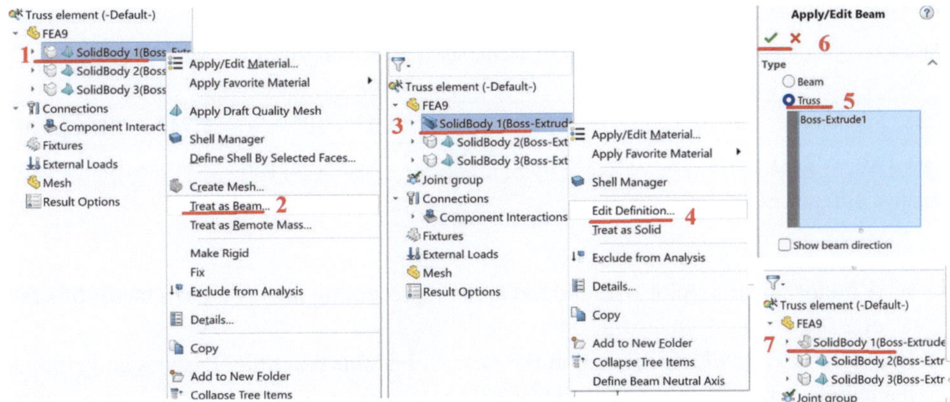

Fig. 3.52 Schematic of defining a truss

Examples of beam and truss elements will be presented in Chap. 4.5 Truss Analysis and Chap. 4.6 Beam Analysis, because they are typically present in an assembly.

Now, the implementation of 3D tetrahedral solid elements and 2D shell elements will be discussed and demonstrated in the following example.

Example 5: 2D shell elements and 3D solid elements (FEA6) Download and open FEA6, which is a sheet-metal model with a thickness of 0.25″. It is fixed on the left end and subjected to a normal force of 200 lbs. on the right end as shown in Fig. 3.53. The material of this model is ASTM36 steel, and the global standard element size is 0.05".

(1) Case One: Run FEA simulation using 2D shell elements. Generate the following plots:
  - von Mises stress plot on the top surface
  - von Mises stress plot on the bottom surface
  - A factor of safety plot with areas under 2.0
(2) Case Two: Run FEA simulation using 3D solid element meshing. Generate the following plots:
  - von Mises stress plot

Fig. 3.53 Schematic of boundary conditions

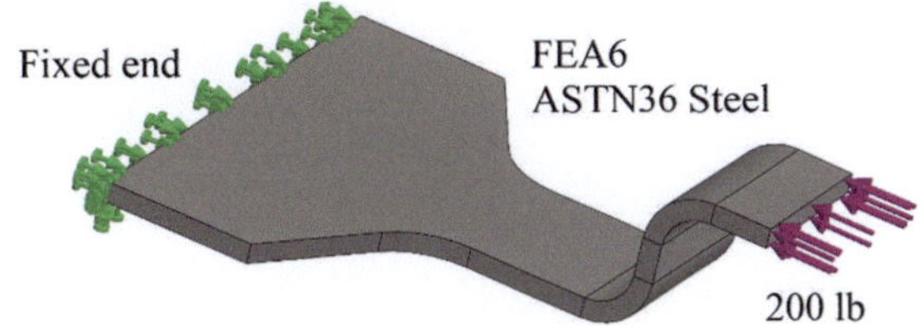

Table 3.4 Some simulation results from Case One and Case Two

Case #	Computing time	Number of elements	Max. von Mises stress	Max. deformation
Case One	3 (s)	20,782	2.005×10^4 (psi)	7.741×10^{-2} (inch)
Case Two	22 (s)	397,335	2.031×10^4 (psi)	7.967×10^{-2} (inch)
Relative differences			1.3%	2.8%

- A factor of safety plot with the red areas representing areas below a factor of safety of 2.0

(3) Fill in Table 3.4, which is listed at the near end of this example. Discuss and compare the results between Case One and Case Two.

Solution

(1) Case One: Meshing by 2D shell elements

Follow Steps 1 to 7 described in Example 1 in Chap. 3.3 to set up this simulation study.

Since the model (FEA6) is created with sheet metal, it is already configured to use 2D shell elements. The FEA simulation setting of Case One is displayed in Fig. 3.54.

Since FEA simulation for 2D shell meshing is focused on computing on the mid-surface, there will be several display options for any stress and any strain as shown in Fig. 3.55. For mechanical component design, the main concerns will be the maximum/minimum stress and strain, which will be on the top surface or the bottom surface. The top and bottom surfaces

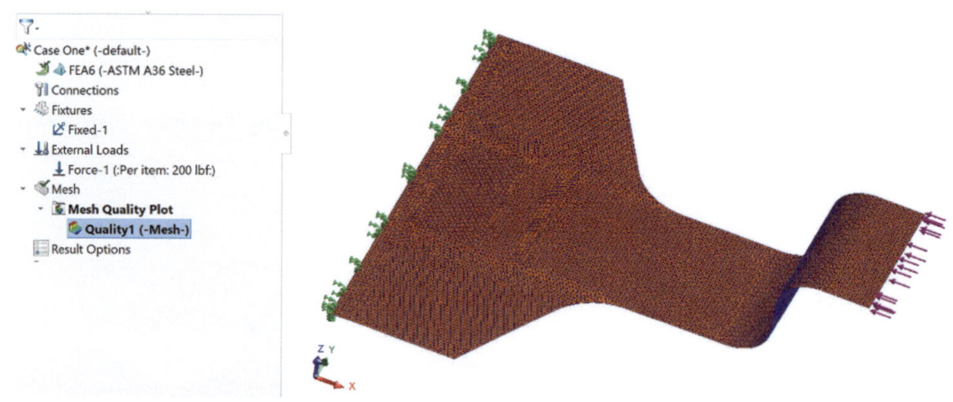

Fig. 3.54 The FEA simulation setting for Case one

are just the outside surfaces of the model for 2D shell meshing. For displacement or factor of safety plots, there is no need to make any surface selections.

The von Mises stress plots for the top and the bottom surfaces are shown in Fig. 3.56. From these figures, we know the maximum von Mises stress is 2.005×10^4 (psi).

The factor of safety plot with red areas below the factor of safety 2.0 is shown in Fig. 3.59a.

(2)   Case Two: Meshing by 3D solid element

Since the model FEA6 is created by using a sheet metal tool, we need to convert it into a model using 3D solid elements. This conversion has been shown in Fig. 3.48 and has been described earlier in this section. The FEA simulation setting for Case Two is shown in

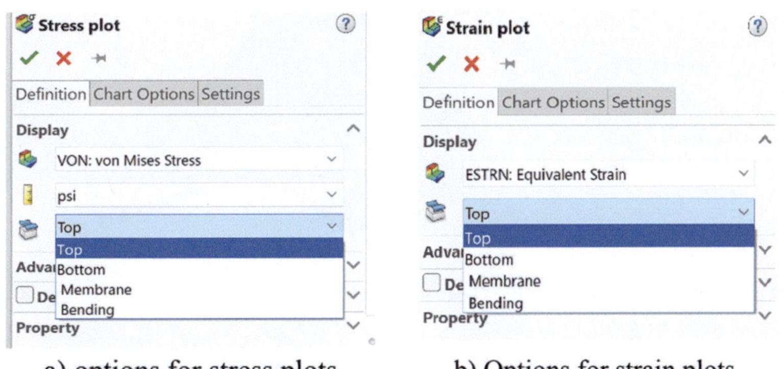

a) options for stress plots   b) Options for strain plots

Fig. 3.55   Options for stress/strain plots

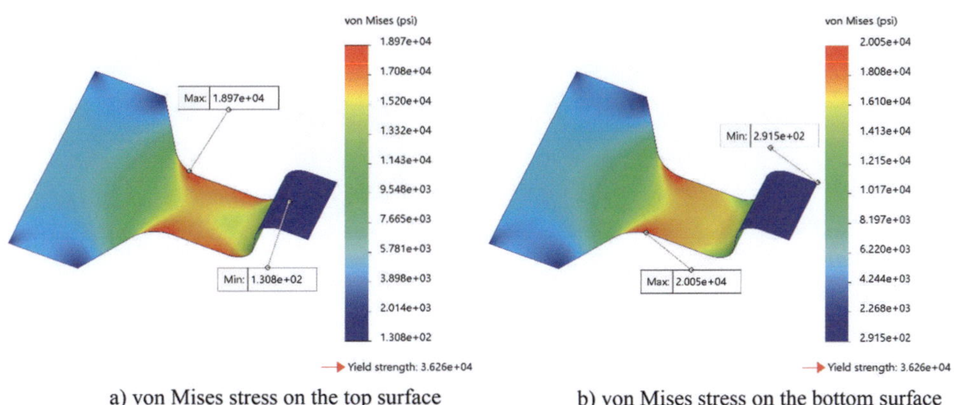

a) von Mises stress on the top surface   b) von Mises stress on the bottom surface

Fig. 3.56   von Mises stress plots on the top and bottom surfaces

Fig. 3.57. The von Mises stress plot for Case Two is displayed in Fig. 3.58. The factor of safety plot with red areas below the factor of safety 2.0 for Case Two is shown in Fig. 3.59b.

(3) Fill in Table 3.4. Discuss and compare the results between Case One and Case Two

After the simulations are completed, we can find the required computing time and total number of elements for each simulation. To do so, right-click "Results" in the simulation

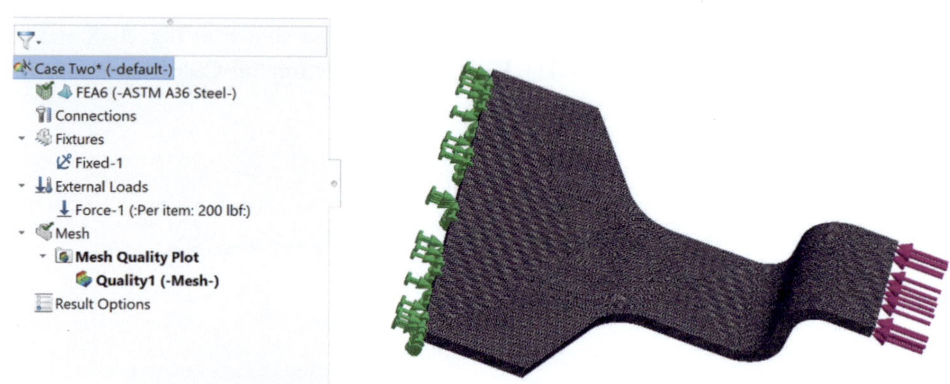

Fig. 3.57 The FEA simulation setting for Case Two

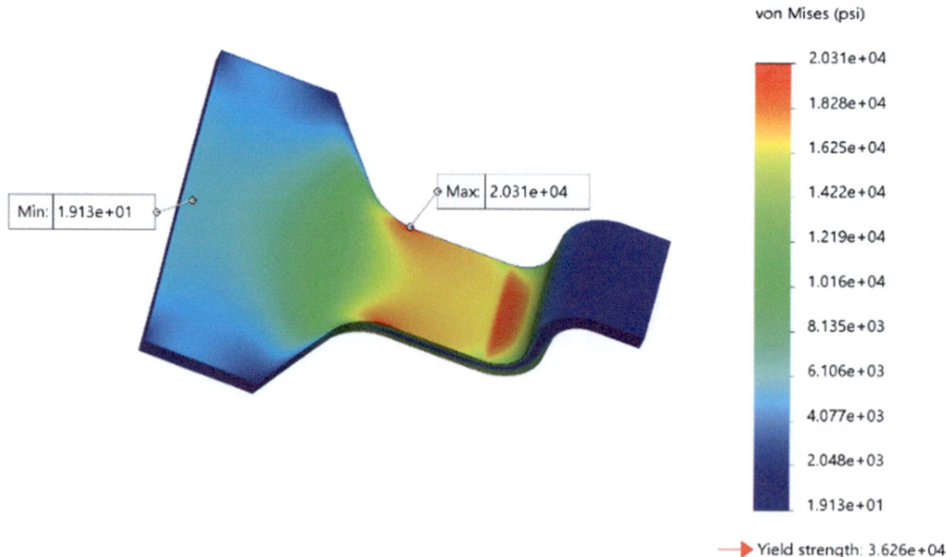

Fig. 3.58 The von Mises stress plot for Case Two

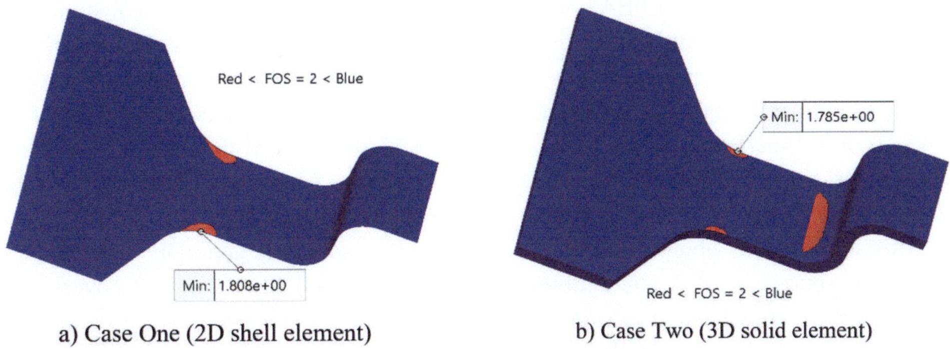

a) Case One (2D shell element)   b) Case Two (3D solid element)

Fig. 3.59 The factor of safety plot with the red area below the factor of safety 2.0

study tree and then select "Solver Messages" in the drop-down menu. Some simulation results from Case One and Case Two are shown in Table 3.4. The relative difference is the ratio of the absolute differences between the two cases to the value obtained from Case Two. Below are some comparisons and discussions about the results.

(a) When the Max. von Mises stresses and Max. deformations are compared between Case One and Case Two, the relative differences are only 1.3% and 2.8%, respectively. Therefore, for mechanical design, we could say that both provide roughly the same simulation results. However, when comparing the computing time and the number of elements between Case One and Case Two, we can see that the simulation by 2D shell meshing in Case One is much faster and simpler.

(b) From Fig. 3.56b and Fig. 3.58, the locations for the maximum von Mises stress in Case One and Case Two are slightly different. However, these locations are still in the same type of locations, the areas near the fillet sections.

(c) If the rated factor of safety is required to be 2.0, Fig. 3.59a and b provide different information. Both figures show that the areas near the fillets will be the failure areas. When redesigning, we will need to pay more attention to these areas. Figure 3.59b also suggests that the second bending area could also fail because its factor of safety is lower than 2.0. But Fig. 3.59a does not provide this information. Even if the simplified model with 2D shell elements is much simpler, runs faster, and provides mostly equivalent simulation results, the original model with 3D solid elements can provide more comprehensive information for mechanical design.

3.7   Fixtures

This section will discuss Step 5: Fixtures of the general FEA simulation procedure and show how to properly define typical supports.

Boundary conditions in FEA simulation are assumptions made about real supporting conditions and load conditions. One of the most difficult tasks in FEA simulation is to properly specify boundary conditions for real products. For the same simulation project, different engineers might use different assumptions based on their understanding of the products, and experiences from testing or previous simulations. If there is a large relative error in the simulation results, the main cause might be from assumptions made about boundary conditions.

After assumptions about boundary conditions are made, it is generally easy to apply these assumptions in FEA simulations, as most commercial FEA software covers most typical types of boundary conditions. Now, let's show how to apply typical supporting conditions in SolidWorks Simulation.

Supporting conditions (Fixtures) in FEA simulation are mathematical descriptions of deformation restraints, which typically include three pieces of information: (1) locations where supports are applied; (2) the type of deformation restraints such as x-, y- and z-direction translation deformation, or rotational deformation about the x-axis, y-axis, and z-axis; (3) magnitude of pre-specified deformation, which is typically set to zero. Supporting conditions are assumptions made about these three pieces of information, through which real supporting conditions of real products can be approximately expressed by mathematical equations or descriptions.

Degrees of freedom are the number of variables needed to fully describe its position and orientation. For example, a point has three degrees of freedom. Its position can be fully defined by x-, y-, and z-coordinates. A surface or an object has six degrees of freedom. Its position can be fully defined by three translational variables (along the x-, y-, and z-axes) and three rotational variables (about the x-, y-, and z-axes).

SolidWorks Simulation uses four types of elements: 1D beam elements, 1D truss elements, 2D shell elements, and 3D solid tetrahedral elements. The nodes (joints) in a 1D beam will have six degrees of freedom because they are simplified as 1D objects. The nodes (joints) of the 1D truss element will only have three degrees of freedom because there is no rotation. Nodes in a 2D shell element have six degrees of freedom because the 2D shell element is the simplified representation of a plate object by the mid-surface. Nodes in 3D solid elements have only three degrees of freedom because they are all points. So, we need to specify six deformation restraints, three translational and three rotational deformation restraints, on supporting places for 1D beam elements, and 2D shell elements. However, 3D solid elements and 1D truss elements will only require three translational deformation restraints.

The schematic of the "Fixture" Property Manager Is shown in Fig. 3.60. Before we discuss each type of fixture in further detail, we will go over some general discussions about fixtures. (1) Right-click the "Fixture" tab in the simulation study tree. A drop-down list will appear and display possible options for fixtures and related functions. (2) Left-click "Fixed Geometry". This will open the property manager for fixed geometry. (3) Left-click "Advanced Fixture". This will open the property manager of the advanced fixtures.

(4) There are six advanced fixtures shown in the "Fixture" Property Manager. (5) In the "Fixture" Property Manager, there is a "Split" tab. This tab can be used to add split lines to the model, to break up one surface into several separate surfaces for applying different fixtures. In this book, this action is covered in Step 1: Pre-processing, as is generally recommended that any necessary split lines be created during the pre-processing step. (6) This section contains the windows used for selecting geometric entities for fixtures and directions. (7) This section is where the values for the deformation on supports can be entered. These values can either be set to zero or set to any specific values needed. Displayed here are values for the translational deformation, but depending on the element type, there can be both translational and rotational deformations specified here. (8) When a fixture is applied, the corresponding symbols here will appear on the selected entities. If we expand "Symbol Settings", we can change the number from 100 to 500 to increase the size of the symbols. Two images with identical fixtures, but different symbol sizes 200 and 400 are shown on the right in Fig. 3.60. (9) This section controls the displaying, hiding, and copying of existing fixtures. "Copy" is used to copy the fixture settings of this simulation project to another simulation project. "Hide All" and "Show All" are used to hide or display the symbols of the fixtures. Typically, when we define a project by adding different types of fixtures, we would like to show these symbols as shown in Fig. 3.57. However, for plots in post-processing, we would like to display clean plots without the symbols of the fixtures. Therefore, we would like to hide all symbols of fixtures as shown in Fig. 3.58. (10) Click the check mark to complete defining a fixture.

Typical fixture tools are "Fixed Geometry"; "Use Reference Geometry"; "On cylindrical Faces"; "On Flat Faces"; and "Symmetry". Using these, we can apply any type of fixture. Now, let's discuss how to use these fixture tools to define fixtures.

"Fixed Geometry" is a tool that is typically used for a connection between a model and the ground. It is the simplest type of fixture and can be applied on surfaces, edges, or points. It will have different settings for deformation restraints based on the different types of elements.

- For a 3D solid model (FEA10), "Fixed Geometry" means that all three translational deformations are zero as shown in Fig. 3.61.
- For a 2D shell model (FEA6), there are two different sets of restraints. "Fixed geometry" means all deformations, including the three translational and three rotational deformations are zero as shown in Fig. 3.62a. "Immovable (No translation)" means all translational deformations will be zero, but there are no restraints on three rotational deformations. The fixture can be applied on the edges or points of 2D shell elements.
- For a 1D beam model (FEA8), there are also two different sets of restraints. The behavior of "Fixed geometry" and "Immovable (No translation)" on 1D beam elements are identical to the behavior with 2D shell elements. The fixture can only be applied on joints as shown in Fig. 3.63. The option "Use Reference Geometry" will be discussed later.

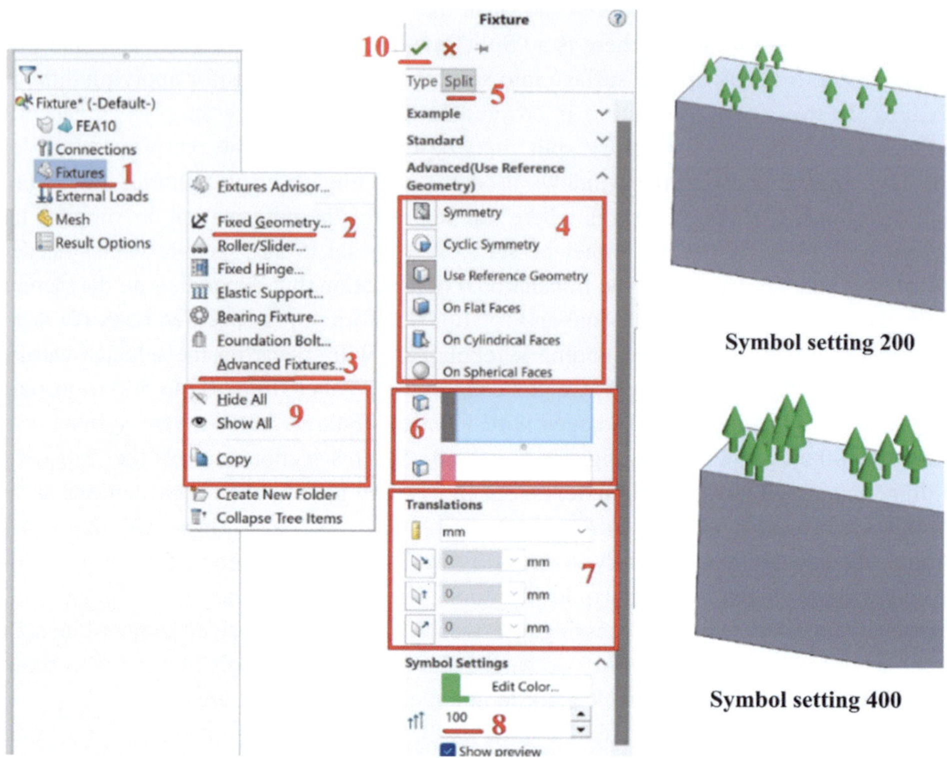

Fig. 3.60 Schematics of Fixture Property Manager

- For a 1D truss model (FEA9), even though there is an option "Fixed Geometry" in the "Fixture" property manager, the option of "Fixed Geometry" is not applicable because no moments can be applied to joints. So "Immovable (No translation)" is the only selection shown in Fig. 3.64. The option "Use Reference Geometry" will be discussed later.

"Use Reference Geometry" is the most powerful tool for defining a fixture, where the directions of deformation restraints are specified by reference geometry such as a line or a surface. It can be used to define any type of fixture.

For this tool, we will explain it in three steps. First, we will show how to access the "Use Reference Geometry" for different types of elements. Then, we will show how to use a line or an axis as a reference geometry for defining a fixture. Finally, we will show how to use a surface as a reference geometry for defining a fixture.

First, let us show how to access the "Use Reference Geometry" option for different types of elements.

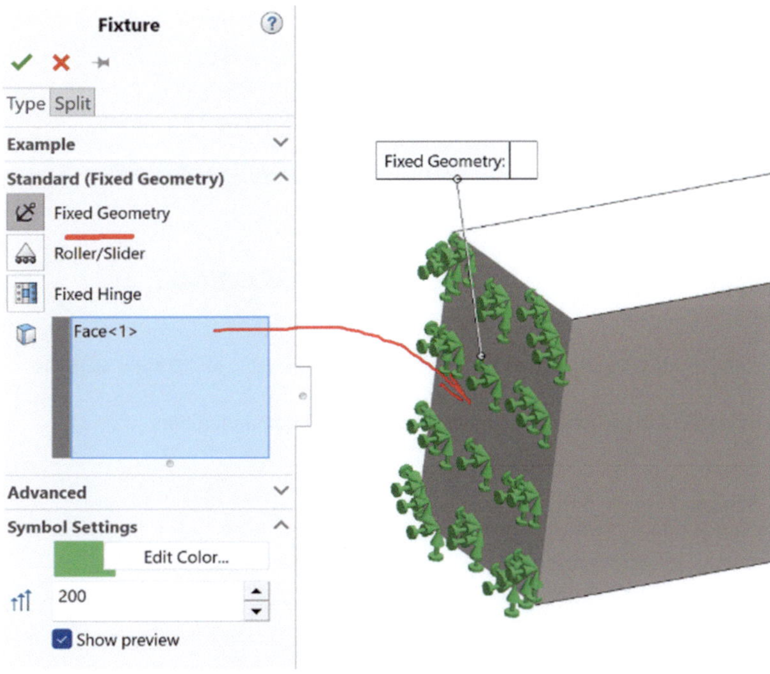

Fig. 3.61 Schematic of defining a fixed geometry for 3D elements

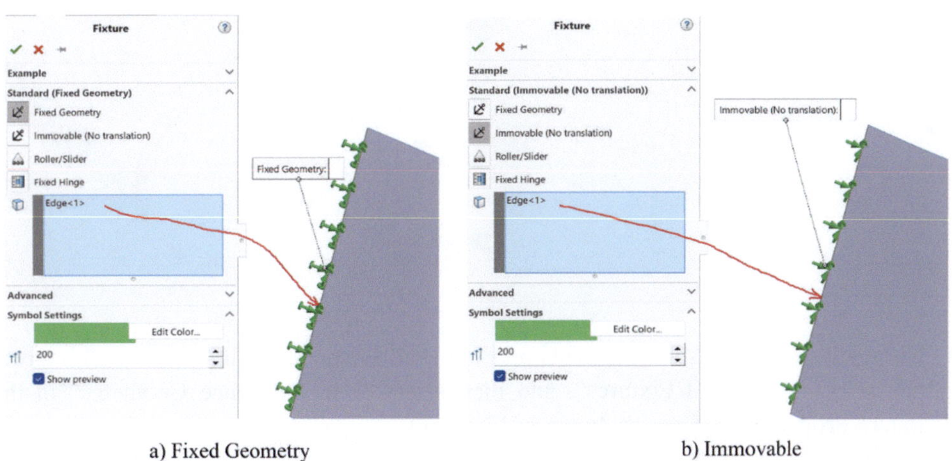

Fig. 3.62 Schematic of defining a fixed geometry for 2D shell elements

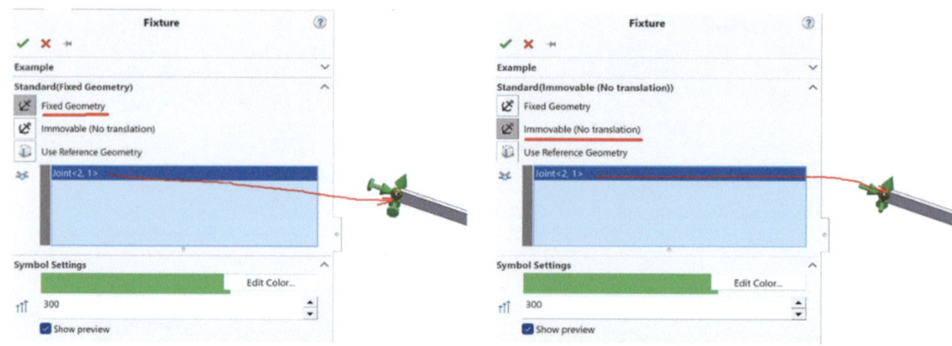

a) Fixed geometry for 1D beam elements   b) Immovable for 1D beam elements

Fig. 3.63 Schematic of defining a fixed geometry for 1D beam elements

Fig. 3.64 Schematic of defining a fixed geometry for 1D truss elements

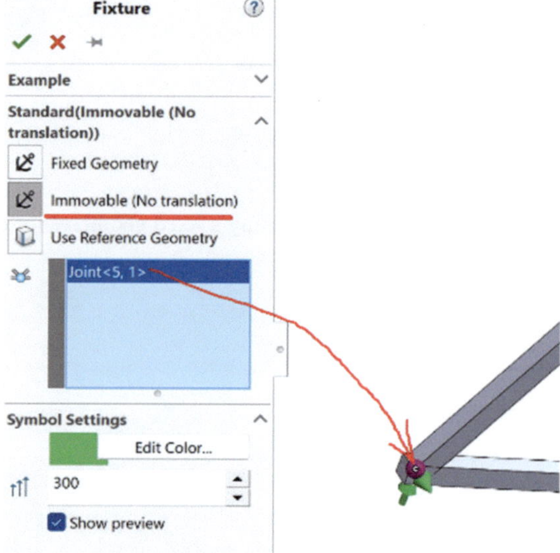

- On models with 3D solid elements and 2D shell elements, right-click the "Fixtures" tab, select "Advanced Fixtures", and then select "Use Reference Geometry" in the fixture Property Manager as shown in Fig. 3.60.
- On models with 1D beam elements and 1D truss elements, right-click the "Fixtures" tab and then select "Fixed Geometry". This will open the fixture Property Manager as shown in Figs. 3.63 and 3.64. Then, we can select "Use Reference Geometry" to update the Fixture Property Manager.

Now let us show how to use a line, edge, or axis as the reference geometry for defining a fixture. When a line, edge, or axis is used as a reference geometry, it will specify the direction of the translational deformation along it and a rotational deformation about it. With this approach, we can select an edge, line, or axis that is parallel to the direction of translational deformation or acts as the axis of the rotational deformation.

- Figure 3.65a is an example of an edge as "Use Reference Geometry" for a 3D solid model (FEA10). Figure 3.65b is an example of an edge as "Use Reference Geometry" for a 2D shell model (FAE6). The first window in the "Fixture" Property Manager is used to select the location for applying the fixture. It can be a face, a line, or a point. The second window is used to select the reference geometry. For 3D elements as shown in Fig. 3.65a, the selected edge is the direction of a translational deformation. But for 2D shell elements as shown in Fig. 3.65b, the selected edge is both the direction of a translational deformation and the axis of a rotational deformation. We need to select "Translations" and/or "Rotation" based on the actual fixture definition.
- Figure 3.66a is an example of an edge selected for "Use Reference Geometry" for the 1D beam model (FEA8). Figure 3.66b is an example of an edge selected for "Use Reference Geometry" for the 1D truss model (FEA9). The first window is used to select the joints on which the fixture is applied. The second window is used to select reference geometry for defining the directions of the deformation restraints. For 1D beam elements, the selected edge is the direction of a translational deformation, and the axis of a rotational deformation as shown in Fig. 3.66a. For the 1D truss elements,

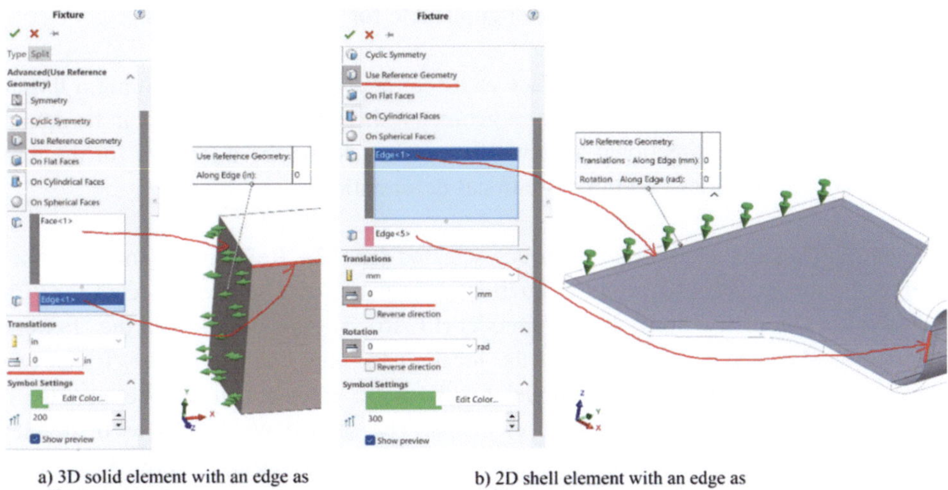

a) 3D solid element with an edge as reference geometry

b) 2D shell element with an edge as reference geometry

Fig. 3.65 An edge used by "Use Reference Geometry" for a 3D solid element and a 2D shell element

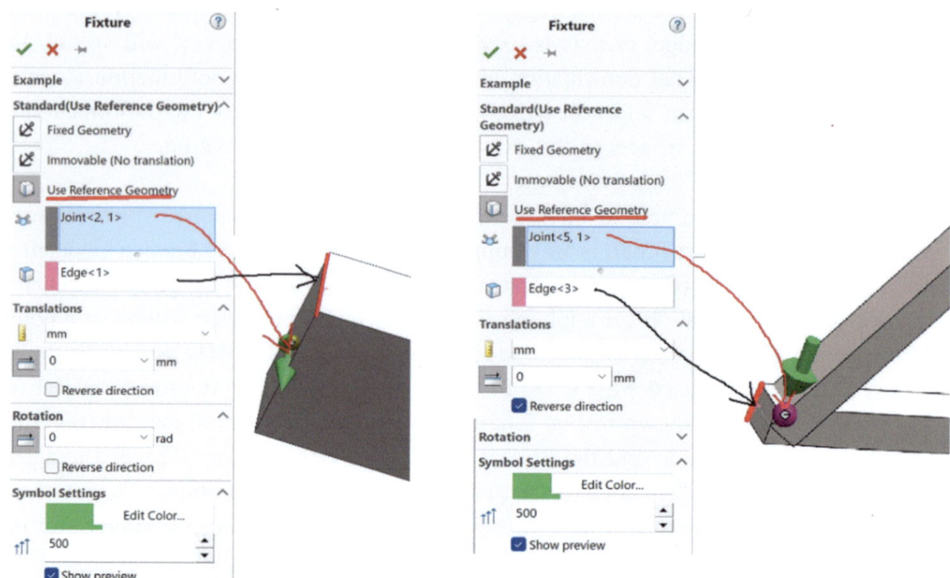

a) 1D beam element with an edge as reference geometry

b) 1D truss element with an edge as reference geometry

Fig. 3.66 An edge as "Use Reference Geometry" for 1D beam and 1D truss elements

the selected edge is only the direction of a translational deformation. We can ignore the "Rotation" option because it is not applicable for 1D truss elements.

Finally, let us show how to use a surface as a reference geometry for defining a fixture. When a flat surface is used as a reference geometry, three directional axes can be defined by this surface. These directional axes will be used to define the directions for the three translational deformations and three rotational deformations.

- Figure 3.67a is an example of a surface selected for "Use Reference Geometry" for a 3D solid model (FEA10). Figure 3.67b is an example of a surface selected for "Use Reference Geometry" for a 2D shell model (FEA6). The first window in the "Fixture" Property Manager is used to select the location for applying the fixture. It can be a face, a line, or a point on the model for 3D solid elements. It can be an edge or a point on the model for 2D shell elements. The second window is used to select a surface as a reference geometry. For 3D elements as shown in Fig. 3.67a, three entries are available for the three directions for translational deformations. But for 2D shell elements as shown in Fig. 3.67b, three entries for translational deformations and three entries for rotational deformations are available. We can select the corresponding deformation entries based on the actual fixtures.

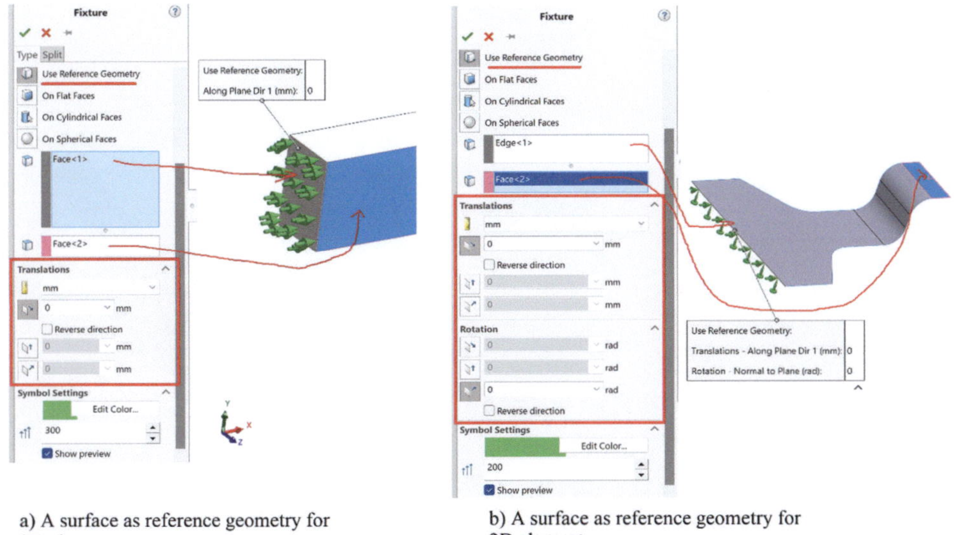

a) A surface as reference geometry for 3D elements

b) A surface as reference geometry for 2D elements

Fig. 3.67 Schematics of a surface as reference geometry for 3D elements and 2D elements

- Figure 3.68a is an example of a surface selected for "Use Reference Geometry" for the 1D beam model (FEA8). Figure 3.68b is an example of a surface selected for "Use Reference Geometry" for the 1D truss model (FEA9). The first window is used to select the joints on which the fixture is applied. The second window is used to select a surface as a reference geometry. For the 1D beam element, three entries for translational deformations and three entries for rotational deformations are available. For the 1D truss element, three entries for translational deformations are available. The entries for rotation are available as well but can be ignored because they do not apply to 1D truss elements. It is recommended to use a surface as the reference geometry when defining fixtures for 1D beam elements and 1D truss elements because it can specify all types of fixtures.

"On cylindrical faces" is typically used to define fixtures on cylindrical surfaces that are supported or connected with pins or shafts.

For mechanical design, when an FEA simulation on a component is run, connections between cylindrical surfaces and supporting pins or shafts can typically be defined by the fixture tool "On cylindrical faces".

- Figure 3.69a is an example of "On Cylindrical Faces" on a model (FEA11) using 3D solid elements. There are three entries for three possible deformation restraints. These entries are for deformations along the radial direction, axial direction, and circumferential direction.

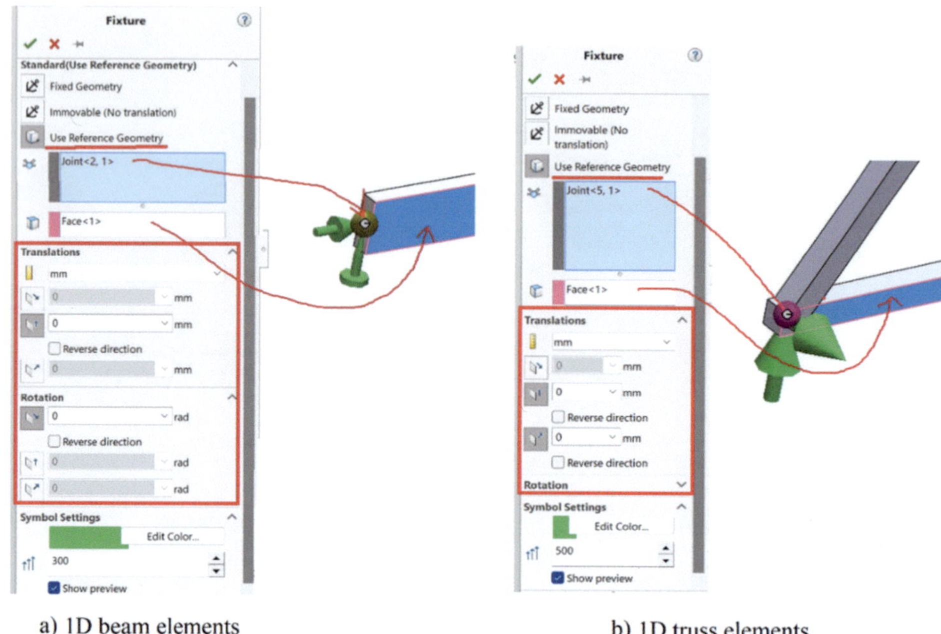

a) 1D beam elements   b) 1D truss elements

Fig. 3.68 Schematics of a surface as reference geometry for 1D beam and truss elements

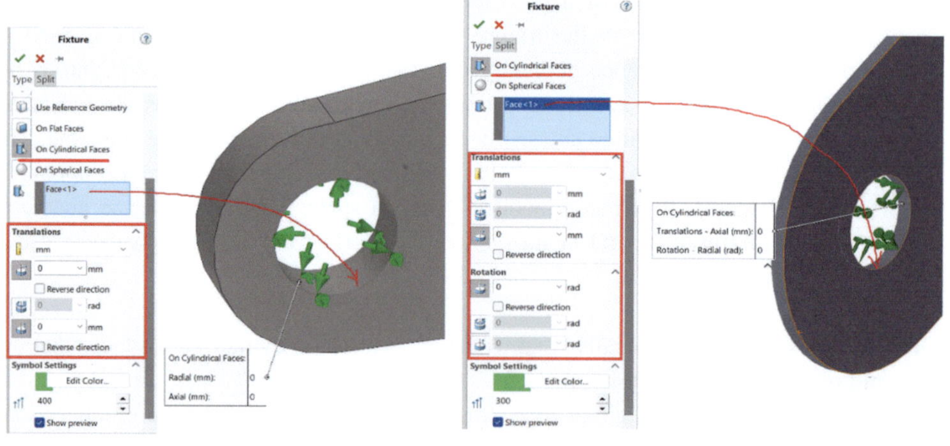

a) "On Cylindrical faces" for 3D solid elements   b) "On Cylindrical faces" for 2D shell elements

Fig. 3.69 Schematic of defining a fixture by "On Cylindrical Faces"

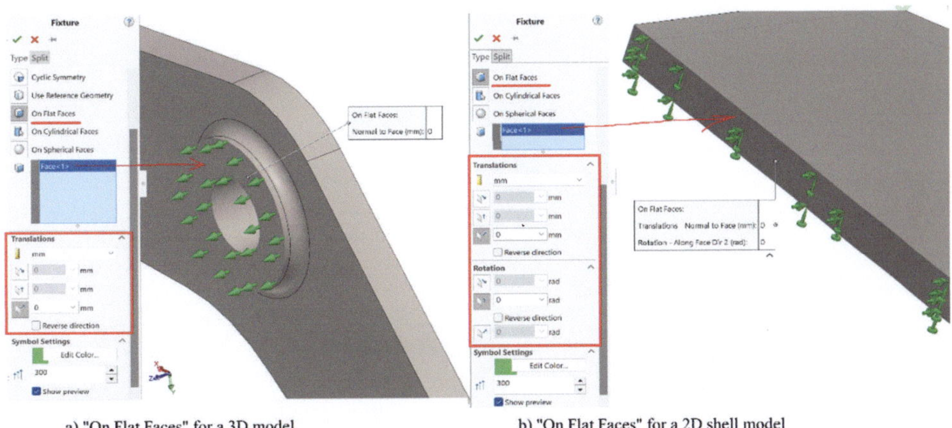

a) "On Flat Faces" for a 3D model   b) "On Flat Faces" for a 2D shell model

Fig. 3.70 Schematic of defining a fixture by using "On Flat Faces"

- Figure 3.69b is an example of "On Cylindrical Faces" on a model (FEA12) with 2D shell elements. There are six entries for three possible translational deformations and three possible rotational deformations.

"On Flat Face" is the same as "Use Reference Geometry" except that the reference geometry and the selected flat face for the fixture are the same. This can be used to specify a fixture on a flat face for 3D models and an edge face for 2D shell models.

- Figure 3.70a is an example of "On Flat Face" on a 3D solid model (FEA11). It provides three entries for the three possible translational deformation restraints. It is typically used to define a fixture without normal deformation.
- Figure 3.70b is an example of "On Flat Face" on a 2D shell model (FEA6), which was created by sheet metal tools such as base flange, edge flange, jog, and sketched bend. The edge faces have a total of six possible deformations including three translational deformations and three rotational deformations.

"Symmetry" is used to simulate a portion of the model instead of the full model. The implementation of "Symmetry" requires that geometry, loads, and materials are all symmetrical. The main purpose of this tool is to significantly reduce the simulation computing time.

- Figure 3.71a is the original model (FEA13) in which one end is fixed and another end is subjected to tensile force. For this case, the loading, fixture, and geometry of the model are symmetrical about the middle plane in the height direction. So a half model with "Symmetry" can be used for the simulation, as shown in Fig. 3.71b. On

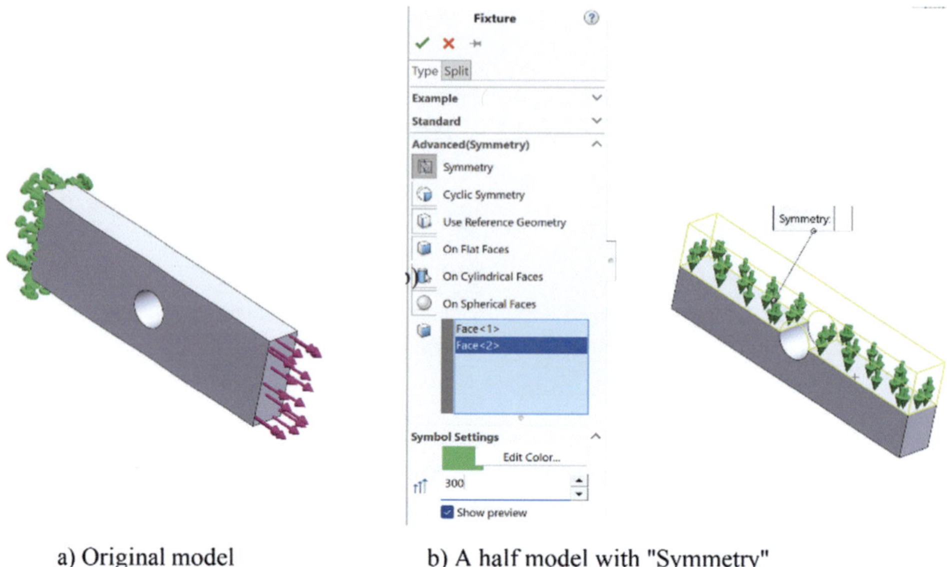

a) Original model   b) A half model with "Symmetry"

Fig. 3.71 Schematic of using "Symmetry" on a solid model

the surfaces where "symmetry" is applied, there is no normal deformation for the solid model.

- Figure 3.72a is an original model called FEA7, which is a thin tube with a fixed bottom end and an internal expanding force. This model is symmetric about the center axis. Figure 3.72b is the quarter model with two "Symmetry" surfaces. Since it is a 2D shell model, the normal deformation and two rotation deformations about axes on the symmetry surfaces will be zero.

SolidWorks Simulation provides many tools for defining a fixture, but we can use the fixture tools discussed above to define any type of fixture on 1D beam elements, 1D truss elements, 2D shell elements, and 3D solid elements. Now, we will use Example 6 to demonstrate that the FEA simulation results might be completely different when different fixtures are assumed.

Example 6: Different assumptions of supports (FEA15) Download and open FEA15, which is a pivot beam model. The device consists of supporter 1, supporter 2, two pins, and a pivot beam (FEA15) as shown in Fig. 3.73. The supporters are fixed on their bottom surfaces. A load of 1750 (lb) is applied on the shaded area of the top surface. There are relatively big gaps between the side surfaces of the pivot beam and the inner side surfaces of supporter 1. However, there is a very small gap between the side surfaces of the pivot beam and the inner side surfaces of the supporter 2. The design concern is the pivot beam. The pivot

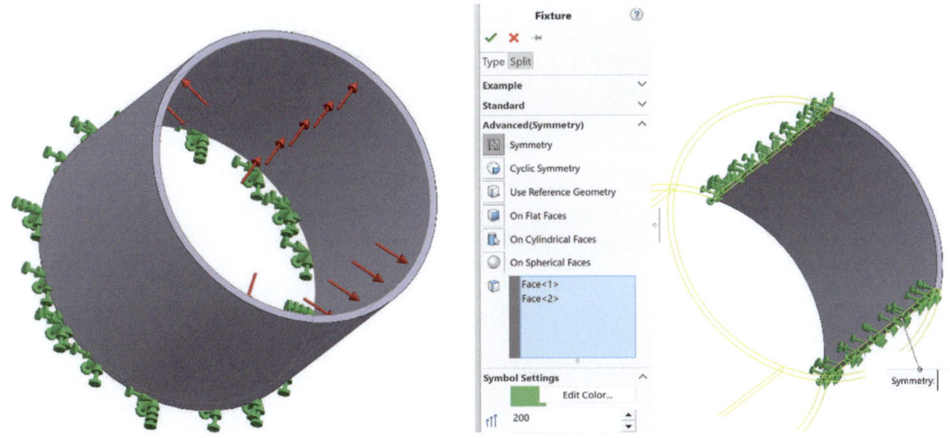

a) Shell model with a fixed bottom and an internal pressure

b) The quarter model with "symmetry"

Fig. 3.72 Schematic of using "Symmetry" on a shell model

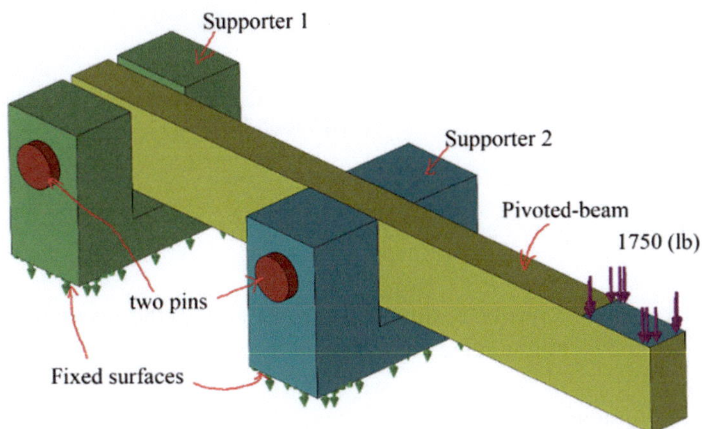

Fig. 3.73 Schematic of an assembly

beam's material is plain carbon steel, and its factor of safety is 1.5. Some possible supporting conditions for the pivot beam are presented as Case One and Case Two in Fig. 3.74.

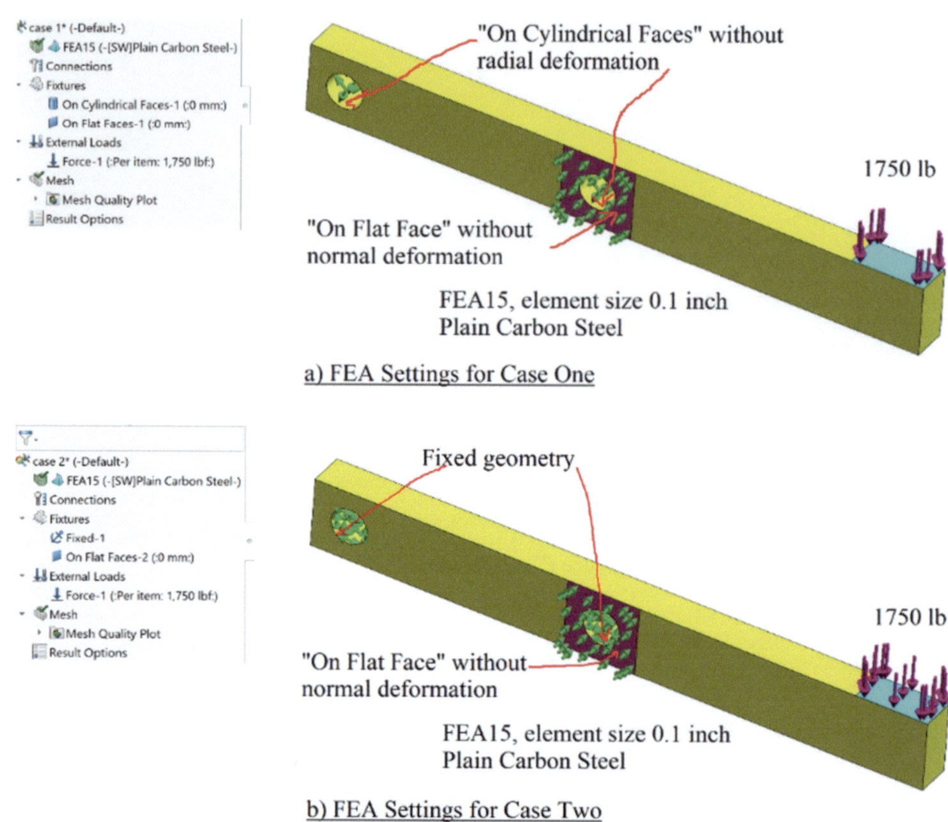

a) FEA Settings for Case One

b) FEA Settings for Case Two

Fig. 3.74 The FEA Settings for Case One and Case Two

Use these two cases to run FEA simulations to determine the maximum deformation, Von Mises stress, and a minimum factor of safety. The global standard element size is 0.1".

(1) Compare the simulation results from two cases. If the component fails, identify how to modify it.
(2) Discuss the two fixture assumptions of both cases.

Solution

(1) **FEA simulation**

Step 1: **Pre-processing**. Since all other components in this device are strong and the pivot beam is the design concern, it is not necessary to run an FEA simulation on the whole device. Instead, we can run a FEA simulation only on the pivot beam component (Note: FEA simulation on the assembly will be discussed in the next chapter).

Therefore, the design issue can be simplified into the design of the pivot beam. Four split lines have been added on side surfaces to define the areas for fixtures near the second cylindrical hole. One split line has been added on the top surface of the pivot beam to define the area for external loading. Now, the model is meshable and ready for FEA simulation.

Step 2: **Set up a project.** This is a static stress analysis by the "Static" module, so we will set up two simulation projects: One for Case One and another for Case Two.

Step 3: **Select the type of element and assign a material.** 3D solid elements with high-quality meshes will be used, so we don't need to make any changes to the type of element. Plain carbon steel is included in the SolidWorks material library so we can directly assign plain carbon steel to the component.

Step 4: Connections. This step is not needed for a component simulation.

Step 5: Fixtures. There are four places for fixtures. Two cylindrical holes are supported by two pivot pins and two side surfaces near the second cylindrical hole are supported by supporter 2. The following sets of fixtures in these two cases will be used for FEA simulation.

Case One: Use "On Cylindrical faces" on two cylindrical holes to define fixtures without radial deformation. This is a typical assumption for connections with pivot pins and support shafts. For this device, there is no loading along the pin's axial direction. Since the gaps between the pivot beam and the inner surfaces of supporter 2 are very small, supporter 2 will prevent the motion of the beam along the pin's axial direction. Given this, we can use "On Flat Face" to define a fixture without normal deformation on any one side surface near the second hole. The fixtures for Case One are shown in Fig. 3.74a.

Case Two: Use "Fixed Geometry" on the two cylindrical holes. Use "On Flat Face" to define a fixture without normal deformation on any one side surface near the second hole. The fixtures for Case Two are shown in Fig. 3.74b.

Step 6: **Loads.** Right-click the tab "External loads" and select "Force". A 1750 (lb) normal force is applied on the right top shaded area.

Step 7: **Meshing.** Right-click the tab "Mesh" and select "Create Mesh". Then check the "Mesh Parameters" and enter $0.1''$ as the global element size for the standard mesh.

Step 8: **Run Simulation.** Right-click the project name and select "Run". This will start the FEA simulation.

Step 9: **Post-processing.** For this example, the resultant deformation plots of both cases are shown in Fig. 3.75a and b. The Von Mises stress plots of both cases are shown in Fig. 3.76a and b. Figure 3.77a and b shows the factor of safety plots for both cases, with red areas indicating areas below a factor of safety of 1.5.

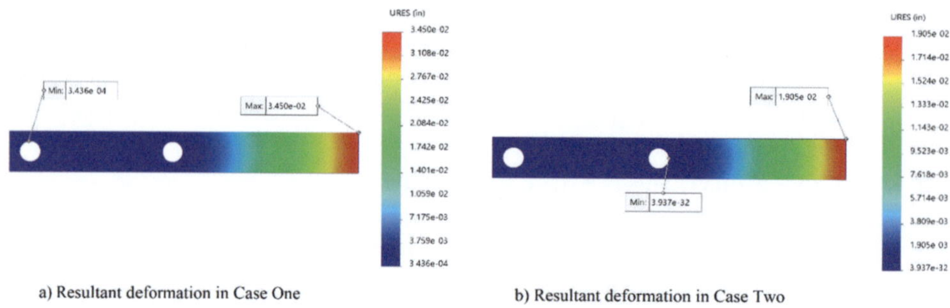

a) Resultant deformation in Case One

b) Resultant deformation in Case Two

Fig. 3.75 The resultant deformation plots

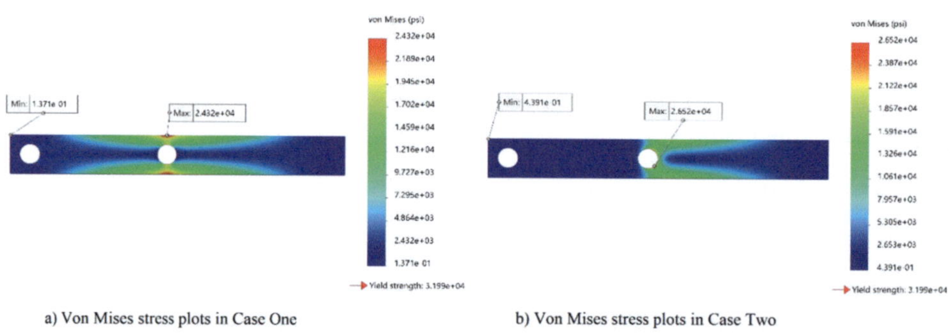

a) Von Mises stress plots in Case One

b) Von Mises stress plots in Case Two

Fig. 3.76 The Von Mises stress plots

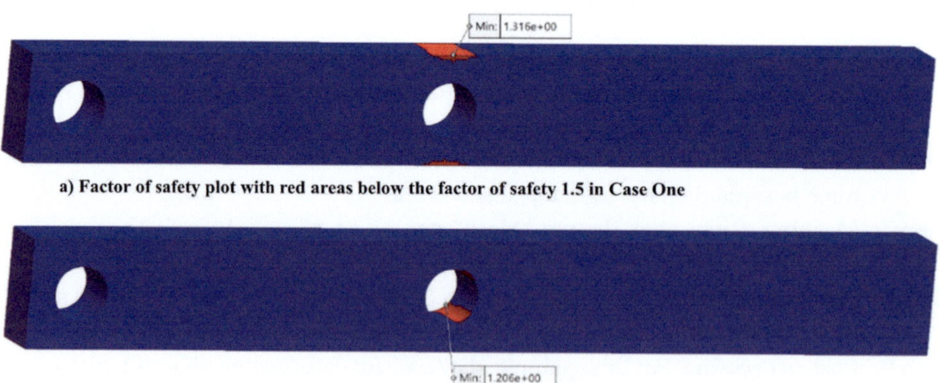

a) Factor of safety plot with red areas below the factor of safety 1.5 in Case One

b) Factor of safety plot with red areas below the factor of safety 1.5 in Case Two

Fig. 3.77 The factor of safety plots with red areas below the factor of safety 1.5

(2) **Simulation results and comparisons**

From Fig. 3.75, the maximum resultant deformations for Case One and Case Two are 0.0345 (inch) and 0.0191 (inch). These two values are quite different. However, the locations of the maximum deformations for both cases are the same, at the right-top edge of the pivoted beam.

From Fig. 3.76, the maximum Von Mises stresses for Case One and Case Two are 24,320 (psi) and 26,520 (psi). The difference between these two values is 9%. However, the locations of the maximum Von Mises stress are quite different. From Fig. 3.76a for Case One, the location of the maximum Von Mises stress is at the top portion of the second cylindrical hole. From Fig. 3.76b for Case Two, the location of the maximum Von Mises stress is on the inner surface of the second cylindrical hole.

In Fig. 3.77, the failure areas where the factor of safety is below 1.5 are quite different for both cases. For Case One, the failure areas are the top and bottom portions of the second cylindrical hole. This result indicates that possible modifications for a redesign could be to increase the moment of inertia of the cross-section on the second hole, as this is a type of failure due to bending stress. For Case Two, the failure areas are on the inner surface of the second cylindrical hole. This result indicates that possible modifications for a redesign could be to increase the hole contact surface area because this is a typical type of failure due to contact stress. The simulation results of these different cases give different directions for possible modifications and redesign when different sets of fixtures are used for the same design problem.

When compared with the simulation results on the assembly shown in Fig. 3.73, Case One provides better results. The predictions of locations for maximum deformation, maximum Von Mises Stress, and possible failure areas for Case One are the same as those from the simulation on the assembly. It should be noted that the fixtures without radial deformation on two cylindrical holes indicate that the supporting pins are assumed to be rigid. In the simulation of the assembly, the pins are deformable and are not treated as rigid. Using fixed fixtures on the two cylindrical holes like in Case Two is not correct because relative deformation or rotation between pins and the holes is allowed.

3.8 Loading

This section will discuss how to apply loads such as force, pressure, bending moment/torque, centrifugal force, and remote load, which are typical loadings for mechanical design. This is Step 6: Loading of the general FEA simulation procedure which specifies all the external loads applied on models as part of defining the boundary conditions.

External Loading on models means the mathematical description of every external load on the models, which are some assumptions made about the real loads on real products or devices. Descriptions of external loads on models include four pieces of information: (1) the type of loading such as force, pressure, or moment; (2) the location where the loadings are applied; (3) the directions along which loadings are applied; (4) the magnitudes of the loadings.

One of the most difficult tasks in FEA simulation is to specify the external loadings on models to represent real loadings on products. After what external loads on models to apply have been decided upon, it will be easy to apply them in FEA simulations as most commercial FEA simulation software has extensive tools that cover the most typical types of loads. Now let's show how to apply typical loads to models.

Follow the sequence shown in Fig. 3.78 to access the tools for defining loads. (1) Right-click the "External loading" tab; (2) In the section of the drop-down list, the typical types of loads such as force, torque, pressure, gravity, centrifugal load, temperature, and remote load are shown. We will discuss all of these load types except for "temperature" in this chapter. "Temperature" will be discussed in Chap. 6 Thermal analysis and thermal stress analysis. (3) This section contains tools for displaying and copying loads. "Copy" is used to copy the loading settings of this simulation project to another simulation project. "Hide All" and "Show All" are used to hide or display the symbols of the loading. Typically, when we define a project by adding different types of loading, we would like to display these loadings like in Fig. 3.74. However, for plots in post-processing, we would like to display clean plots without the loadings displayed, like the plots shown in Fig. 3.75.

Fig. 3.78 The loading property manager

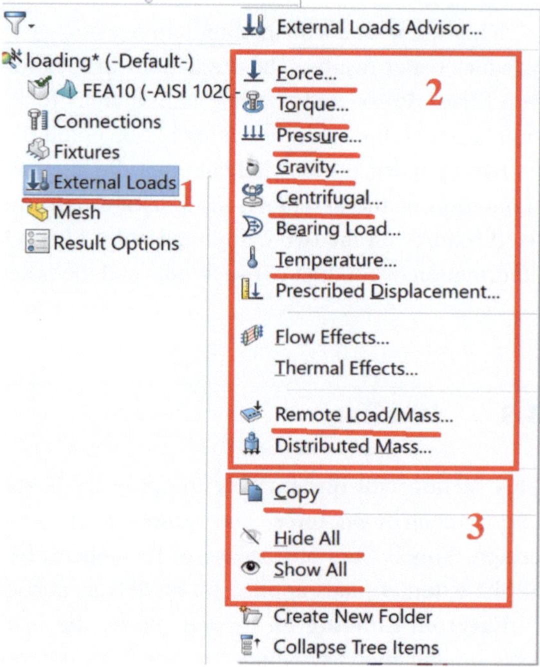

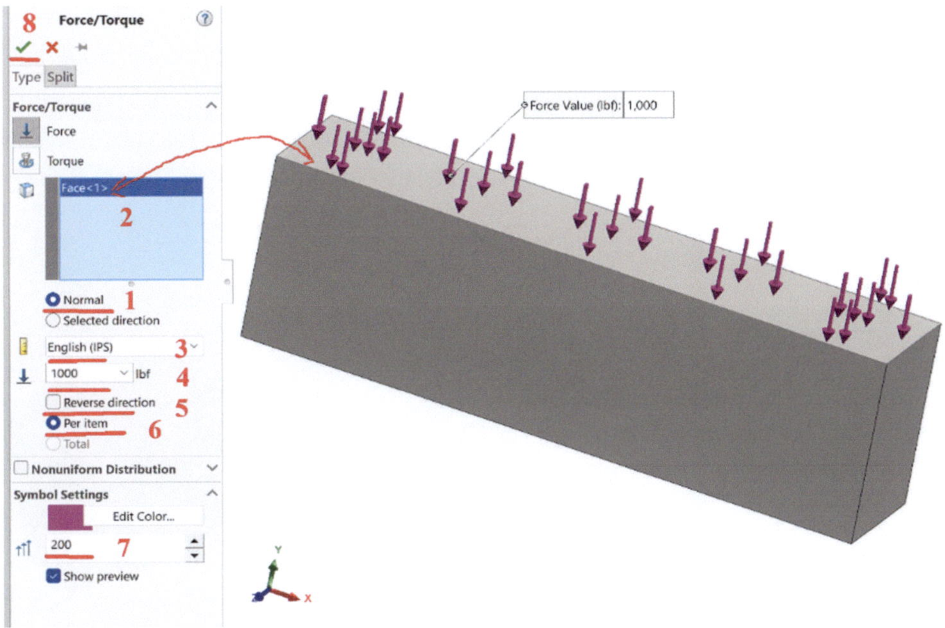

Fig. 3.79 Force normal to a surface (FEA10)

Force is an external load that can be applied along a line, on a point, or a surface. Figures 3.79, 3.80, and 3.81 display applying force normal to the surface, along an edge, and in a direction specified by a surface.

After a simulation project has been set up, right-click the "External loads" tab and select "Force" as shown in Fig. 3.78. Then, the "Force/Torque" Property Manager will appear as shown in Fig. 3.79. There are three ways to define the direction of the force: (1) "Normal" to the surface; (2) "Selected direction" by an edge or line; and "Selected direction" by a surface. Now, we will explain these three approaches.

Force normal to a surface: Follow the sequence in Fig. 3.79 to define a force normal to a surface. (1) Set the direction of the force to be normal to the surfaces where the load is applied; (2) Select a surface for applying load. This surface can be flat or curved; (3) Select the unit system for the load; (4) Enter the magnitude of the load; (5) If desired, we can reverse the force direction, but it will still be normal to the selected surface; (6) Selecting the "Per item" option means that the same magnitude of load will apply on each of the selected surfaces; (7) Here we can change the value between 100 to 500, to scale the size of the displayed loading symbols; (8) When done defining the load, click the check mark to apply the load.

Force along an edge: When the "Selected direction" beneath "Normal" as shown in Fig. 3.79 is selected, the "Force/Torque" Property Manager will become the image as

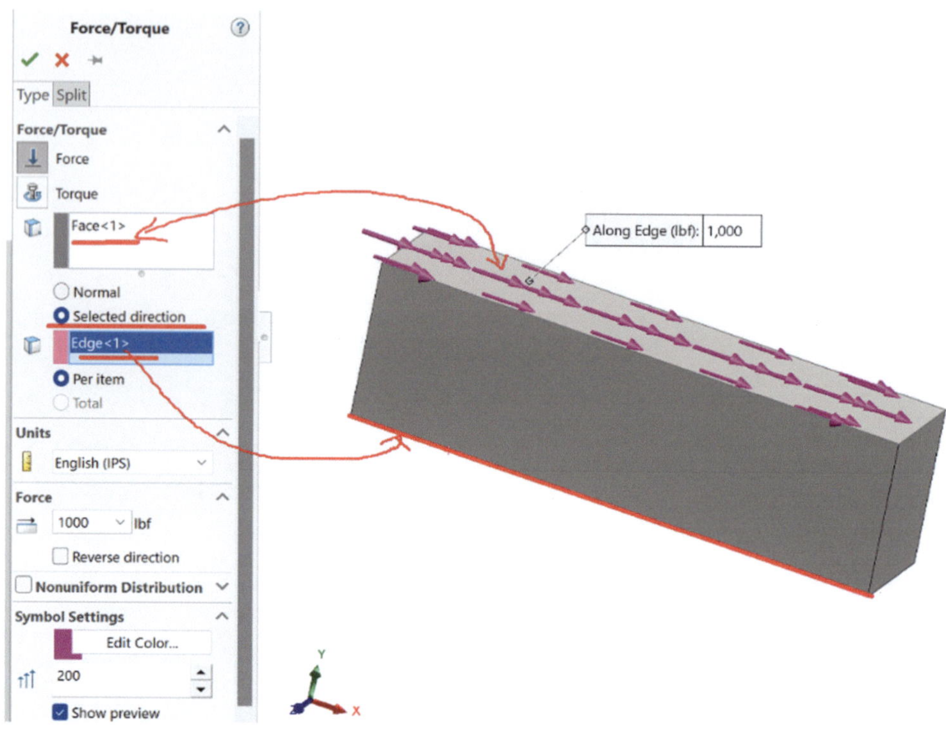

Fig. 3.80 Load along an edge

shown in Fig. 3.80. The first window is for the location where loading is applied. The second window is for reference geometry. If a straight line or edge is selected, the load will be applied along the direction of this edge as shown in Fig. 3.80. The rest of the selections can be the same as the selection shown in Fig. 3.79.

Force along directions by a surface: When a surface is selected to define the "Selected direction", the "Force/Torque" Property Manager will look like the image shown in Fig. 3.81. The first window is for the location where loading is applied. The second window is for a surface as the reference geometry. Since a flat surface can define three directions: two directions along the surface and one direction normal to the surface, three possible directions will appear for selections. Forces with specified magnitude can be applied along any one direction, or all three directions.

A torque is a moment about an axis. A bending moment is also a moment about an axis and can be treated as a torque. To define a torque, we need to specify the location, axis, and magnitude of a torque. Typically, we need to create a reference axis to be used as the axis of the torque. When the center axis of a cylindrical surface is the same as the axis of the torque, we can select the cylindrical surface to replace the axis.

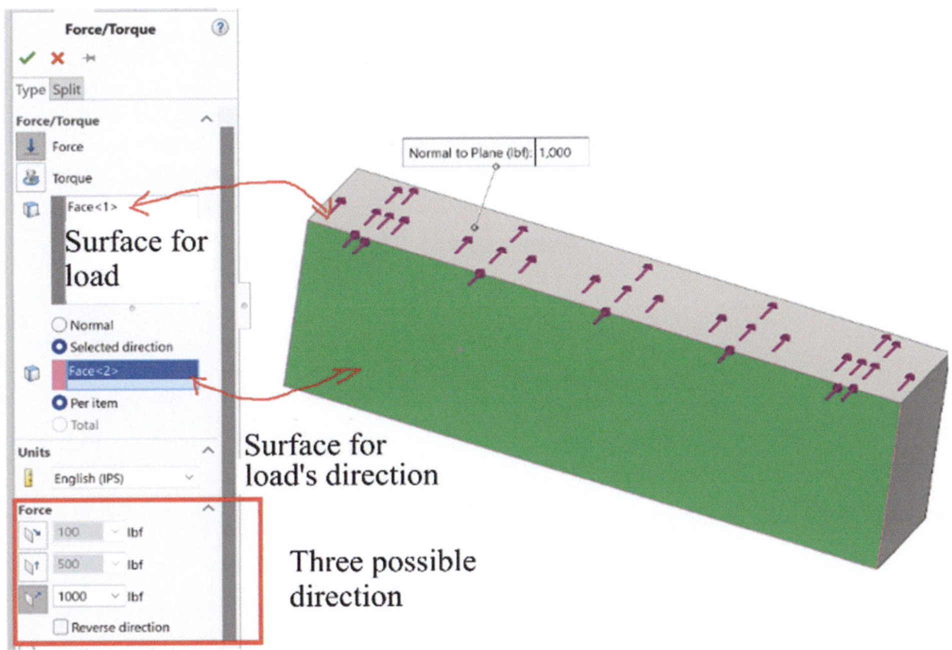

Surface for load

Surface for load's direction

Three possible direction

Fig. 3.81 Force along directions specified by a surface

After a project is set up, right-click the "External loads" tab and select "Torque" as the type of load as shown in Fig. 3.78. The "Force/Torque" Property Manager will appear as shown in Fig. 3.82. In the "Force/Torque" Property Manager, the first window is for the location where torque/moment is applied. The second window is for the reference axis. We can use the same sequence to define a bending moment as shown in Fig. 3.83.

Pressure is the force per unit area on a surface. For defining a pressure, we need to specify the location, direction, and magnitude of the pressure. The procedure for defining a pressure is almost identical to defining a force. There are three ways we can specify the direction of a pressure: normal to surface as shown in Fig. 3.84; along a selected edge as shown in Fig. 3.85; and along one of three possible directions of a reference geometry as shown in Fig. 3.86.

Gravity is the weight of a component. Typically, the weight of a component is negligible because the stress caused by the weight is much smaller when compared with the stress caused by working loads. For defining gravity, we just need to specify a surface or plane that is normal to the gravity direction as shown in Fig. 3.87a.

Centrifugal force is an inertial force about a rotating axis. For defining a centrifugal force, we need to specify a rotating axis and a rotating speed such as RPM (Revolution Per Minute) as shown in Fig. 3.87b. The rotating axis can be a pre-created reference axis.

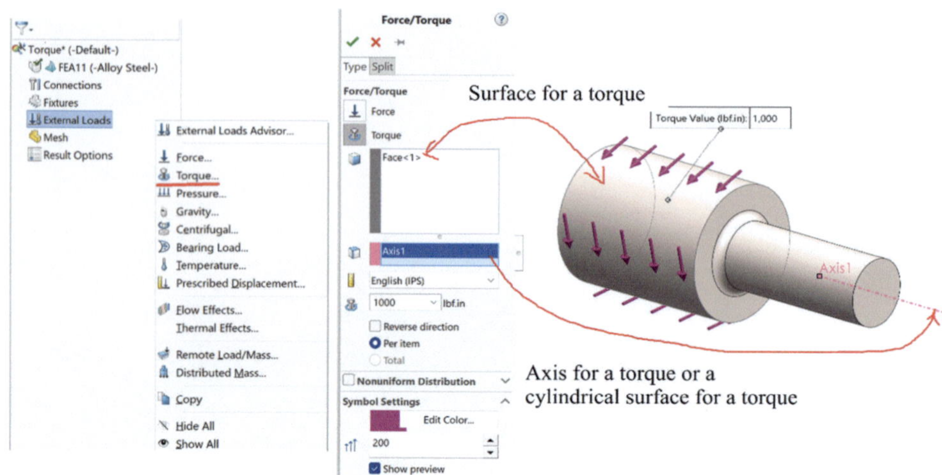

Fig. 3.82 Schematic of applying a torque

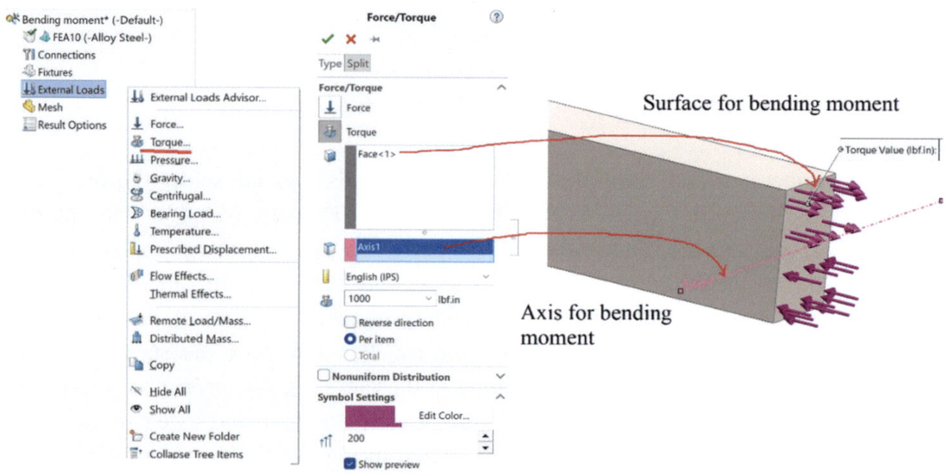

Fig. 3.83 Schematic of defining a bending moment

The rotating axis can also be replaced by a cylindrical surface when the center axis of the cylindrical surface is the same as the rotation axis.

Remote Loads are loads applied at a point outside of the models in the FEA simulation. The purpose of FEA simulation is to obtain the stress/strain/deformation of concerned components without needing to run the FEA simulation on the whole assembly. So, we generally only include the concerned components in the simulation and omit

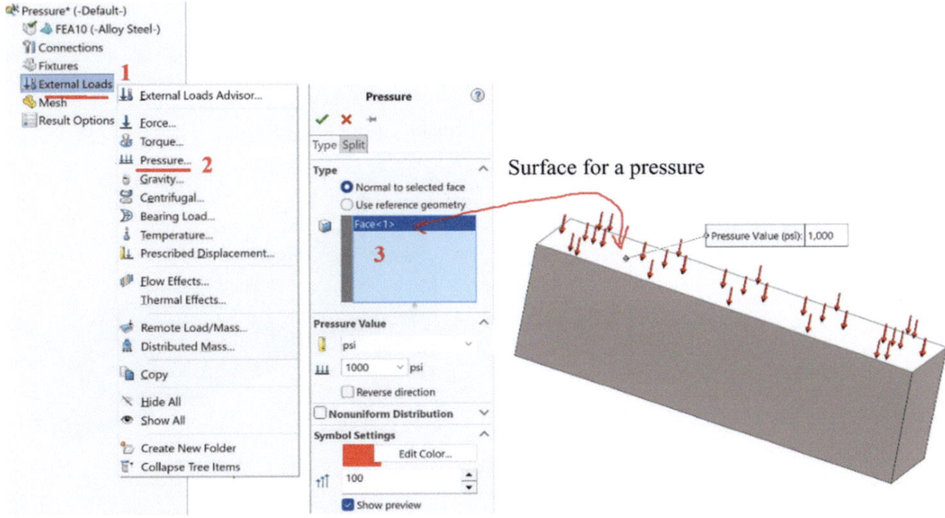

Fig. 3.84 Schematic of defining a pressure normal to a surface

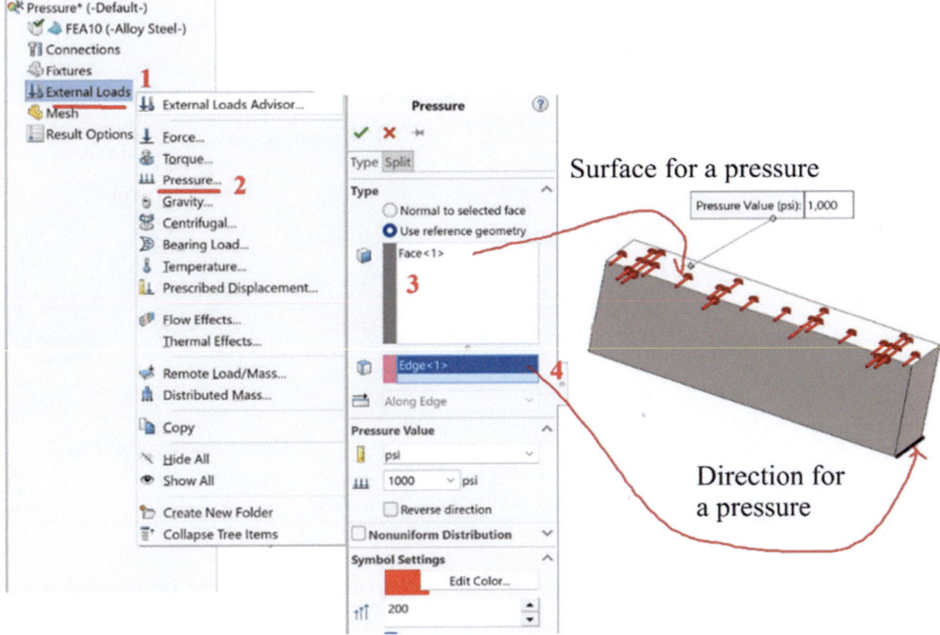

Fig. 3.85 Schematic of defining a pressure along an edge

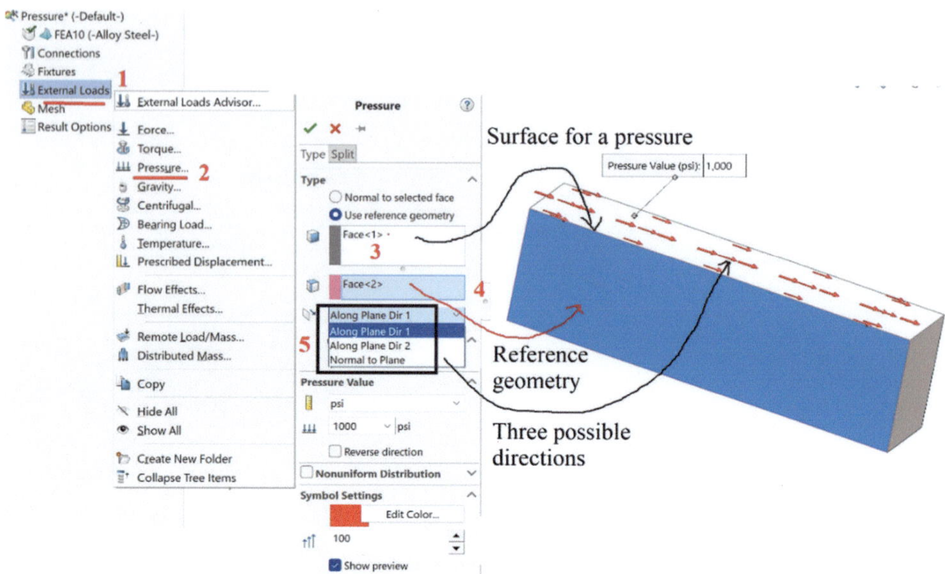

Fig. 3.86 Schematic of defining a pressure by reference geometry

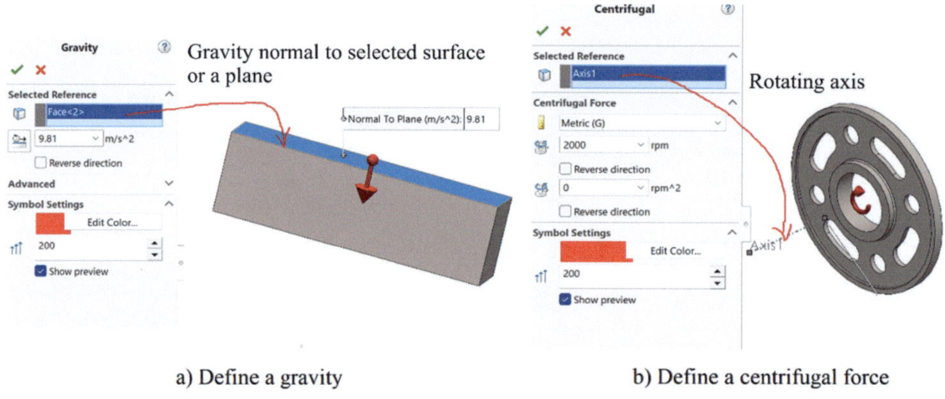

a) Define a gravity   b) Define a centrifugal force

Fig. 3.87 Schematics of defining a gravity and a centrifugal force

all other non-concerned components. On a remote point, there might be three possible forces and three possible moments.

Based on assumptions made about the connection between the omitted components and the component under consideration, there are two different types of connections for remote loads.

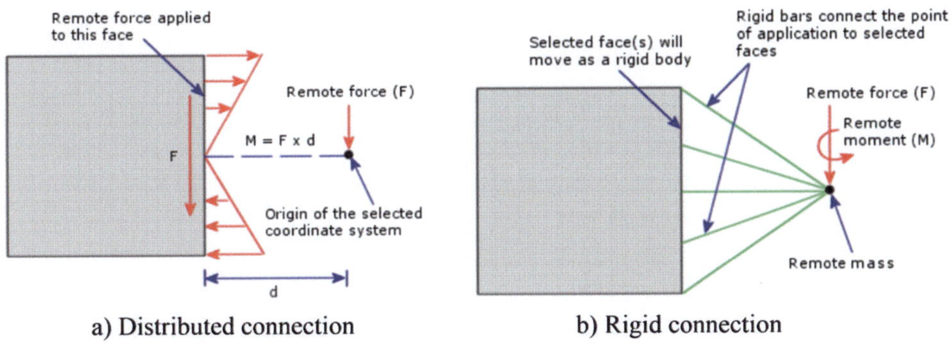

a) Distributed connection   b) Rigid connection

Fig. 3.88 Two types of remote loads

Remote Load with distributed connection: when the omitted component is adequately flexible, but its displacements are still within the small displacement assumption. A force applied at the remote location can be transferred as an equivalent force-moment system applied to the selected face(s) as shown in Fig. 3.88a.

Remote Load with rigid connection: When the omitted component is very strong and has very small deformation, the omitted component is treated as a rigid connection on the place for the remote load as shown in Fig. 3.88b.

For defining remote loads, we need to specify these three types of information: (1) the location for the converted remote loads; (2) the location of the remote point; and (3) the remote loads at the remote point. After the simulation project is set up, right-click the "External load" tab and select "Remote load". The "Remote Loads/Mass" Property Manager will appear as shown in Fig. 3.89. Follow the sequence shown in Fig. 3.89 to define remote loads. (1) This window is used to select the location for the remote load, which can be a surface, an edge, or a point; (2) This section is used to select a coordinate system for defining the remote point. Typically, a global coordinate system is used, but a user-defined coordinate system can be used too; (3) The units and coordinate values of the remote point can be entered here; (4) The "Translational Components" section is where we can select the proper units for forces, and contains the three entries used to define the three forces along three coordinate axes at the remote point; (5) The "Rotational Components" section is where we can select the proper unit system for the moments. The section also contains the three entries used to define moments about the coordinate axes at the remote point; (6) "Connection Type" is where we can select between using remote loads with a distributed connection or remote loads with a rigid connection.

All the typical types of loads and the general approaches for defining these loads have been discussed in detail. Next, we will present three examples of projects applying different loadings. The first example, Example 7, covers a beam under bending. The second example, Example 8, covers a round shaft under torsion. The third example, Example 9, covers a bar under remote loads.

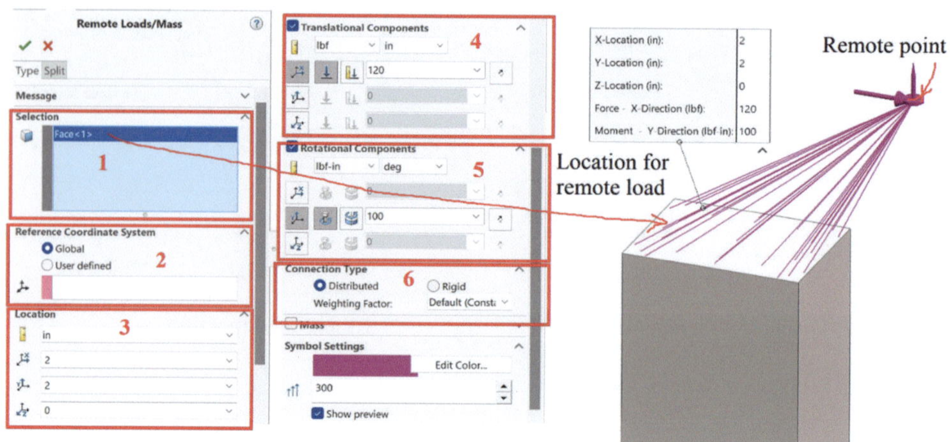

Fig. 3.89 Schematic of defining a remote load

Example 7: A beam under pure bending (FEA17) Download and open FEA17. FEA17 consists of a beam that is subjected to a bending moment of 400 lb-in at the right end and is fixed at the left end as shown in Fig. 3.90. Its material is AISI 1020 steel. Run the FEA simulation with a global standard element size of 0.1 (inch) and compare the simulation results obtained on edge cd and split line ab with theoretical results obtained using beam theory. Discuss the findings.

Solution

(1) **FEA simulation**

Step 1: **Pre-processing**. A reference axis on the right end, axis 1, has been created for defining a bending moment. The stresses on this beam will be x-normal stress according to the loading condition shown in Fig. 3.90. The x-normal stresses from the FEA simulation on line ab and the edge cd will be compared with the theoretical calculation results. The line ab is an added split line in the middle of the beam.

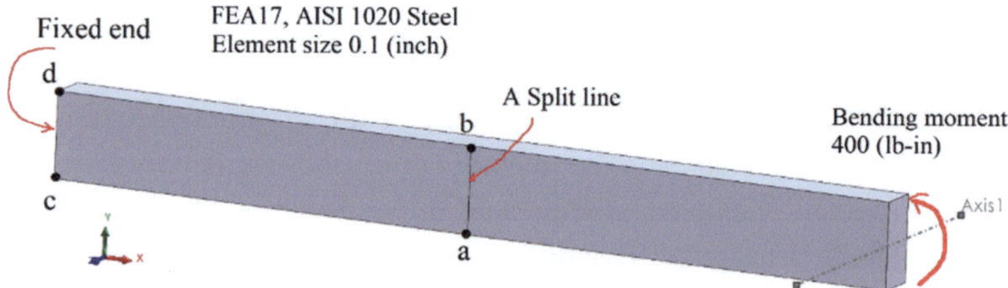

Fig. 3.90 The FEA settings of a beam under a pure bending moment

The model is ready for a FEA simulation, so no further actions are needed in Pre-processing for this example.

Step 2: **Set up a project**. This is a static stress analysis.

Step 3: **Select the type of element and assign a material to each component**. We will use 3D solid elements with a high-quality mesh, so no changes are required to the type of element. AISI 1020 steel is already included in the SolidWorks material library, so we can directly apply AISI 1020 to the component.

Step 4: **Connections**. This step is not needed for a component simulation.

Step 5: **Fixtures**. Right-click the "Fixture" tab and select "Fixture geometry" to define the fixed end on the left end of the beam.

Step 6: **Loads**. Right-click the "External loads" tab and select "Torque" to define a 400 lb-in bending moment on the right end about the reference axis 1.

Step 7: **Meshing**. Right-click the "Mesh" tab and select "Create Mesh". Check the "Mesh Parameters" and enter $0.1''$ as the global element size for standard mesh.

Step 8: **Run Simulation**. Right-click the project name and select "Run". This will start the FEA simulation.

Step 9: **Post-processing**. The x-normal stress plot of the beam is shown in Fig. 3.91. The x-normal stress graphs along the split line ab and the edge cd are shown in Fig. 3.92a and b, respectively.

(2) **Comparison and discussions**

The beam has a height of 1 inch and a width of 0.5 inch. Since the beam is under pure bending moment, the top of the beam will have a positive constant x-normal stress and the bottom of the beam will have a negative constant x-normal stress per beam theory. The magnitude of the x-normal stress on the top or the bottom of the beam is:

Fig. 3.91 The normal stress along the x-axis direction

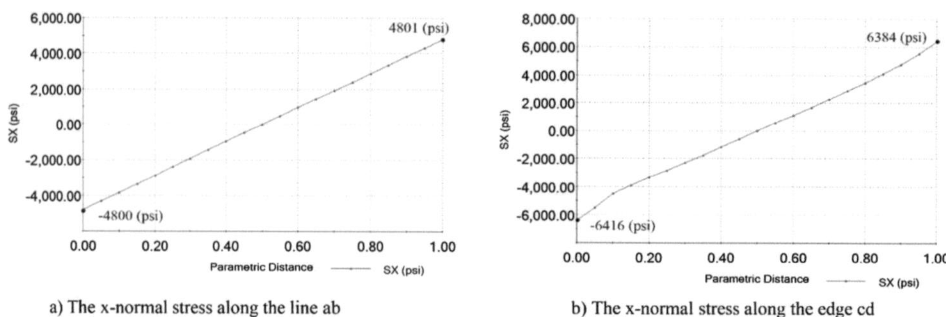

a) The x-normal stress along the line ab

b) The x-normal stress along the edge cd

Fig. 3.92 The x-normal stresses along the line ab and the edge cd

$$\sigma_x = \frac{M_z(h/2)}{\left(h^3 b/12\right)} = \frac{400 \times (1/2)}{\left(1^3 \times 0.5/12\right)} = 4800 \ (psi)$$

In beam theory, theoretical bending stresses will vary linearly on any vertical line such as the split line ab and the edge cd.

From running the FEA simulation, the FEA simulation results at points a and b are $-$ 4800 (psi) and 4801 (psi) as shown in Fig. 3.92a. The relative error of the simulation results when compared with theoretical calculation is less than 0.02%. The x-normal stress along the split line ab as shown in Fig. 3.92a is a linear line. So, the FEA results on the split line ab are the same as the theoretical results. In this example, the split line is in the middle of the beam, but the line can be at any position between two ends, except on the two edges of the two ends.

The x-normal stresses at points c and d per Figs. 3.91 or Fig. 3.92b are -6000 (psi) and 6384 (psi) respectively. The relative errors of FEA simulation results on these two points when compared with the theoretical results are 25% and 33%, respectively. These are big relative errors. Based on the graph in Fig. 3.92b, the x-normal stress on the edge cd is not a linear straight line. Therefore, the x-normal stresses on the edge cd from the FEA simulation don't match the expected theoretical results.

From the same example, we have two completely different conclusions: (1) X-normal stresses on the split line ab from FEA simulation do have the same results as theoretical calculations, and (2) X-normal stresses on the edge cd from FEA simulation do not have the same results as theoretical calculations. For beam theory, one assumption made is that the cross-section for the stress calculation can be freely deformed. The cross-section with the split line ab satisfies this assumption. However, the cross-section with the edge cd is a fixed face, which doesn't satisfy this assumption. This is the main reason why there is such a large relative error. We can say that we should not compare the FEA x-normal stresses on edge cd with theoretical calculation results, because the stress conditions in the FEA simulation are not the same as the stress conditions in theoretical calculations.

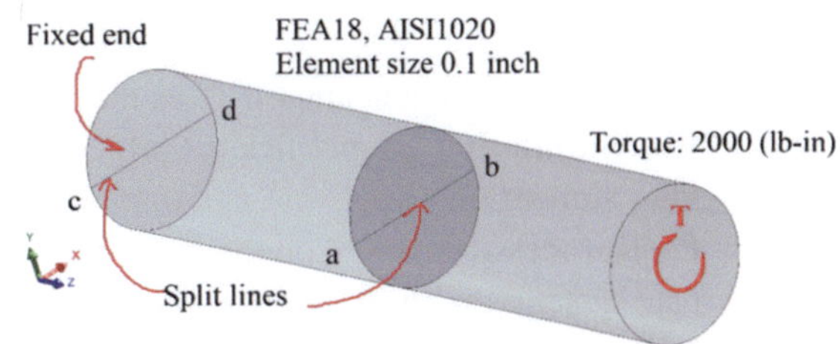

Fig. 3.93 A solid circular shaft under a torsion

Example 8: A shaft under torsion (FEA18-ASM) Download, unzip, and open FEA18-ASM. The project consists of a solid circular shaft, fixed on the left end, and subjected to a torque of 2000 (lb-in) on the right end as shown in Fig. 3.93. The shaft material is AISI 1020 steel. Run an FEA simulation with a standard element size of $0.1''$ and compare the simulation results on line cd and line ab with theoretical results from the torsion formula. Discuss the findings.

Solution

(1) **FEA simulation**

Step 1: **Pre-processing**. Based on Eq. (3.1–3.6), the torsion formula of a solid circular shaft is $\tau = (T\rho)/J$. The shear stress at a point is linearly proportional to its radius and is normal to the line between this point and the center of the circular cross-section. Two split lines ab and cd are created through the center of the cross-section and added to display shear stress on them. To create the split line ab, two half shafts are created. On the left haft shaft, two splits are added as shown in Fig. 3.93. Then two half shafts are connected through a bonded connection, which will be discussed in Chap. 4: FEA simulation on assemblies. In such a way, the split line ab is accessible and the assembly of two half shafts can be treated as a component because they are bonded together. The model is ready for a FEA simulation.

Step 2: **Set up a project**. This is a static stress analysis simulation.

Step 3: **Select the type of element and assign a material to each component**. The 3D solid elements with high-quality mesh will be used so we don't need to make any changes to the type of elements. AISI 1020 steel is included in the SolidWorks material library, and we can directly assign AISI 10,120 to the components.

Step 4: **Connections**. The connection between two half shafts is bonded together, which will be discussed in Chap. 4. No action is needed for this example.

Step 5: **Fixtures**. Right-click the "Fixture" tab and select "Fixture geometry" to define the fixed end on the right end of the shaft.

Step 6: **Loads**. Right-click the "External loads" tab and select "Torque" to define a 2000 lb-in torque on the right end surface of the shaft. The cylindrical surface in this example can be used as the axis for torque.

Step 7: **Meshing**. Right-click the "Mesh" tab and select "Create Mesh". Check "Mesh Parameters" and then enter $0.1''$ as the element size for standard mesh.

Step 8: **Run Simulation**. Right-click the project name and select "Run". This will start the FEA simulation.

Step 9: **Post-processing**. Based on the orientation of the shaft relative to the global coordinate system, the shear stress τ_{zy} is chosen for the stress plot distribution as shown in Fig. 3.94. The shear stress τ_{zy} on the split lines ab and cd are shown in Fig. 3.95a and b respectively.

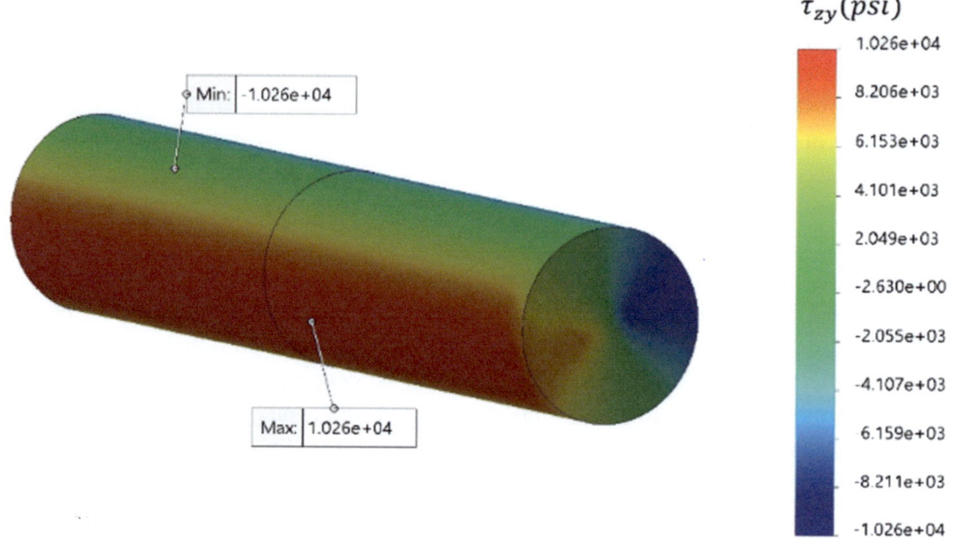

Fig. 3.94 The shear stress τ_{zy} plot of the shaft

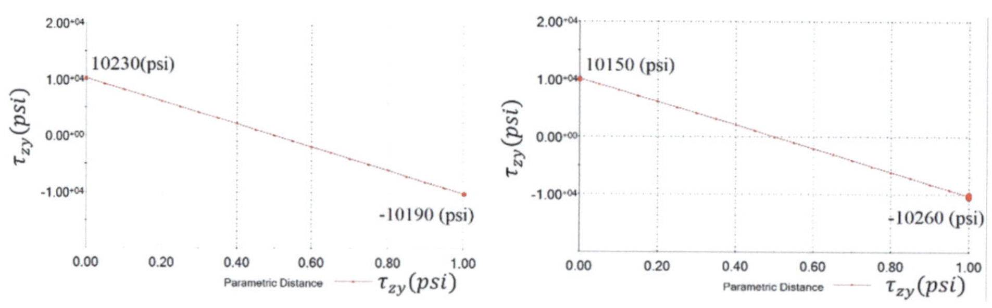

a) Shear stress τ_{zy} graph on the split line ab　　　b) Shear stress τ_{zy} graph on the split line cd

Fig. 3.95 Shear stress τ_{zy} graphs on the split lines ab and cd

(2) **Comparison and discussions**

The solid circular shaft has a diameter of 1 inch. Using the torsion formula, the magnitude of shear stress of this shaft on the outer surface under a constant torque of 2000 (lb-in) is

$$\tau = (T\rho)/J = (2000 \times 0.5)/(\pi \times 1^4/32) = 10186 \ (psi)$$

The graph of shear stress τ_{zy} on the split line ab is shown in Fig. 3.95a. The shear stress τ_{zy} at the two end points of the split line ab are $-10{,}190$ (psi) and $10{,}230$ (psi) respectively. The relative errors of the FEA simulation results at point a, and point b are 0.04% and 0.4%, respectively. The shear stress τ_{zy} on the split-line ab (a diametric line) from FEA simulation varies linearly with its radius of the point for the shear stress. These lead to the conclusion that shear stresses τ_{zy} on line ab from FEA simulation are aligned with theoretical calculation results.

The split line cd is on the fixed surface of the shaft. The graph of shear stress τ_{zy} on the line cd from the FEA simulation is shown in Fig. 3.95b. Shear stresses τ_{zy} at the two endpoints of the line cd are $-10{,}260$ (psi) and $-10{,}150$ (psi) respectively. The relative errors of the FEA simulation results at point c and point d are 0.7% and 0.03% respectively. The shear stress τ_{zy} on the split-line cd from FEA simulation varies linearly with the radius of the stress point. Therefore, the shear stresses τ_{zy} from FEA simulation agree with the theoretical calculation results.

Why does the fixed surface not affect the shear stress calculation? In the torsion formula, the shear deformation is the relative angle of twist between two cross-sections. One assumption is that the cross-section is still a flat surface. Therefore, the shear stress conditions in this FEA simulation on the fixed end surface satisfy all assumptions made for the theoretical shear stress formula, so it will not affect shear stress calculations.

Example 9: Remote loads with distributed connection and equivalent loads (FAE19) Download and open FEA19. It is a vertical column, fixed at the bottom end, and subjected to a remote load with a distributed connection on its top surface. The global coordinate system has an origin in the center on the top surface of the column. The coordinates of the remote point in the global coordinate system are (4, 2, 0) inches. The remote loads as shown in Fig. 3.96a are $F_y = 1500 \ (lb)$ and $M_z = 2000 \ (lb - in)$. When the remote loads are converted into loads directly applied on the top surface of the column, they become a vertical tensile force $F_y = 1500 \ (lb)$ and a bending moment $M_z = 4 \times 1500 + 2000 = 8000$ (lb. in) as shown in Fig. 3.96b. The material of this column is AISI 1020 steel. Run FEA simulations of the column under (1) the remote loads with distributed connection and (2) under the equivalent loads. Compare the FEA simulation results with these two loading conditions. Use 0.1 inch for the global standard element size for both simulations.

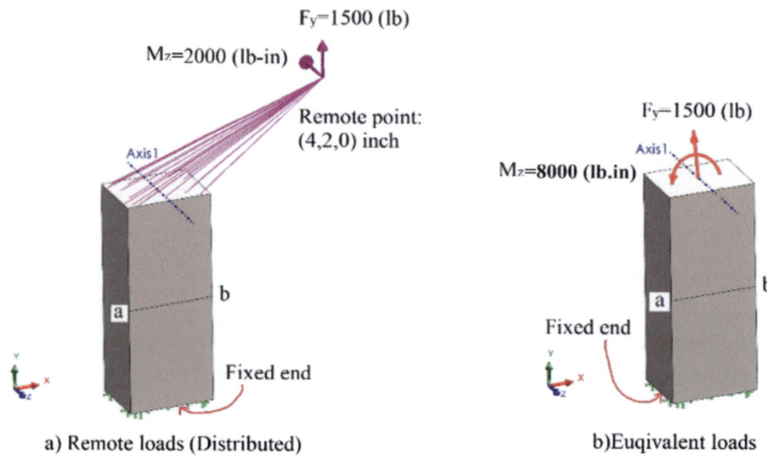

Fig. 3.96 Two sets of loading conditions

Solution

(1) FEA simulation

Step 1: **Pre-processing**. To compare FEA simulation results, a split line ab is added to the model. To define a bending moment on the top surface of the column, a reference axis is added as shown in Fig. 3.96. This prepares the model for FEA simulation.

Step 2: **Set up a project**. Both simulations are static stress analysis simulations.

Step 3: **Select the type of element and assign a material to each component**. 3D solid elements with high-quality mesh will be used, so no changes to the type of elements are needed. AISI 1020 steel is already included in the SolidWorks material library, so we can directly assign AISI 1020 steel as the component material.

Step 4: **Connections**. This is an FEA simulation on a component. There are no connections required.

Step 5: **Fixtures**. For both simulation cases, right-click the "Fixture" tab and select "Fixture geometry" to define the fixed end on the bottom surface of the column.

Step 6: **Loads**. For the FEA simulation with remote loads, Right-click the "External loads" tab and select "Remote Loads" to define remote loads with a distributed connection at the remote point.

For the FEA simulation with equivalent loads, right-click the tab "External loads" and select "Force" to define a 1500 lb normal tensile load on the top surface. Right-click the "External loads" tab again and select "Torque" to define a bending moment of 8000 lb-in about reference axis 1 on the top surface of the column.

Step 7: **Meshing**. For both FEA simulations, right-click the "Mesh" tab and select "Create Mesh". Check the "Mesh Parameters" and then enter $0.1''$ as the element size for the standard mesh.

Step 8: **Run Simulation**. Right-click the project name and select "Run". This will start the FEA simulation.

Step 9: **Post-processing**. Four different types of plots are selected to compare the FEA simulation results on these two loading cases.

The resultant deformation plots for the case with remote loads with distributed connections and the case with equivalent loads are displayed in Fig. 3.97a and b, respectively. Von Mises stress plots for the simulation with remote loads with distributed connections and the simulation with the equivalent loads are displayed in Fig. 3.98a and b, respectively. The y-normal stress plots for the two cases are displayed in Fig. 3.99. Y-normal stress graphs on the split line ab for the two cases are displayed in Fig. 3.100.

(2) **Comparisons and discussions**

From Fig. 3.97, the maximum resultant deformations for the remote loads with distributed connection is 3.562×10^{-3} inches, while the deformation with the equivalent loads is 3.512×10^{-3} inches. The relative difference between these two cases is only 1.4%, and the locations for the maximum resultant deformations are the same. As a result, the resultant deformation plots appear almost identical.

In Fig. 3.98, the maximum von Mises stress of both cases is 1.275×10^4 (psi). The locations of the maximum von Mises stresses are also the same for both cases.

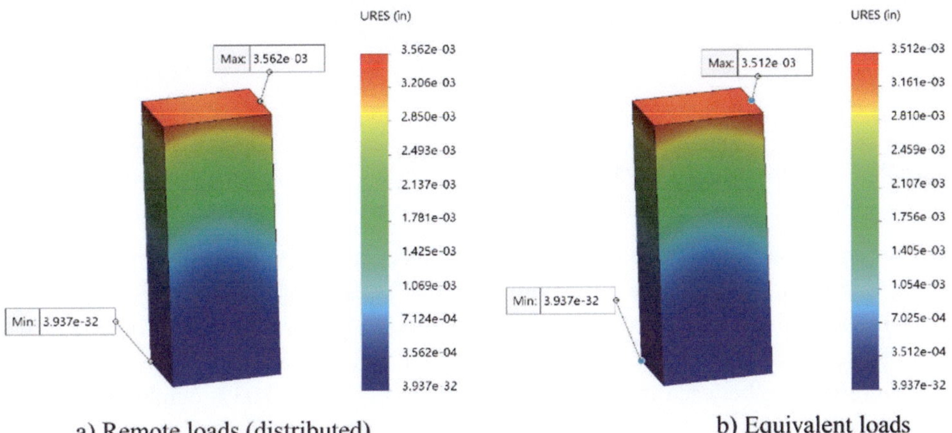

a) Remote loads (distributed)   b) Equivalent loads

Fig. 3.97 Resultant deformation plots

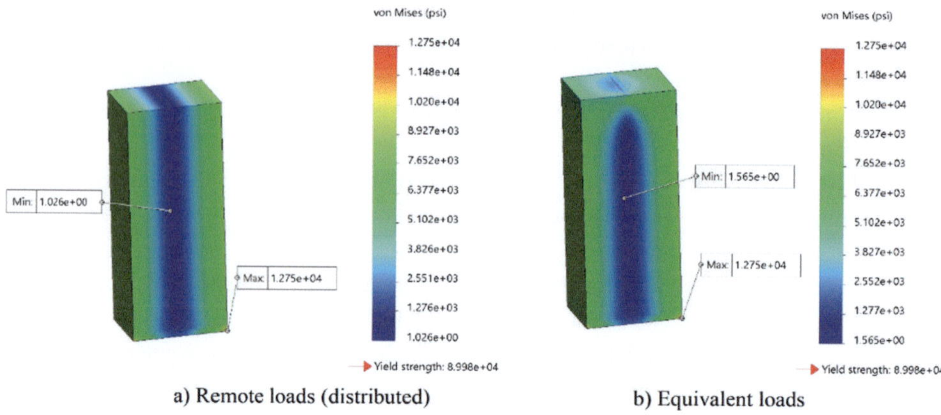

a) Remote loads (distributed) b) Equivalent loads

Fig. 3.98 Von Mises stress plots

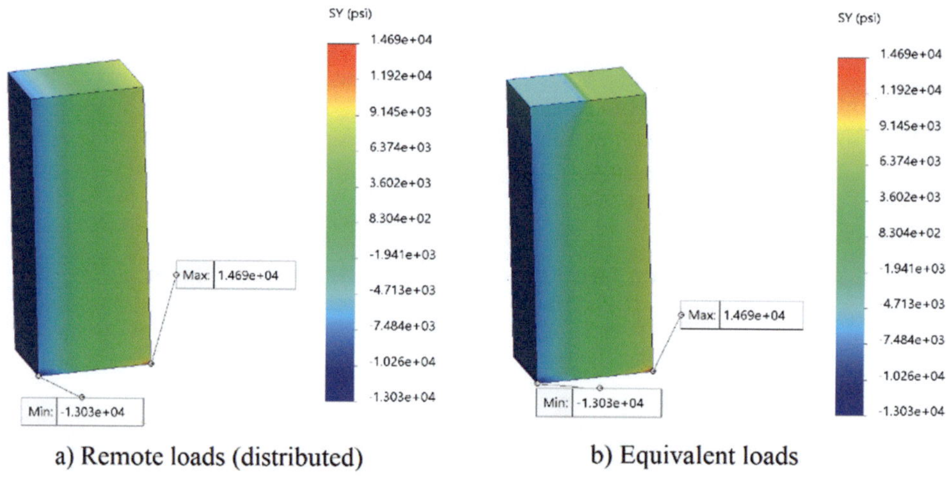

a) Remote loads (distributed) b) Equivalent loads

Fig. 3.99 The y-normal stress plots

The remote loads and the equivalent loads in this example will mainly cause y-normal stress. Based on Fig. 3.99, the maximum y-normal stresses of both cases is 1.469×10^4 (psi). The locations for the maximum y-normal stress are the same as well. From Fig. 3.100, it can be seen that the y-normal stress graphs on the split line ab are almost identical. The relative difference between the y-normal stresses at point a is 0.03%, while the values at point b are the same.

From the comparison results above, it can be concluded that the FEA simulation results of the column have no significant differences when using remote loads with distributed connections compared to using equivalent loads.

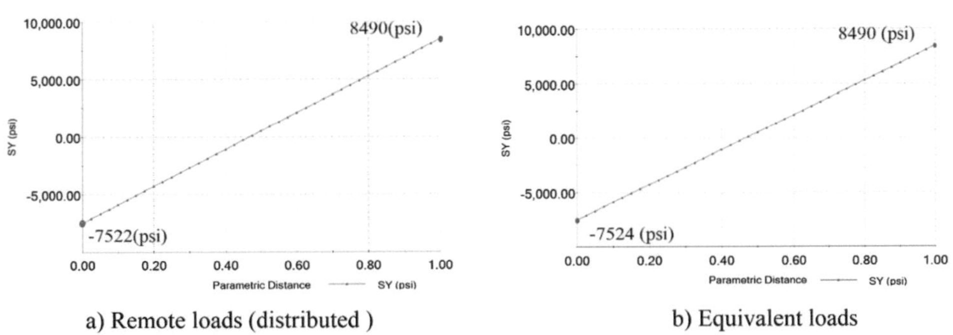

a) Remote loads (distributed)   b) Equivalent loads

Fig. 3.100 The y-normal stress graphs along the split line ab

3.9 Meshing

One of the core techniques for FEA simulation is called Meshing, where an object is broken into lots of small elements. The key concerns for meshing are to make the component or assembly in question meshable and to get acceptable FEA simulation results. Skill with meshing can be accumulated by running many simulations. This section will discuss some recommendations for meshing.

Meshing is a process of dividing models into thousands of small elements. Recall that FEA simulation studies the assembly of thousands of elements, which can be used as a virtual representation of actual physical components. In FEA simulation, the assembly of small deformable elements is used to represent the original mechanical component.

Meshing is a very crucial step in FEA simulation. So, making components "meshable" is a must for FEA simulation, as without meshing, there is no FEA simulation.

Typically, smaller element sizes mean more accurate simulation results. However, more elements also mean more computing time. For mechanical component design via FEA simulation, the main purpose is not to obtain the true stress/strain values at every point, but to obtain the maximum stress and maximum deformation values as well as their locations. This implies that for mechanical component design, finer element sizes can be used in possible weak spots, while larger element sizes can be used in low-stress/strain/deformation locations, to maintain a balance between accuracy and computation times.

Element size: To properly capture the trend of stress/strain variations and find corresponding possible maximum values, there should be at least five elements along the shortest geometrical dimension in possible weak locations or critical stress concentration areas. For example, the component shown in Fig. 3.101 is subjected to a 500 lb load on the top surface of the right end. The left end of this component is fixed. Based on the load condition and supporting condition of this component, the failure of this component

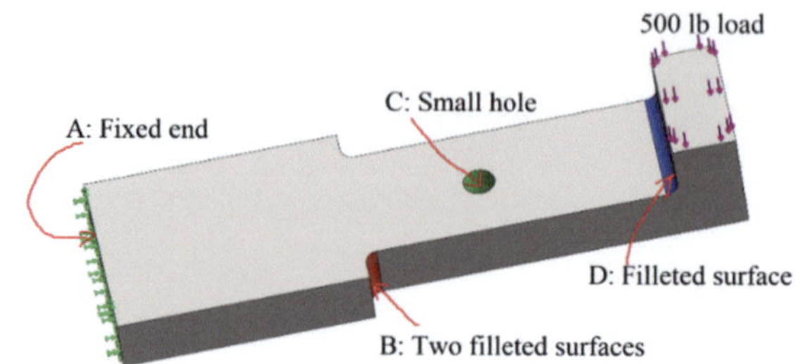

Fig. 3.101 Schematic of possible weak locations (FEA22)

is likely to be caused by bending stress. The possible weak spots and critical stress concentration locations for this example are four areas: the fixed end (A); The two filleted surfaces (B); the small hole (C); and a filleted surface near the load (D).

The filleted surface (D) has some stress concentration, but the maximum stress will not happen here because the bending moments applied to this filleted surface will be relatively small. For the fixed end (A), the shortest dimension will be the height ($0.5''$), and therefore the maximum element size will be $0.5/5 = 0.1''$. For the two filleted surfaces, the shortest geometric dimension will be the arc length ($0.16''$) of the fillet, and therefore the maximum element size will be $0.16/5 = 0.032''$. For the small hole, the perimeter and the length are $0.79''$ and $0.5''$ respectively, so the maximum element size will be $0.5/5 = 0.1''$. In this example, the element size $0.1''$ can be used as the global element size for the whole model and the element size $0.032''$ can be used as the local element size (mesh control) on the two filleted areas.

SolidWorks simulation offers a variety of different user-friendly tools for meshing. It can be summarized as three methods and their combinations for meshing: (1) Use mesh controls to define some areas with smaller local element sizes; (2) Create meshes using global element size; (3) Use adaptive methods (h-adaptive and p-adaptive). We will discuss these three methods and their applications. Typically, we need to apply mesh control first and then create meshing.

Create a local element size: "Apply mesh control" is a tool for defining element size in some areas with large stresses or high-stress concentrations. This element size is also called the local element size. For mechanical component design, our main concern is the maximum stress/deformation on any possible weak locations. In these areas, we want to use smaller element sizes to obtain more accurate results. The local element size can be determined by having at least five elements along the shortest geometrical dimension in the targeted mesh control area. Another approach is that the local element size could be at least one-fifth of the global element size. The mesh controls can be applied on edges and surfaces. We can even apply mesh control on parts in an assembly. For the same simulation, several different mesh controls can be applied.

Right-click the "Mesh" tab in the simulation study tree and select "Apply Mesh Control". The "Mesh Control" Property Manager will appear as shown in Fig. 3.102. Follow this sequence to apply a mesh control for a local element size. (1) This box can be used if you want to apply mesh control on a whole component of an assembly. This will be discussed in further detail in Chap. 4: FEA simulation on assemblies. (2) This window can be used to select the faces or edges that could be possible weak locations or high-stress concentration areas and might need a smaller element size. (3) The slider bar "mesh factor" can be used to change the local element size, also known as mesh density. Typically, we don't need to do anything here. (4) This section is used to specify a local element size. We need to select the unit for the element size in the first box. We can enter a local element size in the second box. The third and fourth boxes are the minimum element size and the minimum number of elements in a circle for the curvature-based meshes only, which will be discussed later. (5) If you want to create meshing on the selected areas for mesh control, we can click "Create mesh". But we typically create meshing using both global and local meshing. (6) click the check mark to complete the definition of "Apply Mesh Control".

Create a global element size: "**Create mesh**" is a tool for defining a global element size for the bulk part of a component. This element size is usually called global element size. Right-click the "Mesh" tab in the simulation study tree and then select "Create

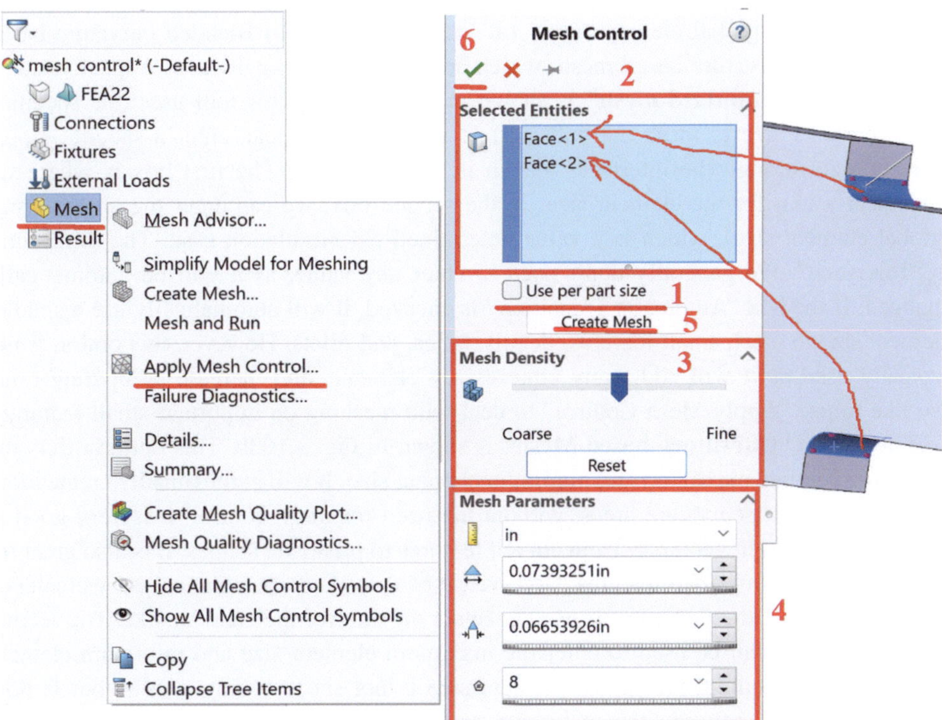

Fig. 3.102 The "Mesh Control" Property Manager

a) Standard mesh   b) Curvature-based mesh   c) Blended curvature-based mesh

Fig. 3.103 Create a mesh using global element size

mesh". The "Mesh" Property Manager will appear as shown in Fig. 3.103. There are three options for global element sizes: (a) Standard mesh; (b) Blended curvature-based mesh, and (c) Curvature-based mesh, which are shown in Fig. 3.103a–c, respectively.

The option "Standard mesh" is shown in Fig. 3.103a. This tool uses one specified global element size to mesh the bulk part of the model. We can enter a specific global element size through the interfaces shown in the red block. The first box is where we can select a unit for the element size. In the second box, we can enter the element size (global element size), which is a value determined per simulation case. The third entry is "Tolerance". We generally don't need to enter any value, as it will be automatically changed. If the box "Automatic Transition" is checked, it will automatically use a smaller element size to mesh small features, details, holes, and fillets. However, this option is not typically used as it will add many unnecessary elements and increase computing time. We like to use "Apply Mesh Control" to deal with meshing on important small features.

The option "Curvature-based Mesh" is shown in Fig. 3.103b. This option offers the ability to specify a maximum and minimum element size. It will automatically create more elements in higher-curvature areas, without the need for mesh control. It is very good at capturing changes in geometry from curved features to prismatic shapes. It is also great for geometry with many small features. However, it can also add many unnecessary elements. In the red section, the first entry is used to select the unit for the element size. The second and third entries can be used to enter the maximum element size and minimum element size. With this method, the global element size is not set to a single value but is now a range of values between the maximum and minimum element sizes. The maximum element size is used to mesh the bulk parts of the components, and the minimum element size is used to mesh smaller and more detailed features such as small fillets or small holes.

The program will automatically provide default values for the maximum and minimum element sizes, but we can change these values depending on the simulation cases. The fourth entry is where we enter the minimum number of elements in a circle. This number can be changed and increased depending on the simulation case. In this case, we will have a minimum of 8 elements for any circles in the mesh. The fifth box is the default element size growth ratio. This value is used to smoothly transition between the minimum and maximum element size. Typically, we will keep this value at the default of 1.4.

The option "Blended Curvature-based Mesh" is shown in Fig. 3.103c. It is an extension of the Curvature-based mesh and offers more advanced options to capture very small geometric features. It automatically adapts the element size to the local curvature of the geometry to create a smooth mesh pattern. The interface for this tool option is the same as that of "Curvature-based mesh".

The main difference between the standard mesh and the curvature-based mesh is how the tool will mesh small features such as fillets or small holes. For preliminary runs of a simulation project, we might use the default option "Blended curvature-based mesh". But for the final simulation, it is recommended to use "Standard mesh" and "Apply mesh controls" only, on any possible weak spots in the design. This approach gives us more control about the meshing. This book will use this approach for meshing.

The h-adaptive and p-adaptive methods: These methods will automatically run several FEA simulations with different settings on the same simulation problem and display a convergence graph. These adaptive methods are only available for FEA simulation with solid elements.

Right-click the "Project Name" tab in the simulation study tree and then select "Properties". The "Static" Property Manager will appear. Select "Adaptive" and then the "h-adaptive" and "p-adaptive" adaptive methods will appear as shown in Fig. 3.104a and b.

The "h-adaptive" method will automatically refine the mesh element size in critical regions with higher stress. When a pre-specified target accuracy is satisfied, the h-adaptive method will end. Otherwise, the h-adaptive method will continuously run until the pre-set maximum number of loops is reached. The interface for the "h-adaptive" option is shown in Fig. 3.104a.

- Use the "Target accuracy" slider to set the desired level of "Target accuracy", which is used as a convergence condition. The target accuracy is the accuracy level for the strain energy norm. We can use 98% or 99% here, which is equivalent to a 2% or 1% relative error.
- "Accuracy bias" is an indicator of focus between the local and the global. The default position is used.
- "Maximum no. of loops" is the maximum number of loops allowed when we run the study. The maximum possible number of loops with this method is 5.

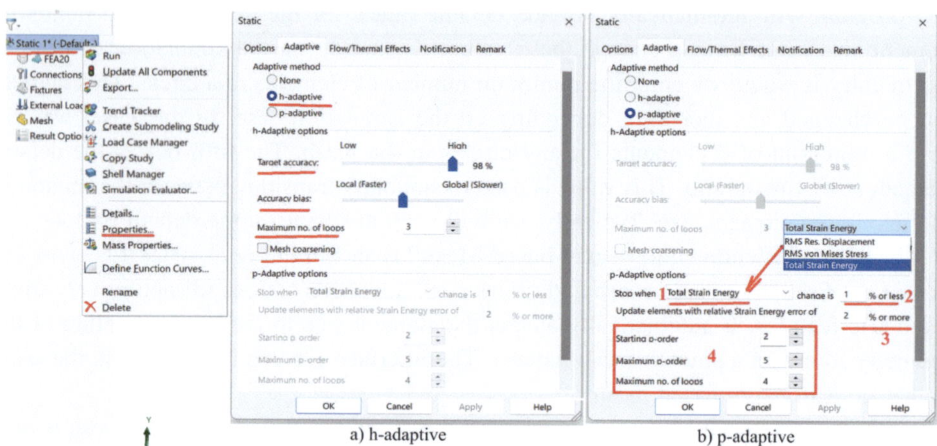

a) h-adaptive   b) p-adaptive

Fig. 3.104 The h- and p-adaptive methods

- "Mesh coarsening" allows the program to coarsen the mesh in regions with a low error during the adaptive loops. However, this option is not typically selected.

The "p-adaptive" method does not refine the element size, and instead uses progressively higher polynomial order deformation functions to improve the results. When a pre-specified target accuracy is satisfied, the p-adaptive method will end. Otherwise, the p-adaptive method will continuously run until the maximum number of loops is reached. The interface for the "p-adaptive" option is shown in Fig. 3.104b.

(1) First select a physical variable for checking the convergence conditions from three options: "Total Strain Energy", "RMS Von Mises stress" or "RMS Res. Displacement". Typically, we use "Total Strain Energy".
(2) Then enter the level of convergence. For this example, "Total Strain Energy" and a level of convergence of 1% are selected. This means that when the relative change in the total strain energy is less than 1%, the p-adaptive method will be terminated. Otherwise, the program continuously runs until the maximum number of loops is reached.
(3) Set the maximum allowable relative error in the strain energy of each element. When the relative error in the strain energy of each element is more than the maximum allowable relative error in the strain energy, the program will increase the polynomial order for those elements' deformation functions.
(4) Enter "Starting p-order", which is the element order for the first loop. Typically, the starting p-order is 2. Then enter "Maximum p-order". The highest possible order is 5. Finally, enter "Maximum no. of loops", which is the maximum number of loops

allowed in this analysis. The maximum possible number of loops with this setting is 4.

The h-adaptive and p-adaptive methods are a time-consuming process. They are typically used for the final detailed FEA simulation. Generally, these methods are used together with "Apply Mesh control" and "Create mesh" by global element size to obtain more accurate stress/strain in the possible weak locations.

Define Adaptive Convergence Graph: After the h-adaptive or p-adaptive method is completed, we can create a convergence graph to show the convergence level. Right-click the "Results" tab in the simulation study tree and select "Define Adaptive Convergence Graph" as shown in Fig. 3.105. The "Convergence Graph" Property Manager will appear as shown in Fig. 3.105a. We can select at least one physical variable for the convergence graph. The five possible variables for convergence graphs are "Target accuracy"; "Maximum von Mises Stress"; "Maximum Resultant Displacement"; "Total Strain Energy"; and "Number of Nodes". An example of a convergence graph with target accuracy and maximum von Mises stress is shown in Fig. 3.105b. For the example shown, the target accuracy for the strain energy norm is set as 1% from this graph. From this graph, the h-adaptive method completed its maximum loops and the total relative strain energy norm error at the end of the h-adaptive method is 1.60682%, which does not satisfy the convergence condition (less than 1%). The element sizes for meshing need to be reduced and then we can rerun the h-adaptive method until the convergence condition is satisfied.

Meshing is a crucial step for FEA simulation. The primary purpose of FEA simulation is not to pursue accurate results for the entire component, but to find the maximum stress/deformation and the possible weak locations for mechanical design. So, the approach of standard meshing for global element size and specifying local mesh controls for possible weak locations or high-stress concentration areas can be suitable for mechanical design.

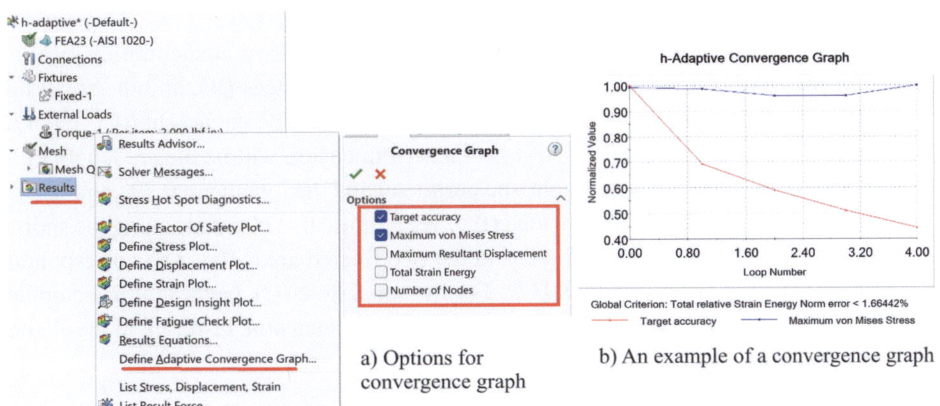

a) Options for convergence graph

b) An example of a convergence graph

Fig. 3.105 An example of an adaptive convergence graph

Now that each step of the general FEA simulation procedure has been discussed and explained. Now we will present three examples for meshing, applying the adaptive methods, and a design case.

Example 10: Local and global element sizes (FEA22) Download and open FEA22. It is a supporter, fixed at the left end and subjected to 500 lb force on the top surface of the right end as shown in Fig. 3.101. The material of this component is AISI 1020 steel. The factor of safety is required to be 1.5.

(1) Explain and set up mesh parameters and show the meshing
(2) Show the von Mises stress and resultant displacement distribution plots.
(3) Display the factor of safety plot with red areas below 1.5. Is this component safe?

Solution

(1) **FEA simulation**

Step 1: **Pre-processing**. The model FEA22 is ready for FEA simulation.
Step 2: **Set up a project**. This is a static stress analysis.
Step 3: **Select the type of element and assign a material to each component**. For this example, 3D solid elements with high-quality mesh will be used so no element type changes are required. AISI 1020 steel is included in the SolidWorks material library and can be directly assigned to the component.
Step 4: **Connections**. This is an FEA simulation on a component, so there are no connections.
Step 5: **Fixture**. Right-click the "Fixture" tab and select "Fixture geometry" to apply the fixed fixture on the left end of the component.
Step 6: **Loads**. Right-click the "External loads" tab and select "Force" to define a 500 lb force on the top surface of the right end.
Step 7: **Meshing**. The possible weak locations and important stress concentration locations for this setting are the fixed end (A), two filleted surfaces (B), and the small hole (C). For the fixed end (A), the shortest dimension will be the height (0.5″). For the two filleted surfaces, the shortest geometric dimension will be the arc length of the fillet (0.16″). For the small hole, the perimeter and the length are 0.79″ and 0.5″, so the shortest geometric dimension is 0.5. We will use five elements along the shortest dimensions on the fixed end (A) and the two filleted areas (B). The corresponding element sizes are 0.1″ and 0.032″. For the small hole (C), because it is a complete hole, we can use ten elements along the shortest dimension, resulting in the element size of 0.05″.

We will apply mesh controls on the two filleted surfaces (B) with an element size of 0.032″ and on the small hole (C) with an element size of 0.05″. We will use the

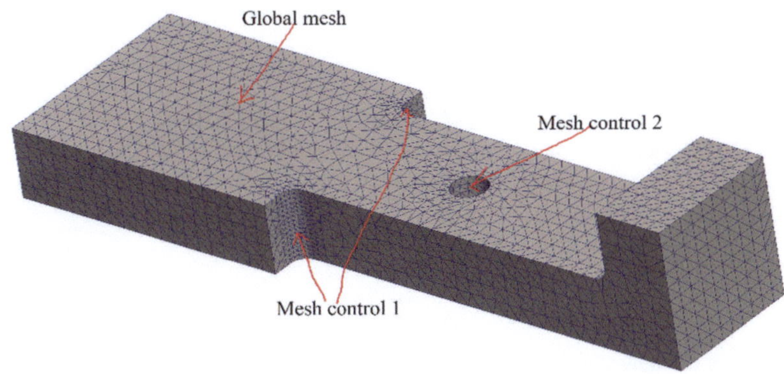

Fig. 3.106 Meshing of this example

standard mesh with a global element size of 0.1″. The meshing of this component is displayed in Fig. 3.106.

Step 8: **Run Simulation**. Right-click the project name and select "Run". Alternatively, we can click "Run this study" in the toolbars. This will start the FEA simulation.

Step 9: **Post-processing**. The von Mises stress plot and the resultant displacement plot are shown in Fig. 3.107a and b. The factor of safety plot and the factor of safety plot with red areas below the factor of safety 1.5 are shown respectively in Fig. 3.108a and b.

(2) **Interpretation and Discussions**

From Fig. 3.107a, the maximum von Mises stress is at the two filleted surfaces (B). This is expected. The maximum bending moment will happen at the left fixed end. But at the two

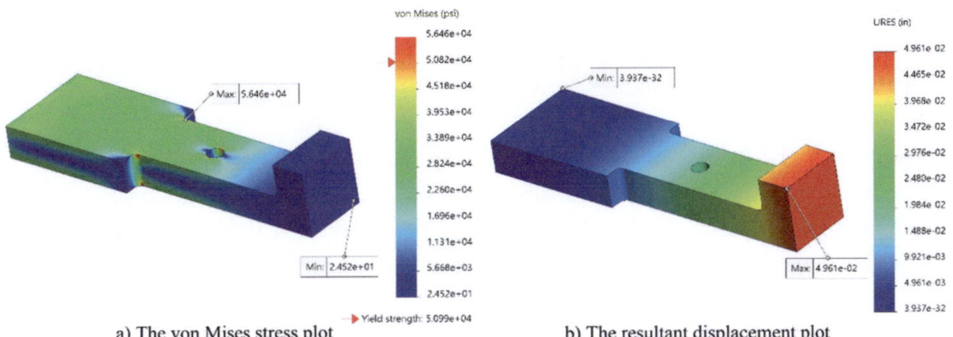

a) The von Mises stress plot   b) The resultant displacement plot

Fig. 3.107 The Von Mises stress and the resultant displacement plots

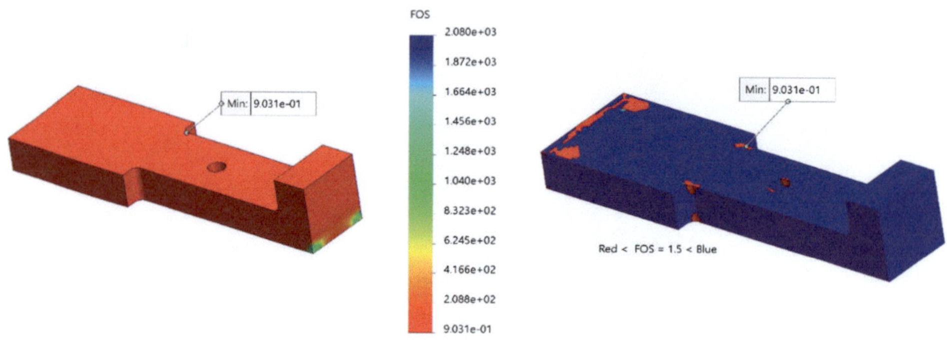

a) The factor of safety plot   b) The factor of safety with the areas below 1.5

Fig. 3.108 The factor of safety plots

filleted surfaces (B), the effective cross-section area is smaller than that of the left fixed end. Additionally, there are also stress concentrations due to the fillets.

From Fig. 3.107b, the maximum displacement is at the top surface of the right end. This is also expected because the part is similar to a cantilever beam.

From Fig. 3.108a, the minimum factor of safety is only 0.9, therefore this component needs to be redesigned. Since the factor of safety is required to be more than 1.5, knowing where the factor of safety is lower than 1.5 will be useful. Figure 3.108b is the factor of safety plot with red areas below the factor of safety 1.5. This information is what we need to improve the mechanical component design because it provides possible directions for further modification or redesign. From Fig. 3.108b, the weak locations are near the left fixed end, the two filleted surfaces, and the small hole. It also shows that the failure areas in the weak locations are near the top and the bottom surfaces, which indicates that these failures are mainly caused by the bending stress. These results are the same as what we predict using mechanical design theory. So, FEA simulation is a useful tool for mechanical component design because it can help identify possible weak locations, and determine the maximum stress, maximum deformation, factor of safety, and provide some insights for further modification or redesign.

Example 11: The h- and p- methods (FEA23) Download and open FEA23, which is a square bar, fixed at the left end, and subjected to a torque moment of 2000 (lb. in) at the right end as shown in Fig. 3.109. The material of this bar is AISI 1020. The target accuracy is 1%. Use the h-adaptive method and the p-adaptive method to run FEA simulations.

(1) Show the meshing before and after the adaptive runs for both cases
(2) Show the adaptive convergence graphs for both cases.
(3) Compare the FEA simulation results with theoretical calculations and interpret the results

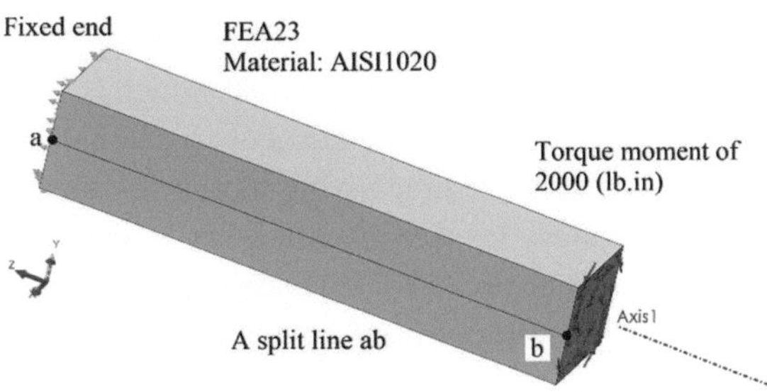

Fig. 3.109 Schematic of FEA settings

Solution

Step 1: **Pre-processing**. A center axis is added for applying torque on the right end. A split line ab is added on the middle of one surface as shown in Fig. 3.109. This split line will be used to display the maximum shear stress τ_{zy}. After these adjustments, the model is ready for the FEA simulation.

Step 2: **Set up a project**. This is a static stress analysis.

Step 3: **Select the type of element and assign a material to each component**. 3D solid elements with high-quality mesh will be used, so the type of element can remain unchanged. AISI 1020 steel is already included in the SolidWorks material library, so it can be directly assigned to the component.

Step 4: **Connections**. This is an FEA simulation on a component, so there are no connections.

Step 5: **Fixture**. Right-click the "Fixture" tab and select "Fixture geometry" to apply the "Fixed" fixture on the left end of the component.

Step 6: **Loads**. Right-click the "External loads" tab and select "Torque" to define a 2000 lb-in torque on the right end of the bar around Axis 1.

Step 7: **Meshing**. There are no small features for this component. The shortest geometrical dimension of this component is the side length of $2''$. We will use "Standard mesh" with an element size of $0.4''$.

Right-click the "project name" tab in the simulation study tree and select "Properties". In the prompted "Static" Property Manager, we can select "h-adaptive" or "p-adaptive" methods as shown in Fig. 3.104. For the h-adaptive method, the target accuracy is 1% and the maximum number of loops is 5. For the p-adaptive method, the target accuracy is 1% and the maximum order number is 4. The rest of the settings can remain as the default.

Step 8: **Run Simulation**. Right-click the project name and select "Run". This will start the FEA simulation.

Step 9: **Post-processing**.

For the h-adaptive method, the meshing before and after applying the h-adaptive method are displayed in Fig. 3.110a and b. The adaptive convergence graph with the element size $0.4''$ is shown in Fig. 3.111a. The accuracy level at the end of the h-adaptive method is 1.54%, which is larger than the target accuracy of 1%. Given this, we change the element size into $0.25''$ and rerun the simulation. The adaptive convergence graph with the element size of $0.25''$ is shown in Fig. 3.111b. Now the accuracy level at the end of the h-adaptive method is 0.89%, which is less than the target accuracy of 1% and satisfies the convergence condition. The shear stress τ_{zy} plot and shear stress τ_{zy} along the split line ab are shown in Fig. 3.112a and b.

For the p-adaptive method, the meshing is displayed in Fig. 3.113a and the p-adaptive convergence graph with element size $0.4''$ is displayed in Fig. 3.113b.

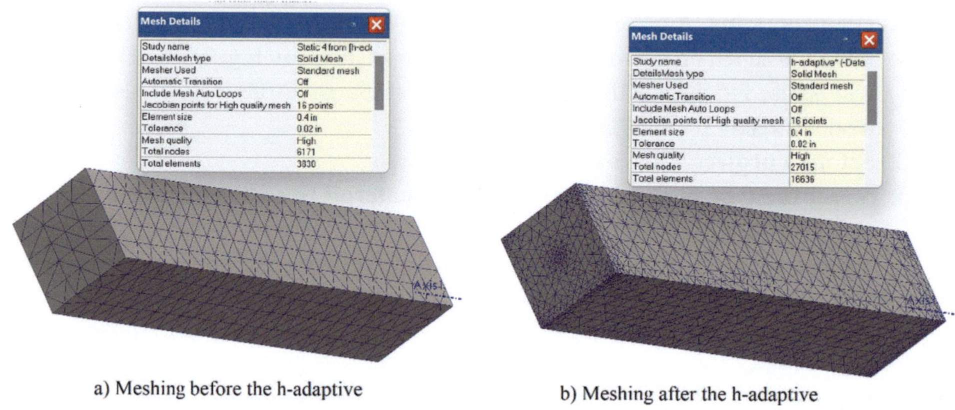

a) Meshing before the h-adaptive

b) Meshing after the h-adaptive

Fig. 3.110 The meshing before and after the h-adaptive method

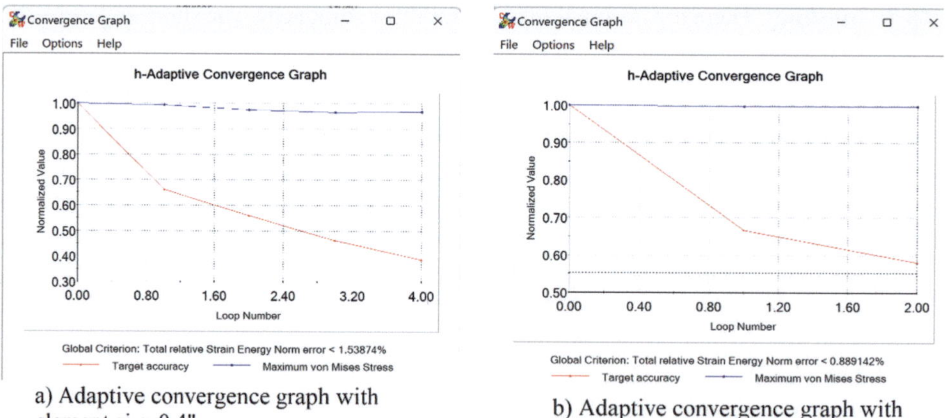

a) Adaptive convergence graph with element size 0.4"

b) Adaptive convergence graph with element size 0.25"

Fig. 3.111 Adaptive convergence graphs

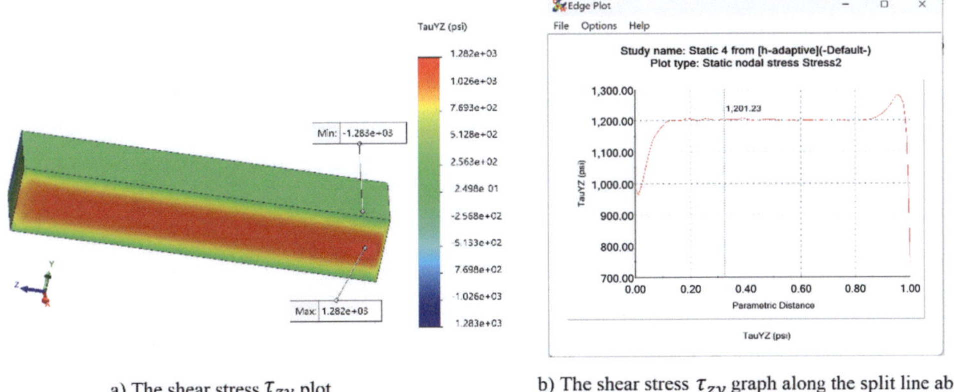

a) The shear stress τ_{zy} plot

b) The shear stress τ_{zy} graph along the split line ab

Fig. 3.112   The shear stress τ_{zy} plot and the shear stress graph along the split line ab in h-method

From this graph, we can see that the convergence condition ($<1\%$) is satisfied after the second loop. The shear stress τ_{zy} plot and shear stress τ_{zy} along the split line ab are shown in Fig. 3.114a and b.

(2)   Interpretation and Discussions

The h-adaptive method will change the meshing as shown in Fig. 3.110. The number of total elements before the h-adaptive method is 3830, but after the h-adaptive method is applied, the number of elements becomes 16,636. For the p-adaptive method, the meshing does not

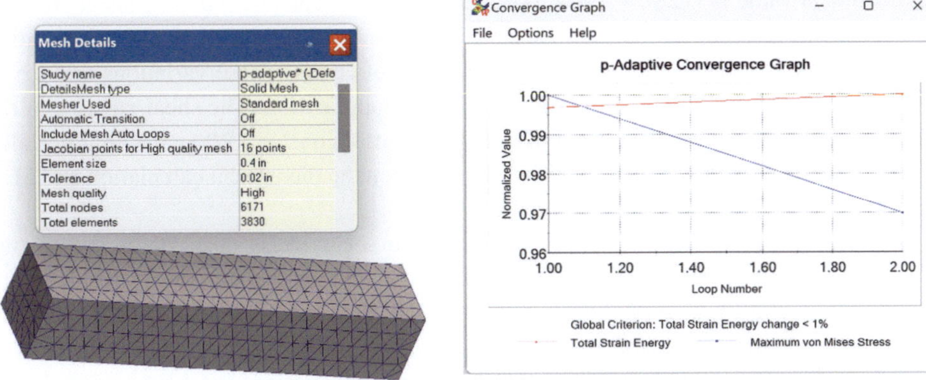

a) The meshing with element size 0.4"

b) The p-adaptive convergence graph with element size 0.4"

Fig. 3.113   The meshing and the p-adaptive graph

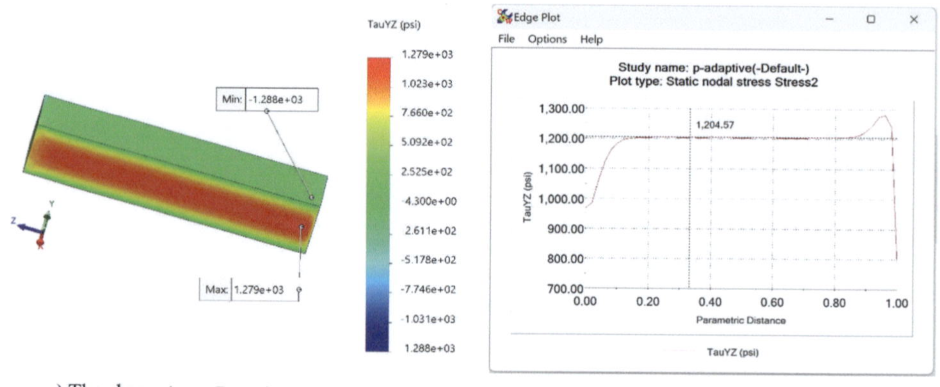

a) The shear stress τ_{zy} plot   b) The shear stress τ_{zy} graph along the split line ab

Fig. 3.114 The shear stress τ_{zy} plot and the shear stress graph along the split line ab in p-method

change as shown in Fig. 3.113a. The numbers of total elements are the same before and after applying the p-adaptive method.

The p-adaptive method is more effective than the h-adaptive method. The h-adaptive method does not satisfy the convergence condition with the element size $0.4''$ as shown in Fig. 3.111a. But the p-adaptive method does satisfy the convergence condition with the same element size $0.4''$ as shown in Fig. 3.113b.

The theoretical result of the maximum shear stress τ_{zy} for a square bar with a side length of $2''$ under a torque of 2000 (lb. in) is 1202.8 (psi).

For the h-adaptive method, the FEA simulation value for the maximum shear stress τ_{zy} from Fig. 3.112 a) is 1282 (psi). The relative error of the FEA simulation result compared to the theoretical value is $(1282 - 1202.8)/1202.8 = 6.6\%$. Theoretically, the shear stress at any point on the split line ab will be constant. However, from Fig. 3.112b, the FEA simulation results indicate that there are large variations near both ends. These variations are due to the left fixed-end condition and the right-end torque application. The shear stresses in the middle section without the effects of both ends are almost constant. When this middle value of 1201.23 (psi) is used, the relative difference between the theoretical calculation and FEA simulation is $abs(1201.23 - 1202.8)/1202.8 = 0.13\%$.

For the p-adaptive method, when the maximum shear stress of 1279 (psi) from the shear stress τ_{zy} the plot in Fig. 3.114a is used, the relative error between the theoretical calculation and FEA simulation is $(1279 - 1202.8)/1202.8 = 6.3\%$. But if the shear stress value 1204.57 (psi) from the middle area without the effects from both ends is used, then the relative error between the theoretical calculation and FEA simulation is $(1204.57 - 1202.8)/1202.8 = 0.15\%$.

From the discussions above, we can conclude that the FEA simulation results in this example match the theoretical calculation results. However, when we compare the FEA

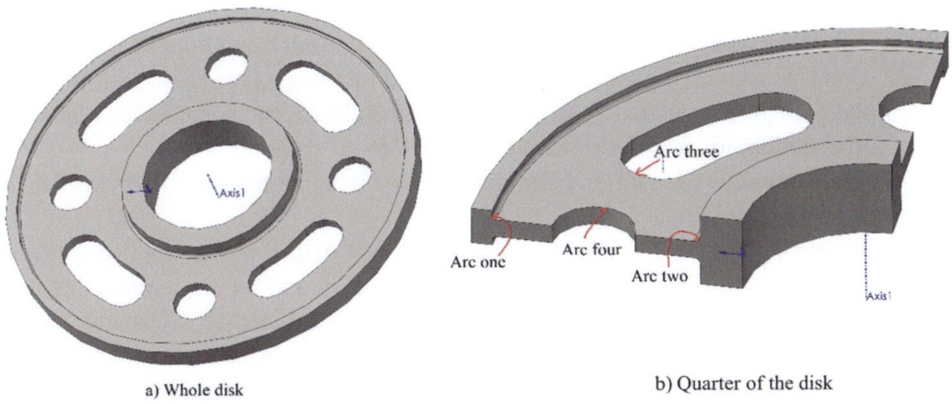

Fig. 3.115 The flywheel and a quarter of the flywheel

simulation results with theoretical calculation results, we must pay attention to the assumptions made and conditions required for the theoretical formulas. When the conditions in the theoretical formula are different from the conditions in the FEA simulation, the results should not be compared.

Example 12: A design case- maximum allowable speed (FEA24) Download and open FEA24, which is a flywheel. It is made of cast iron steel as shown in Fig. 3.115a. Determine the maximum allowable RPM (revolution per minute) for this flywheel when the design specification for the factor of safety is 2.0.

Solution

The external force for a flywheel is centrifugal force. It will be difficult to run theoretical calculations to obtain the stress, and then calculate the factor of safety. However, FEA simulation can obtain the maximum stress and the minimum factor of safety. So, for this design case, several FEA simulations with different RPMs will be run to obtain a table or a curve of the factor of safety vs. RPM. From the table or curve, the maximum allowable RPM with a factor of safety 2.0 can be identified.

(1) **FEA Simulation**

Step 1: **Pre-processing**. Based on the geometrical shape of the flywheel, a quarter of the flywheel with symmetry restraints as shown in Fig. 3.115b can be used for FEA simulation. The flywheel is mounted on a shaft with a mechanical key. But in this design case, the main force is the centrifugal force, and failure will not happen in the key seat. Therefore, the mechanical key seat can be ignored and is not included in this model.

Step 2: **Set up a project**. This is a static stress analysis.

Step 3: **Select the type of element and assign a material to each component**. The 3D solid elements with high-quality mesh will be used for this project, so no adjustments to the element type are required. Cast iron steel is already included in the SolidWorks material library and can be directly assigned to the component.

Step 4: **Connections**. This is an FEA simulation on a component, so there are no connections.

Step 5: **Fixtures**. On the cut-off surfaces, the symmetry fixture will be applied as shown in Fig. 3.116a. The flywheel is mounted on a shaft, and we can assume that there is no radial deformation on the mounting cylindrical surface. Typically, there is one shoulder of the shaft, or some other part such as a spacer in the shaft-flywheel assembly to stop any possible axial movement along the shaft axis. So, it can be assumed that one side of the flywheel hub does not have any normal deformation. For this example, three types of fixtures: "Symmetry", "On Cylindrical faces", and "On Flat Faces" are applied as shown in Fig. 3.116a.

Step 6: **Loads**. We will start the FEA simulation with the flywheel rotating at 1000 RPM. Right-click the "External loads" tab and select "Centrifugal Force" to define a 1000 RPM rotation about the central axis.

Step 7: **Meshing**. There are four possible weak locations or stress concentration areas as shown in Fig. 3.115b. Arc one is a fillet surface. The arc length is the shortest geometric dimension at 0.31″. Arc two is also a fillet surface. In this example, it may not experience high stress, but we can still treat it as a possible weak location. The shortest geometrical dimension for this fillet surface is the arc length 0.47″. For arc three and arc four, the shortest geometric dimensions are their thickness of 1.0″.

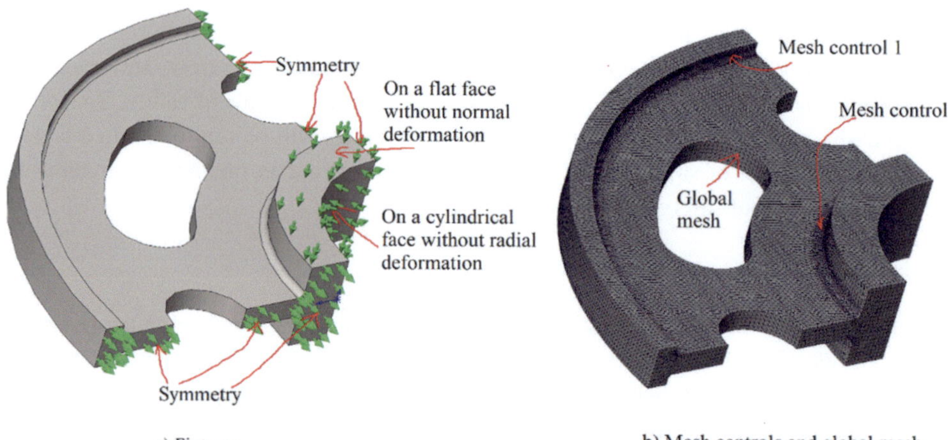

a) Fixtures   b) Mesh controls and global mesh

Fig. 3.116 Fixtures and meshing

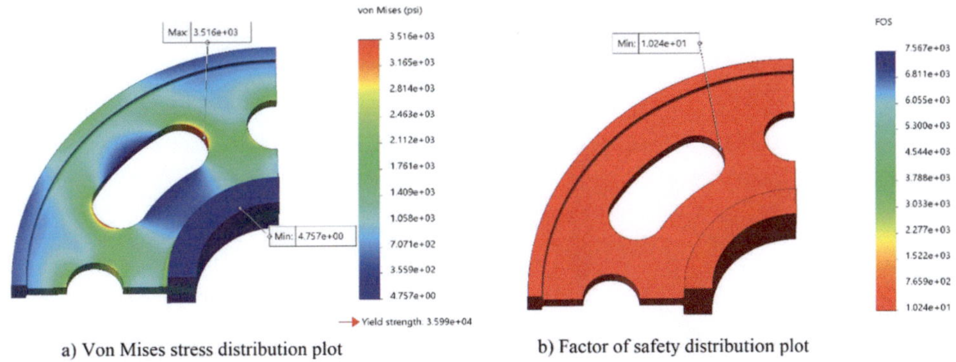

a) Von Mises stress distribution plot   b) Factor of safety distribution plot

Fig. 3.117 Von Mises stress plot and factor of safety plot

For this example, we will use a standard mesh with a global element size of $1.0/5 = 0.2''$. Mesh control one is applied on the arc one's fillet areas with a local element size of $0.31/5 = 0.062''$. Mesh control two is applied on the arc two' fillet areas with a local element size of $0.47/5 = 0.0814''$. The meshing of the model is shown in Fig. 3.116b.

Step 8: **Run Simulation**. Right-click the project name and select "Run". This will start the FEA simulation.

Step 9: **Post-processing**.

The von Mises stress distribution plot and the factor of safety distribution plot with 1000 RPM are displayed in Figs. 3.117a and b, respectively.

(2) **FEA simulation interpretations and maximum allowable RPM.**

The maximum von Mises stress is 3516 (psi). The minimum factor of safety with 1000 RM is 10.24 as shown in Fig. 3.117. The Iso clipping for the stress of more than 3000 psi is shown in Fig. 3.118a. From Figs. 3.117a,b, and Fig. 3.118a, the weakest location of this flywheel due to centrifugal force is in the areas of arc three. This is an expected result. The design specification for the factor of safety is 2.0. So, we can continuously increase the RPM by 500 and run the FEA simulations until the factor of safety drops below this benchmark. The RPM and corresponding minimum factor of safety are listed in Table 3.5 and displayed as a graph of the factor of safety vs RPM as shown in Fig. 3.118b. Based on either Table 3.5 or the graph in Fig. 3.118b, the maximum allowable RPM with a minimum factor of safety of 2.0 is around 260 RPM.

This example shows that difficult theoretical analysis and design problems can easily be solved by FEA simulation. FEA simulation a powerful tool for mechanical engineers, as mechanical engineers can use concepts and theory of mechanical design to interpret the FEA simulation results, while the bulk of calculations will be done using FEA simulation.

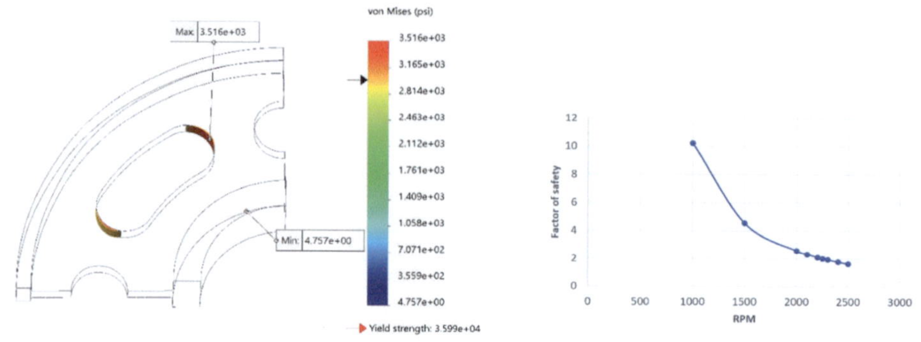

a) Iso clipping for the stress of more than 3000 (psi)   b) Factor of safety vs the RPM (revolution per minute)

Fig. 3.118 Iso clipping and the factor of safety vs. RPM

Table 3.5 The factor of safety vs. the RPM (revolution per minute)

RPM	1000	1500	2000	2100	2200	2250	2300	2400	2500
Factor of safety	10.24	4.543	2.555	2.318	2.112	2.018	1.932	1.775	1.635

3.10 Minor Design Project

So far, every step of a general procedure for FEA simulation on a component has been discussed, explained, and demonstrated. With these discussions, we have accumulated the necessary skills and knowledge about FEA simulation on a component. FEA simulation can be used as a tool to conduct any type of mechanical component design. We can display and identify possible weak locations or find the maximum stress/strain of a component with FEA simulation. However, we still need to accumulate experience in three areas: loads, fixtures, and meshing by running many FEA simulations. The best way to accumulate this experience is through running many simulation design projects. We could run an FEA simulation on any real mechanical component. However, because theoretical or established solutions for the mechanical components might not be available, it is hard to compare the simulation results with known and established solutions. Here are some projects that are good for us to practice running FEA simulations and accumulate experience.

Minor design project: The suggested approach is to run a group of special design projects, which can be called minor design projects. The minor design project for mechanical component design is an FEA simulation on a component, which doesn't have any available explicit theoretical solutions due to complicated loads or geometrical shapes. However, there are some established tables with empirical formulas for the solutions. There are many such components in mechanical component designs. For example, the stress concentration factor of notched components and the stiffness of bolted joints are two examples of minor design projects. For the stress concentration factor of notched components, there are no simple theoretical analytical equations, but various design tables of stress concentration factors are available for different types of typical notches [3–5]. Similarly, for the stiffness of bolted joints, there are no theoretical analytical equations for solutions. However, there are some empirical equations available for the estimated stiffness of typically bolted joints [6].

The following will be one example and a description of one minor design project.

Example 13: Stress concentration factor of a plate with a symmetrical hole under tension (FEA25). Download and open FEA25. The component is a flat plate with a symmetrical hole in the center with a uniform thickness of $1.00''$ and is subjected to tensile forces as shown in Fig. 3.119 a). Use FEA simulation to determine the stress concentration factor. Explain and discuss the process used to obtain the result, then compare it with the value obtained from the curve in Fig. 3.119c [6].

Solution

A uniform plate without a hole under an axial load will have a uniform constant normal stress along the load direction. But when there is a symmetrical hole as shown in Fig. 3.119a, there is a stress concentration phenomenon due to a sudden geometric change as shown in Fig. 3.119b. One of the key issues for mechanical design is to determine the maximum stress.

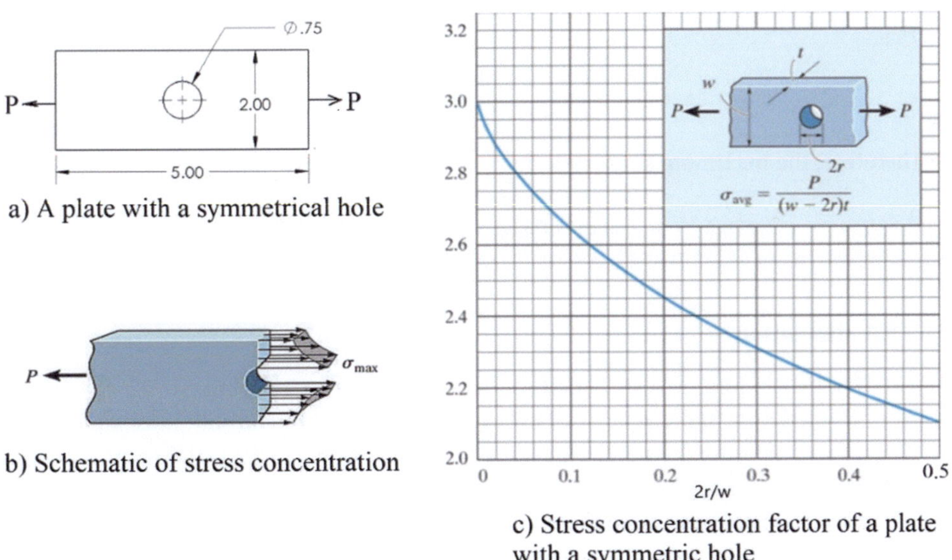

Fig. 3.119 Stress concentration factors of a plate with a symmetrical hole under axial loads

However, there is no simple theoretical analytical formula for the calculation of maximum stress in this scenario. The concept of stress concentration factor is introduced for estimating the maximum stress in these stress concentration areas.

$$k_t = \frac{\sigma_{max}}{\sigma_{nom}} \qquad (3.3)$$

where k_t is stress concentration factor; σ_{max} is the maximum stress in the stress concentration area; and σ_{nom} is the nominal stress, which is calculated by the effective cross-section area without any consideration of stress concentration.

Design handbooks typically provide tables or curves of the stress concentration factors of typical notched geometrical shapes. These tables or curves are determined by experiments or other more advanced techniques. The stress concentration factor of a plate with a symmetrical hole under axial load is shown in Fig. 3.119c. So, design engineers frequently use the following equations to determine the maximum stress.

$$\sigma_{max} = k_t \sigma_{nom} \qquad (3.4)$$

The stress concentration factor is independent of the type of materials and external load. For this example, let's assume that the material is plain carbon steel, and the external axial load is 1000 lb. The σ_{nom} is:

$$\sigma_{nom} = \frac{Axial\ load}{Effective\ area} = \frac{P}{(w - 2r) \times t} = \frac{1000}{(2 - 0.75) \times 1} = 800\ (psi)$$

From the table shown in Fig. 3.119 c), the stress concentration factor at the value $2r/w = 0.375$ is

$$k_{t-graph} = 2.24$$

Therefore, the maximum stress will be:

$$\sigma_{max} = k_{t-graph} \times \sigma_{nom} = 2.24 \times 800 = 1792(psi)$$

(1) **FEA simulation**

Step 1: **Pre-processing.** The provided model (FEA25) is not too big, so we can use the full model for static analysis. However, we will need to add split lines along which we can measure the maximum stress as shown in Fig. 3.120a. If the full model is used for the FEA simulation, the "Use inertial relief" option will be used because there is no restraint applied and the component is theoretically in equilibrium. For this example, the model has geometric symmetry and load symmetry, we can use a one-eighth model for FEA simulation as shown in Fig. 3.120b.

Step 2: **Set up a project.** This is a static stress analysis.

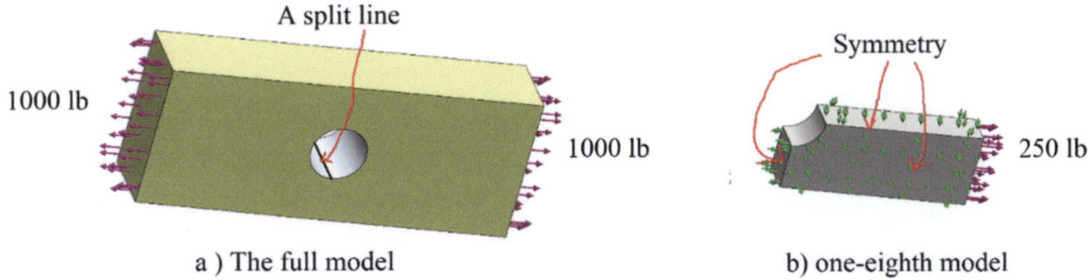

a) The full model

b) one-eighth model

Fig. 3.120 Choices of pre-process

Step 3: **Select the type of element and assign a material to each component**. 3D solid elements with high-quality mesh will be used so no element type changes are needed. Additionally, plain carbon steel is included in the SolidWorks material library so we can directly assign plain carbon steel to the component.

Step 4: **Connections**. This is an FEA simulation on a component, so there are no connections.

Step 5: **Fixture**. There are three symmetrical planes in the one-eighth model as shown in Fig. 3.120b. On each symmetrical plane, there is no normal displacement.

Step 6: **Loads**. Since the one-eighth model is used, the external load will be 250 lbs of normal force since the loading area is a quarter of the original end surface.

Step 7: **Meshing**. In this simulation case as shown in Fig. 3.121a the shortest dimension is $0.5''$. So, the global element size with a standard mesh will be $0.5/5 = 0.1''$. Stress concentration will happen on the inner cylindrical surface and line AB will have the maximum stress. We will use mesh control on the cylindrical surface and the surface that contains the line AB. For this example, the local element size can be one-fifth of the global element size, $0.1''/5 = 0.02''$. The meshing is shown in Fig. 3.121b.

Step 8: **Run Simulation**. Right-click the project name and select "Run". This will start the FEA simulation.

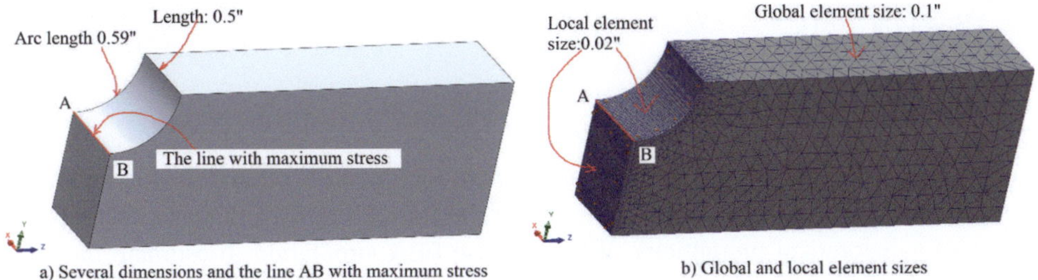

a) Several dimensions and the line AB with maximum stress

b) Global and local element sizes

Fig. 3.121 Meshing

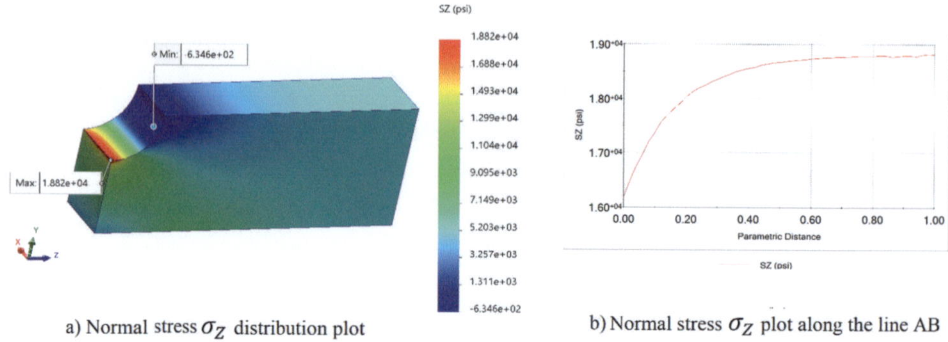

a) Normal stress σ_Z distribution plot   b) Normal stress σ_Z plot along the line AB

Fig. 3.122 The normal stress σ_Z distribution plot and its graph along the line AB

Step 9: **Post-processing**. To estimate the stress concentration factor by FEA simulation, we need to find the maximum stress. In this example, the axial load will generate normal stress σ_Z which is the normal stress along the axial load direction. In the stress concentration area, the stress conditions will be more complicated, but we can also use von Mises stress to calculate the maximum stress. So, we need to find both the maximum normal stress σ_Z and von Mises stress σ_{von}.

The normal stress σ_Z distribution plot is shown in Fig. 3.122a. The maximum normal stress σ_Z in the stress concentration area is 1882 psi, on split line AB. Theoretically, the maximum normal stress along the AB line is constant. But in FEA simulation, it isn't a constant. By using the post-process tool called "List selected", the normal stress σ_Z along the line AB is displayed in Fig. 3.122b. The average normal stress σ_Z along the line AB is 1835 (psi).

The von Mises stress σ_{von} distribution plot is shown in Fig. 3.123a. The maximum von Mises σ_{von} in the stress concentration area is 1777 (psi), on the split line AB. Theoretically, the maximum von Mises stress along the AB line is constant, but in the FEA simulation, it isn't a constant. By using a post-process tool called "List selected", the von Mises σ_{von} along the line AB is displayed in Fig. 3.123b. The average von Mises stress σ_{von} along the line AB is 1756 (psi).

(2) **Interpretation and Discussions**

From Figs. 3.122a and Figs. 3.123a, the maximum normal stress σ_Z and the maximum von Mises stress σ_{von} from FEA simulation are on the line AB, which matches the expected location for theoretical predictions. When we use FEA simulation to estimate the stress concentration factor K_t, we might get four different values based on the approaches we can use. If the normal stress σ_Z is used to calculate the stress concentration factor, we can use the maximum normal stress σ_{Z-max} or the average maximum normal stress along the line AB $\sigma_{Z-averagemax}$. If the von Mises stress σ_{von} is used to calculate the stress concentration factor,

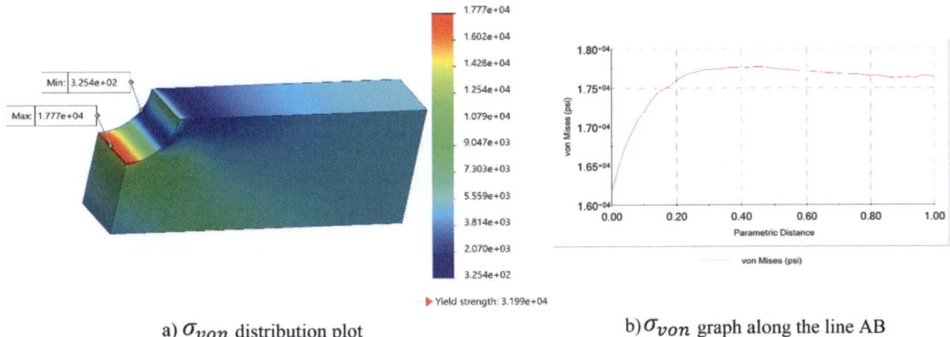

a) σ_{von} distribution plot   b) σ_{von} graph along the line AB

Fig. 3.123 The von Mises σ_{von} distribution plot and its graph along the line AB

we can use the maximum von Mises stress $\sigma_{von-max}$ or the average maximum von Mises stress along the line AB $\sigma_{von-averagemax}$. The estimated stress concentration factors and corresponding data are listed in Table 3.6. In this table, the relative difference is calculated by the following equation:

$$Relative\ difference = \frac{abs(K_{t-FEA} - K_{t-graph})}{K_{t-graph}}$$

From Table 3.6, the maximum relative difference between the value obtained from the graph in Fig. 3.119c and values obtained by using σ_{Z-max} is 4.91%. The minimum relative difference between the value obtained from the graph and values obtained by using $\sigma_{von-max}$ is 0.89%. But all of these relative differences are less than 5%. This implies that any of the four approaches mentioned above can provide a reasonable estimation for the stress concentration factor for mechanical component design.

Now, we will describe one case as a Minor design project for a group of 2 to 4 team members.

Table 3.6 Stress concentration factors and relative difference

Maximum stress	σ_{Z-max} 1882 (psi)	$\sigma_{Z-average\,max}$ 1835 (psi)	$\sigma_{von-max}$ 1777 (psi)	$\sigma_{von-average\,max}$ 1756 (psi)
Nominal stress	800 (psi)			
K_{t-FEA}	2.35	2.29	2.22	2.20
$K_{t-graph}$	2.24			
Relative difference	4.91%	2.23%	0.89%	1.79%

The minor design project one: Stress concentration factors of typical notched components

<u>Introduction</u>

Stresses at or near a discontinuity, such as a hole in a plane, are higher than the stresses if such a discontinuity does not exist. This phenomenon is called stress concentration. The actual stress in the stress concentration area is very complex, so it is very difficult to get the simple theoretical solution/expression to determine the actual maximum stress in that area. A general way to determine the maximum stress in the stress concentration area is to use the stress concentration factor. The stress concentration factor is the factor used to relate the actual maximum stress at the discontinuity to the nominal stress without the discontinuity:

$$K_t = \frac{\text{actual maximum stress}}{\text{average stress}} = \frac{\sigma_{max}}{\sigma_{nom}}$$

Some observations and experimental results about the stress concentration factors confirm:

(1) The stress concentration factor is independent of the part's material properties and loads.
(2) The stress concentration factor is significantly affected by geometry.
(3) The stress concentration factor is significantly affected by the type of discontinuity.

The stress concentration factors can be determined by experiments and running the FEA simulation. In most machine design books, there are tables or plots of the various stress concentration factors. In this minor project, FEA simulation is used to estimate stress concentration factors.

<u>Minor project tasks</u>

There are many different typical discontinuities in mechanical components. Any one of these can be selected as a minor project. Let's use a stepped shaft as an example as shown in Fig. 3.124. The shaft dimensions are as follows: D/d = 1.05, d = 1.00″, and the fillet ratio r/d = 0.05, 0.1, 0.15, 0.2, 0.25, and 0.3. The total length of the stepped shaft is 5″. The stepped cross-section is in the middle of the shaft.

(1) Use FEA simulation to provide the stress concentration factor curves of the assigned discontinuous geometry. The relative difference between the values obtained from the FEA simulation and the values obtained from the tables or curves provided in the design handbook should be less than 5%.
(2) Use FEA simulation to confirm that the stress concentration factor is independent of the part's material properties and applied loads.

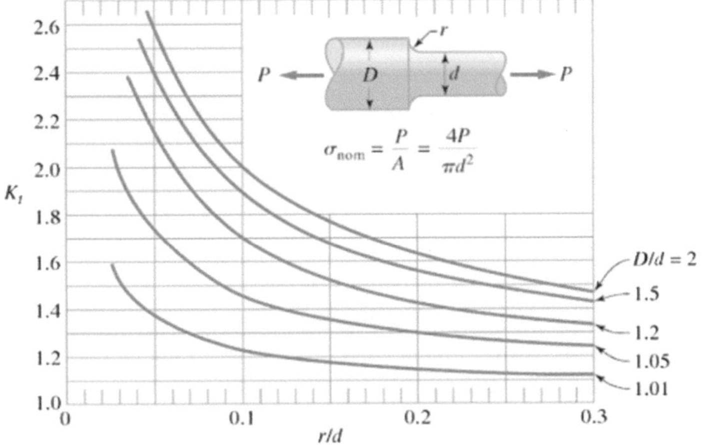

Fig. 3.124 Stress concentration factor of step-shaft under axial loads [7]

<u>Report Format</u>
Write a report for the minor project. The report should include:

- Cover page
- Table of contents
- Introduction
- FEA analysis
- Results and discussions
- Conclusions
- References (at least three references)

Exercises

3-1 List and explain the types of elements used in SolidWorks Simulation. Explain one application for each type of element.

3-2 What is the general procedure for FEA simulation? Explain the main tasks in each step.

3-3 Download and open FEA1. It is made of plain carbon steel, fixed on the bottom surface, and is subjected to 2000 lbs as shown in Fig. 3.125.

    (1) Run FEA simulation with a global standard element size of 0.1″ and provide the following plots

        - The von Mises stress plot with the annotations of maximum and minimum stress

Fig. 3.125 The schematic of
the load and fixture (FEA1)

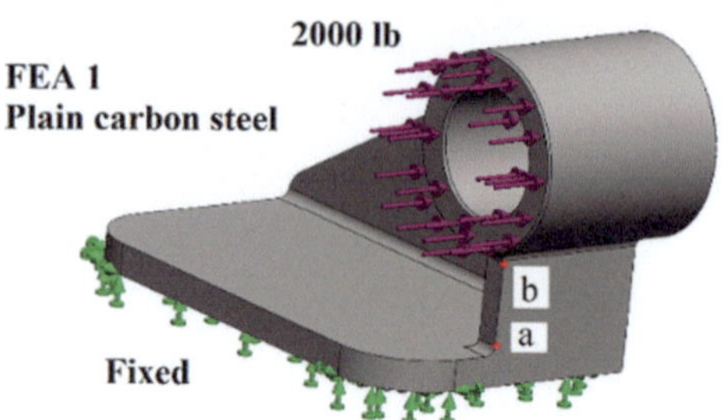

- The von Mises stress graph along the edge ab.
- The resultant displacement plot with the annotations of maximum and minimum displacements.
- The factor of safety plot with red areas below the factor of safety 2.50.

(2) Run FEA simulation with a global element size of 0.4″ and calculate the relative differences between the maximum von Mises stresses obtained with element sizes of 0.1″ and 0.4″.

3-4 Discuss and explain the possible errors in the FEA simulation shown in Fig. 3.30. When there is a relatively big error for a FEA simulation compared with theoretical/experiment results, what should you check first in the FEA simulation setting and why?

3-5 What are the purposes of pre-processing? List and explain several typical actions done during pre-processing.

3-6 Download and open FEA26. It is a round bar made of Alloy steel that is subjected to an axial load of 2000 lbs as shown in Fig. 3.126. There is a split line in the middle of the bar. Run FEA simulations with a global standard element size of 0.1″ on the following two cases.

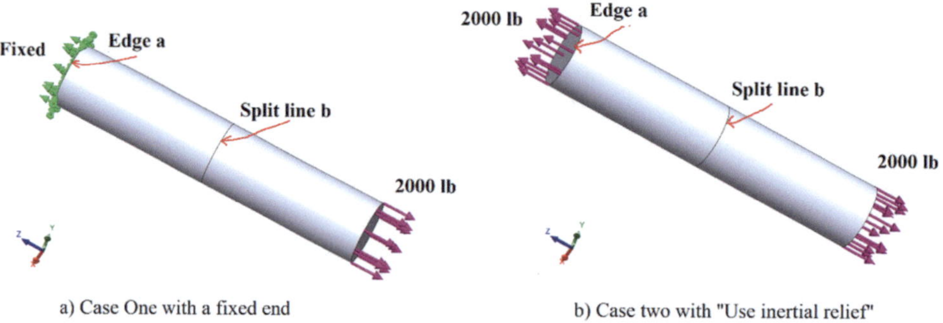

a) Case One with a fixed end   b) Case two with "Use inertial relief"

Fig. 3.126 Schematic of the loading and fixture for Case A and Case B

Fig. 3.127 Schematic of the load and fixture of a support plate

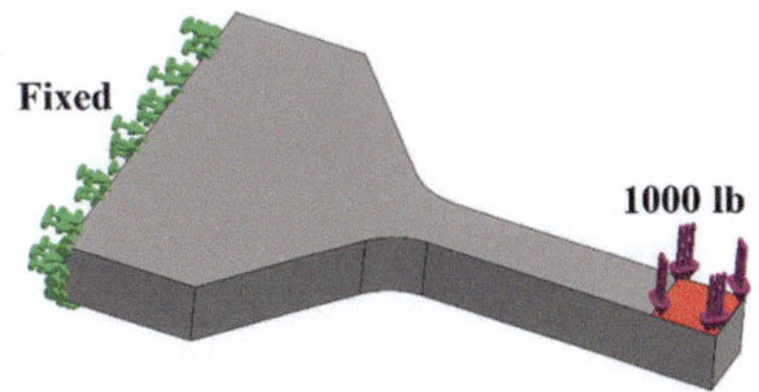

Case One: The one end is fixed and a 2000 lb axial load is applied on the right end.

Case Two: Both ends of the round bar are subjected to a 2000 lb axial load. There is no fixture for this case, but "Use inertial relief" is used to automatically balance the unbalanced forces during the FEA simulation.

- Generate the z-normal stress distribution plot with the annotation of maximum and minimum stresses.
- Display the z-normal stress graph along the edge a and the split line b.
- Compare the FEA simulation results from Case One and Case Two with theoretical results. Discuss the differences.

3-7 Download and open FEA27. The material of this beam is plain carbon steel. One end is fixed and a 1000 lb normal force is applied on the red area of the top surface near the other end as shown in Fig. 3.127.

(1) Run FEA analysis with different global element sizes ($1''$, $0.75''$, $0.5''$, $0.25''$, and $0.125''$) and fill the following table. d_i is the global element size; N is the number of total elements; σ_i is the maximum von Mises stress; $e_{\sigma i}$ is the relative difference between the current von Mises stress with the value in the previous run. The equation for $e_{\sigma i}$ is $e_{\sigma i} = abs(\sigma_i - \sigma_{i-1})/\sigma_{i-1}$; f_i is the maximum resultant displacement; e_{fi} is the relative difference between the current resultant displacement with the value in the previous run and will be calculated using $e_{fi} = abs(f_i - f_{i-1})/f_{i-1}$.

(2) Provide a von Mises stress plot and a resultant displacement plot with the annotation of maximum and minimum when the global element size is 0.125".

(3) Use the FEA results to draw four graphs: d_i vs. σ_i; d_i vs. $e_{\sigma i}$; d_i vs. f_i and d_i vs. e_{fi}

(4) Discuss the trends of the curves in step (3).

d_i (in)	N	σ_i (Psi)	$e_{\sigma i}$	f_i (in)	e_{fi}
1"		$\sigma_1 =$	No value	$f_1 =$	No value
0.75"		$\sigma_2 =$		$f_2 =$	
0.5"		$\sigma_3 =$		$f_3 =$	
0.25"		$\sigma_4 =$		$f_4 =$	
0.125"		$\sigma_5 =$		$f_5 =$	

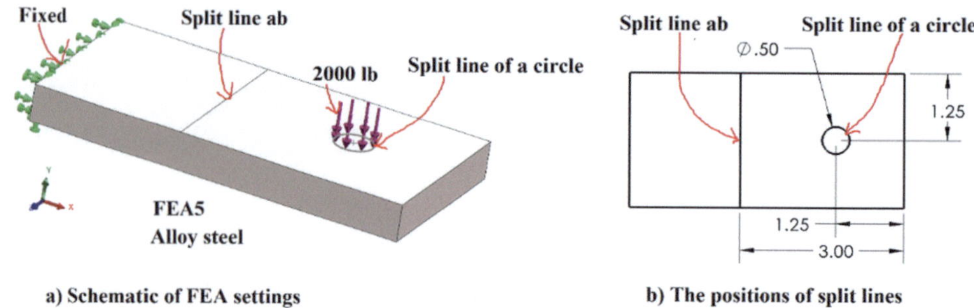

a) Schematic of FEA settings   b) The positions of split lines

Fig. 3.128 Schematics of the FEA setting and dimensions for split lines

3-8 Download and open FEA5. The loads and fixtures are shown in Fig. 3.128a. The 2000 lb normal force is applied to the area circled by the split line of a circle. Create split lines per the dimensions shown in Fig. 3.128b. Run an FEA simulation with a global standard element size of 0.1″ and provide the following plots.

- The von Mises stress distribution plot with the annotations of maximum and minimum stress.
- The x-normal stress graph along the splint line ab.
- The factor of safety distribution plot with an annotation of the minimum factor of safety.
- The factor of safety plot with red areas below 2.00.

3-9 Download and open FEA28. The material of this part is plain carbon steel. The loads and fixtures are shown in Fig. 3.129.
(1) Case A: Run the FEA simulation using solid elements with a global standard element size of 0.05″ and provide the following plots:
  - The von Mises stress distribution plot with the annotation of maximum and minimum stress.
  - The factor of safety plot with red areas below the factor of safety 2.25
(2) Case B: Run the FEA simulation using shell elements with a global standard element size of 0.05″ and provide the following information.
  - The von Mises stress distribution plots on the top and the bottom surface with the annotation of maximum and minimum
  - The factor of safety plot with red areas below the factor of safety 2.25
(3) Compare the maximum von Mises stress and computation time from both cases and discuss the results.

Fig. 3.129 The schematic of FEA settings

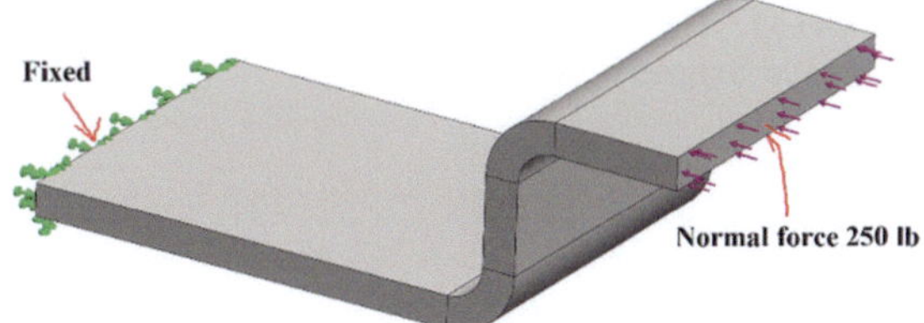

3-10 Download and open FEA11. The material of this component is AISI 1020. The loads and fixtures are displayed in Fig. 3.130. Run FEA simulation with a global standard element size of 0.075".
(1) Generate the following plots and information.
- The von Mises stress distribution plot with the annotation of maximum and minimum stress.
- The resultant displacement distribution plot with the annotation of maximum and minimum displacement.
(2) Create a factor of safety plot with red areas below the factor of safety 2.5. Discuss the results of this factor of safety plot. If the required factor of safety for this question is 2.5, does this design fail? If it fails, what is your suggestion for redesigning?

3-11 Download and open FEA15. Its material is AISI 1020. The load and fixtures are displayed in Fig. 3.131. Run FEA simulation with a global standard element size of 0.1".
(1) Generate the following plots and information.
- The von Mises stress distribution plot with the annotation of maximum and minimum stress.
- The factor of safety plot with red areas below the factor of safety 2.0.
(2) If the required factor of safety for this question is 2.0, what is your suggestion for redesign?

Fig. 3.130 The schematic of loads and fixtures of a v-plate

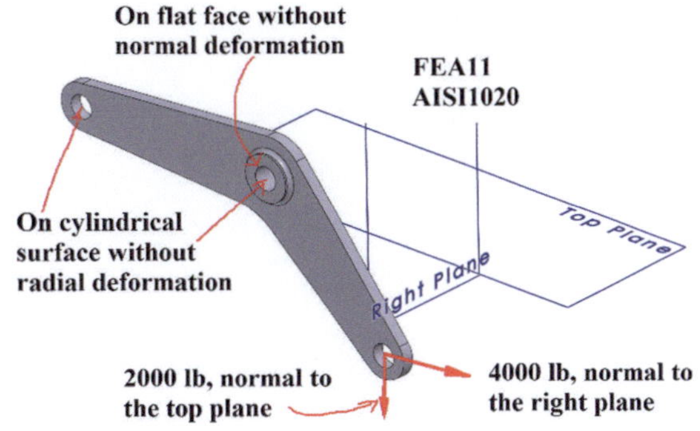

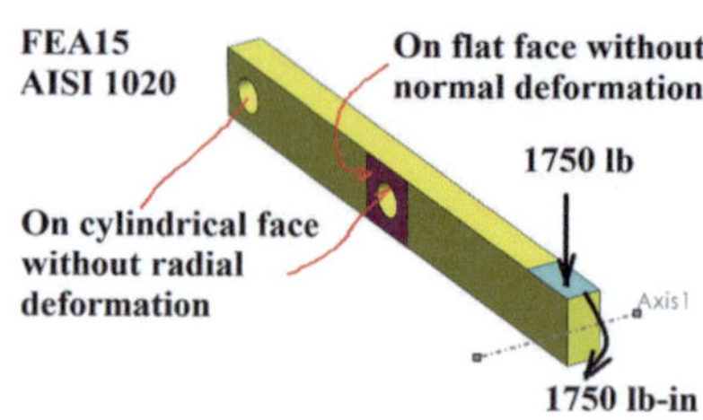

Fig. 3.131 The schematic of loads and fixtures of a beam

3-12 Download and open FEA29. Its material is AISI 1020. The loads and fixtures are shown in Fig. 3.132. The connection type for the remote loads is "Distributed". Run FEA simulation with a global element size of 0.075″. Provide the following plots and information.

- The von Mises stress distribution plot with the annotation of maximum and minimum stress.
- The factor of safety distribution plot with the annotation of the minimum factor of safety.

3-13 Download and open FEA 16. Its material is alloy steel. The loads and fixtures are shown in Fig. 3.133. Run FEA simulation with a global standard element size of 0.2″ and a local element size of 0.08″ on the fillet area. Provide the following plots and information.

- The shear stress τ_{xy} distribution plot with the annotation of maximum and minimum stress. Provide some comments about this plot.
- The shear stress τ_{xy} graph along the arc ab.
- The factor of safety plot with red areas below 1.5. Provide some comments about this plot.

3-14 Download and open FEA7. Its material is Alloy steel. The loads and fixtures of this model are shown in Fig. 3.134. Since this model is symmetrical, a quarter of the model can be used for FEA simulation. Run FEA simulation using shell elements with a global standard element size of 0.1″. Provide the following plots and information.

Fig. 3.132 The schematic of loads and fixture

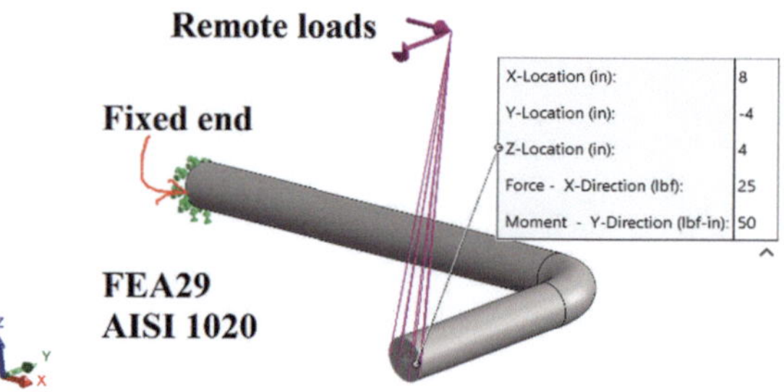

Fig. 3.133 The schematic of the load and the fixture

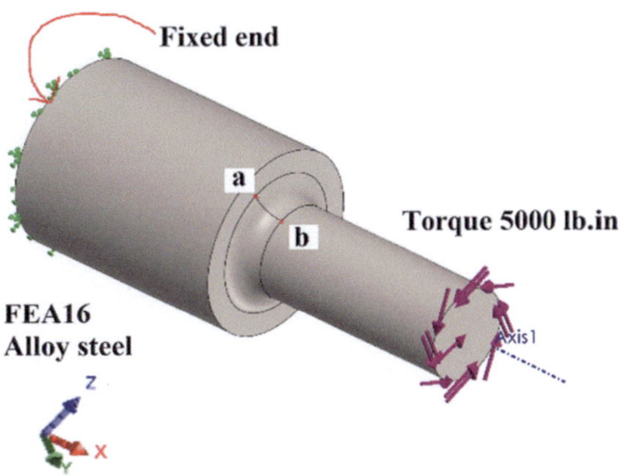

- The von Mises stress distribution plot on the top and bottom surface with the annotation of maximum and minimum stresses.
- The factor of safety distribution plot with an annotation of the minimum factor of safety.
- The factor of safety with red areas below the factor of safety 5. Provide some comments about this plot.

3-15 Download and open FEA22. Its material is plain carbon steel. The loads and fixtures are shown in Fig. 3.135. Use "Apply mesh control" and "Standard mesh" to run the FEA simulation.

- Explain how to determine the local element sizes and the global element size.
- Generate the von Mises stress distribution plot with the annotation of maximum and minimum stress. Comment on the possible weak locations.

Fig. 3.134 Schematic of the load and fixtures of the quarter model

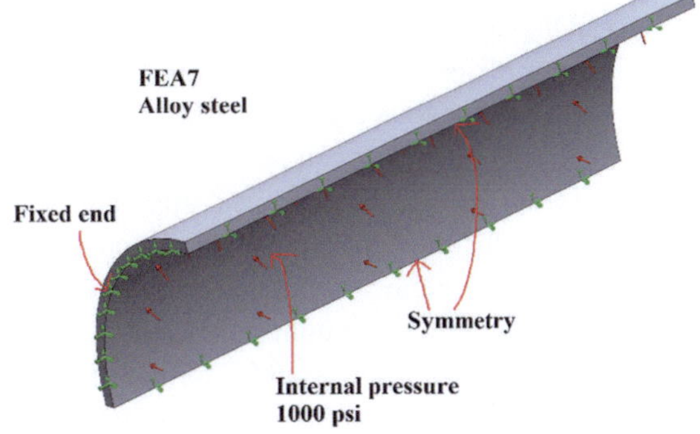

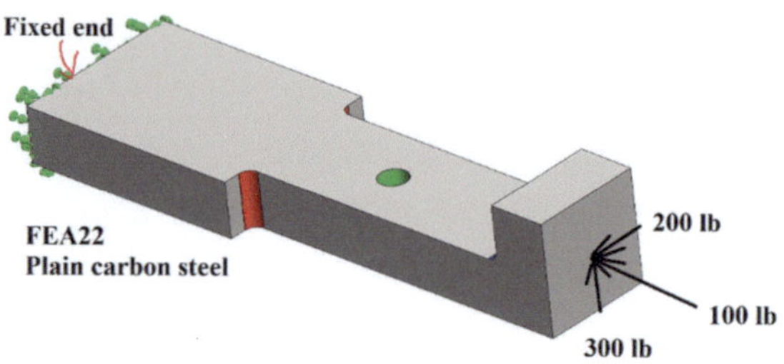

Fig. 3.135 A schematic of loads and fixture

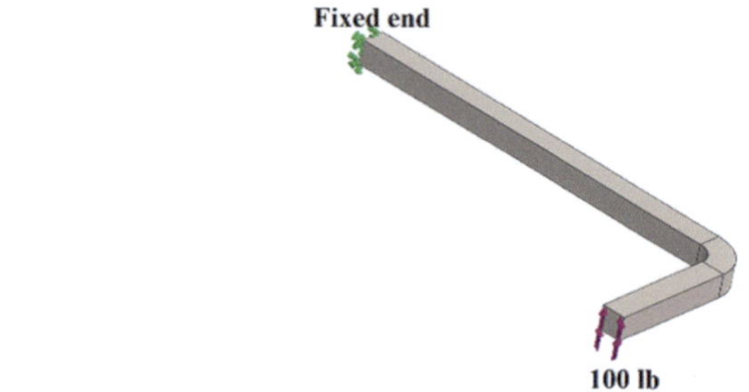

Fig. 3.136 Schematic of the load and fixture of a L-shape bar

- The factor of safety is required to be 1.5. Generate the factor of safety plot with red areas below the factor of safety 1.5. Comment on the failure areas

3-16 Download and open FEA30. Its material is alloy steel. The loads and fixtures are shown in Fig. 3.136.

Case A: Use the "h-adaptive" approach with a global element size of 0.075″. The settings for the "h-adaptive" are 99% target accuracy and maximum number of loops 5. Generate the following distributions and information.
- Display the initial meshing and its mesh detail before the h-adaptive method.
- Display the final meshing and its mesh detail after the h-adaptive method. Comment on the differences between these two meshes.
- Display the convergence graph with the options of target accuracy and maximum von Mises stress. If the target accuracy is not satisfied, change the initial global element size accordingly and rerun the adaptive method.
- Display the factor of safety plot with red areas below 1.5. Comment on the plot.
Case B: Use the "p-adaptive" approach with a global element size of 0.075″. The settings for the "p-adaptive" are: stop when total strain energy change is less than 1%; update element with relative strain energy error of more than 2%; starting p-order of 2 and maximum p-order of 5; 4 maximum number of loops. Generate the following distributions and information.
- Display the initial meshing and its mesh detail before the p-adaptive method.
- Display the final meshing and its mesh detail after the p-adaptive method. Comment on these two meshes.

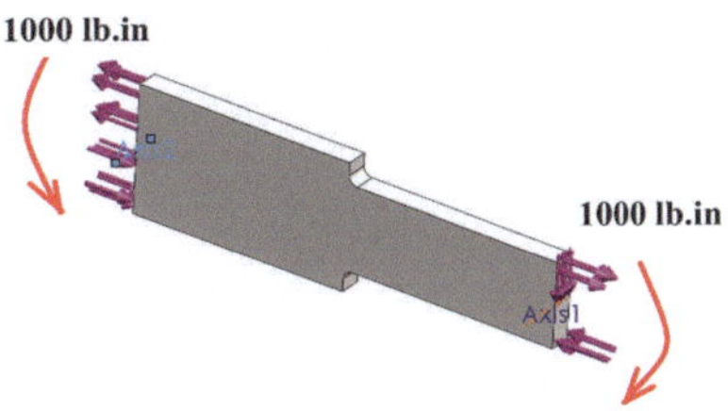

Fig. 3.137 Schematic of the setting for this question

- Display the convergence graph with the options of total strain energy and maximum von Mises stress. If the target accuracy is not satisfied, change the initial global element size accordingly and rerun the adaptive method.
- Display the factor of safety plot with red areas below 1.5. Comment on the plot.

3-17 Download and open FEA14, which is a step-plate experiencing pure bending as shown in Fig. 3.137. Its material is alloy steel. Run an FEA simulation to determine the stress concentration factor and compare it with a stress concentration factor value from any design handbook or textbook. Is the relative difference less than 5%? If not, change the settings to make the relative difference less than 5%.

References

1. Solidworks Web Help, https://help.soliworks.com
2. Paul M. Kurowski (2022) Engineering Analysis with SOLIDWORKS Simulation 2022, SDC Publications
3. Xiaobin Le, Anthony Duva (2020) Facilitate students to integrate FEA simulation skills through a practical simulation project Paper presented at the 2020 ASEE Virtual Annual Conference Content Access, Virtual Online. https://doi.org/10.18260/1-2-34657
4. Xiaobin Le, Ali Moazed, and Anthony Duva (2016) The Design Projects for the Simulation-Based Design Course Paper presented at 2016 ASEE Annual Conference & Exposition, New Orleans, Louisiana. https://doi.org/10.18260/p.26118
5. Robert L. Norton (2006) Machine design: an integrated approach, Pearson Prentice Hall, Upper Saddle River, N.
6. Richard G. Budynas, J. Keith Nisbett (2015) Shrigley's mechanical engineering design, McGraw-Hill Education, New York.
7. Collins, J. (2010). Mechanical design of machine elements and machines. Hoboken, NJ: Wiley

FEA Simulation on Assemblies

4

Abstract

This chapter discusses and demonstrates how to use the "Static" module In Solid-Works Simulation to run an FEA simulation on assemblies. The procedure and skills discussed in Chap. 3 can be directly implemented in this chapter except that there are interfaces on an assembly. First, three types of contact interactions on an interface of assembly in SolidWorks Simulation are discussed and explained. They are "Bonded", "Contact" and "Free". They can be defined by three different approaches which are "Global interaction (default)", "Local interaction" and "Component interaction". Several examples with step-by-step instructions are provided to demonstrate how to define the type of contacts and run FEA simulation on an assembly. Second, if some components in an assembly are designed by other specific approaches such as pins and bolts, they can be simplified as mathematical models (connectors) in the assembly. Several typical connectors such as pin, bolt, and interference fit are presented and demonstrated. Third, examples of truss elements and beam elements will be discussed in this chapter because they are typically assemblies. Fourth, there is a mixture of different elements in an FEA simulation on an assembly because different components can be meshed by different types of elements. One example is presented to demonstrate a mixture of solid, beam, and shell elements in FEA simulation on an assembly. Finally, a major design project in FEA simulation on a real product will be presented.

Supplementary Information The online version contains supplementary material available at https://doi.org/10.1007/978-3-031-64132-9_4.

4.1   Introduction

Mechanical devices or systems typically contain several components. Each component is arranged and assembled according to the designed specific ways on interfaces so that they can complete specified functions or tasks. For mechanical component design by FEA simulation, we start the FEA simulation on an assembly. Strain/stress and the safety of each component can be obtained from the simulation results. Then, we could run the FEA simulation on critical components as required with much finer meshing based on the information on the interface obtained from the assembly simulation to generate a model with a much smaller relative error. This chapter will explore how to run an FEA simulation on an assembly. The concise descriptions of each section of this chapter are listed here.

Ch4.2 Type of contacts: This section will explain three typical types of contacts on the interface of an assembly in SolidWorks simulation.

Ch4.3 Define types of contacts: This section will show how to interpret the default global contact and how to specify the types of contacts. Two examples will be provided as demonstrations.

Ch4.4 Several typical connectors: FEA simulation is mainly used to obtain stress/strain for component design. But if some components in an assembly are not the concern or will be designed by other specific approaches such as pins and bolts, they can be simplified as mathematical models (connectors) in the assembly. This section will explain several typical connectors such as pin, bolt, and shrink fit. One example will be discussed and demonstrated for each type of connector.

Ch4.5 Mixture of different types of elements: This section will use one example to demonstrate a mixture of solid, beam, and shell elements in an FEA simulation on an assembly.

Ch4.6 Major design project: Any FEA simulation on a real product can be used as a Major design project. This section will use one real product as an example for FEA simulation on assemblies.

4.2   Types of Interactions

This section will discuss how a component interacts with another component through interfaces. It will explain three types of contacts: contact interaction, bonded interaction, and free interaction.

Interface in mechanical design can be defined as boundary conditions between components, through which components will interact with each other per specified assembly relationships. Components are assembled through interfaces in such a specific way so that mechanical devices can have the required functions. For example, the boundary conditions between a shaft and a bearing are a typical interface where the assembly relationships

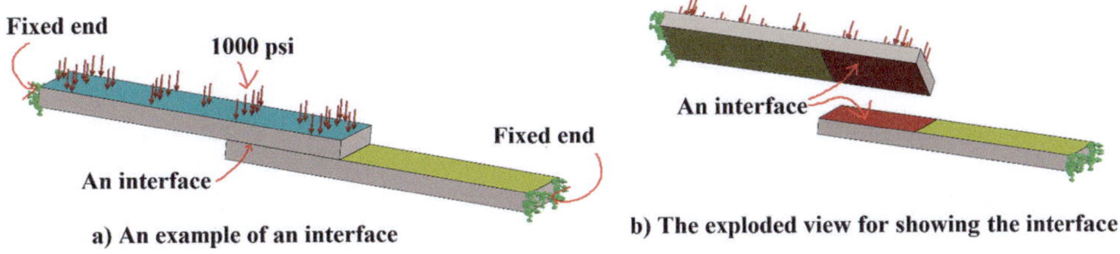

Fig. 4.1 An example of an interface

could be interference fit, transient fit, or clear fit. For another example, the boundary condition between a pivot pin and the pivoted part is another typical interface where the pivoted part is supported by the pin and can have relative rotational motion on the pin.

Interactions for FEA simulation can be defined as the features of interfaces between components, which control the behaviors of interfaces during FEA simulation. Types of Interactions in SolidWorks Simulation can be three different types: contact interaction, bonded interaction, and free interaction. An example as shown in Fig. 4.1 will be used to explain these three types of interactions. The top plate is fixed at the left end and subjected to 1000 psi pressure on the top surface. The bottom plate is fixed at the right end. The interface as shown in Fig. 4.1b is the overlapping area between the top and the bottom plates. Before any load is applied, two surfaces of the interface are overlapping or touching to each other.

Contact interaction is an interface that allows relative motion on a contacting surface or forms a gap but prevents penetration between entities of the interface. Initial contact surfaces or gap-separated surfaces of an interface can be defined as a contact interaction. For a contact interaction, the surfaces can have a relative sliding motion on the interface, or the interface can be partially or completely separated. For mechanical systems, most interfaces except for welded interfaces can be defined as contact interactions.

For example, when the interface shown in Fig. 4.1 is defined as a contact interaction, the resultant displacement distribution plot is shown in Fig. 4.2 (Note: Example 1 in the next section will explain and show how to run the FEA simulation for this case). The 1000 psi pressure on the top surface of the top plate causes deformation on the top plate. The top plate forces the bottom plate to be deformed through the contact interaction of the interface. Because the interface is a contact interaction, the original fully overlapping areas of the interface become only a portion of the contacting area, and gaps are formed in most areas of the interface.

Bonded interaction is an interface that behaves as if entities of the interfaces are welded together. Initial contacting entities or gap-separated entities on an interface can be defined as a bonded interaction. During the FEA simulation, the bonded interaction will automatically make entities of the interface to be contacting and weld them together. Only welded entities can be defined as bonded interactions. For example, the interface

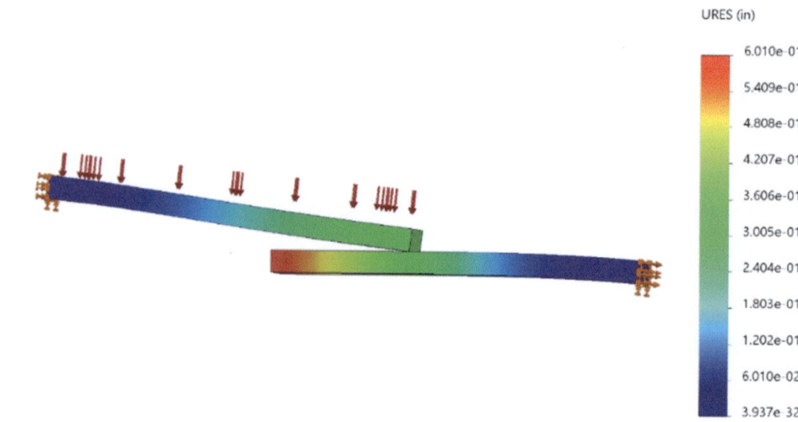

Fig. 4.2 Resultant displacement distribution plot with the type of contact interaction

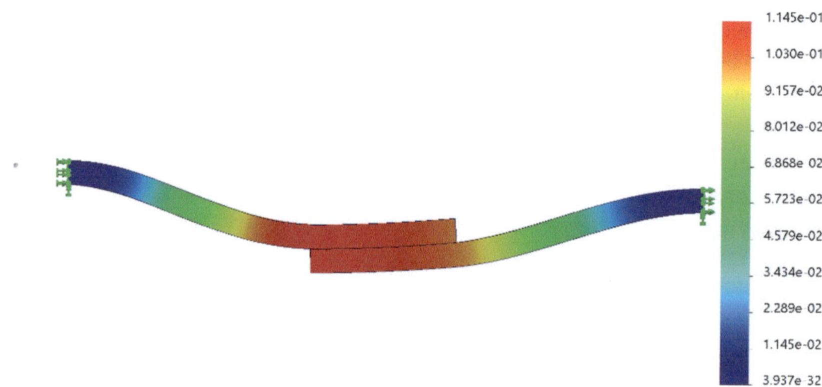

Fig. 4.3 Resultant displacement distribution plot with the type of bonded interaction

of a bolted joint cannot be defined as a bonded interaction and should be defined as a contract interaction.

For example, when the interface shown in Fig. 4.1 is defined as a bonded interaction, the resultant displacement distribution plot is shown in Fig. 4.3. The 1000 psi pressure on the top surface of the top plate causes deformation on the top plate. When the interface is a bonded interaction, the shared contacting points on the interface will have the same deformations because the two surfaces of the interface are bonded or welded together.

Free Interaction is defined as no interactions between entities of different components on the interface. In this type of interface, the entities of one component treat other entities of another component to be non-existent so that interaction due to contact will not happen. Interaction on interfaces will consume lots of computing time. Therefore, free interaction

Fig. 4.4 Resultant displacement distribution plot with the type of free interaction

can save computing time. However, this type of interaction should not be defined unless it is certain that there are no interactions between entities on this interface.

For example, when the interface shown in Fig. 4.1 is defined as a free interaction, the 1000 psi pressure on the top surface of the top plate causes deformation of the top plate. However, since the interface is defined as free interaction, there is no interaction at the interface. So, the bottom plate will have no deformation because the 1000 psi pressure on the top surface of the top plate will not affect the bottom plate due to no interaction. When the free interaction on the interface is defined, the resultant displacement distribution plot is shown in Fig. 4.4. The bottom plate has zero deformation. Of course, the assumption in this example is not correct. This example is only used to explain and demonstrate the effect of a free interaction.

4.3   Define Types of Contacts

In SolidWorks simulation, there are three ways to define types of typical contacts: contact interaction, bonded interaction, and free interaction. They are "Global Interaction", "Local Interactions" and "Component Interactions". This section will explain how to use these three approaches to define types of contacts.

"Global Interactions" is the default setting for contacts of an assembly and defines two types of contacts: (1) If the gap on an interface between any entities is zero or less than 0.01% of the characteristic length of the model, the interface will be defined as a bonded interaction; and (2) If there is a larger gap on an interface between entities, the interface will be defined as a free interaction.

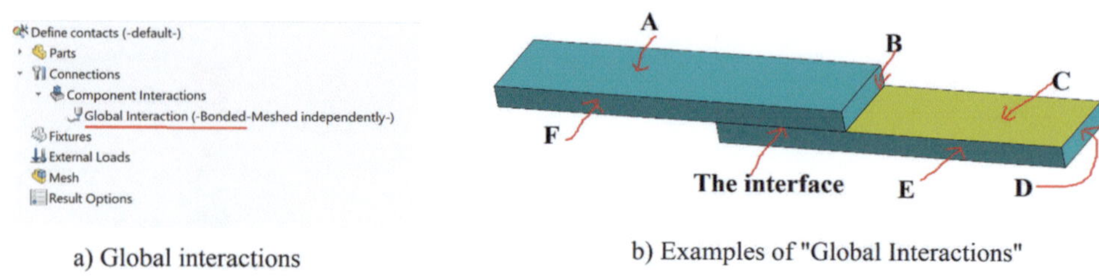

a) Global interactions   b) Examples of "Global Interactions"

Fig. 4.5 Global interactions and their explanations

When any FEA simulation study of an assembly is created, the global interaction is automatically formed as shown in Fig. 4.5a. In the simulation study tree, expand "connections", then expand "Component Interactions" and "Global interaction" will appear. The default "global interactions" does not need to be redefined and it is recommended that "Global interactions" not be modified or deleted. In an FEA simulation, calculations associated with contacting interactions are time-consuming. Free interaction treats other entities as nonexistent, so it will not consume any computing time. In Fig. 4.5b, the surfaces A, B, and F belong to the top plate, and the surfaces C, D, and E belong to the bottom plate. The contact between surface A and surface C, D, or E will be defined as free interactions through global interaction. In the same way, the contact between surface C and surface A or surface B or surface F will also be defined as free interactions through global interaction. Since the overlapping interface shown in Fig. 4.5b is contacting, it will be defined as a bonded interaction through the global interaction.

The global interactions define default types of contacts for all possible interfaces of the assembly: bonded interaction if they are touching each other or free interaction if they are not touching each other. These default types of contacts can be overwritten by the definitions from "Local Interactions" and "Component Interactions".

"Local Interactions" is used to manually define a type of contact at an interface. This can be used to define any type of contact at the interface such as free interaction, bonded interaction, and contact interaction. The type of contact defined by "Local Interactions" will overwrite the type of contact defined by "Global Interactions".

It is better to have an exploded view of an interface to show entities when the "local interaction" is used for defining the type of contact so that corresponding surfaces can be easily selected as shown in Fig. 4.6. The exploded view of the assembly will not affect simulation results.

As shown in Fig. 4.6, in the simulation study tree, right-click the tab "Connections", select "Local Interactions" on the prompt drop-list menu, and then the local interaction Property Manager will appear. Follow this sequence to define a local interaction as shown in Fig. 4.6. (1) First choose the type of contact: Contact, Bonded, and Free, which represent "Contact Interactions", "Bonded Interactions" and "Free Interactions". (2) Select the first of the entities for an interface, which can be faces, edges, and vertices of either

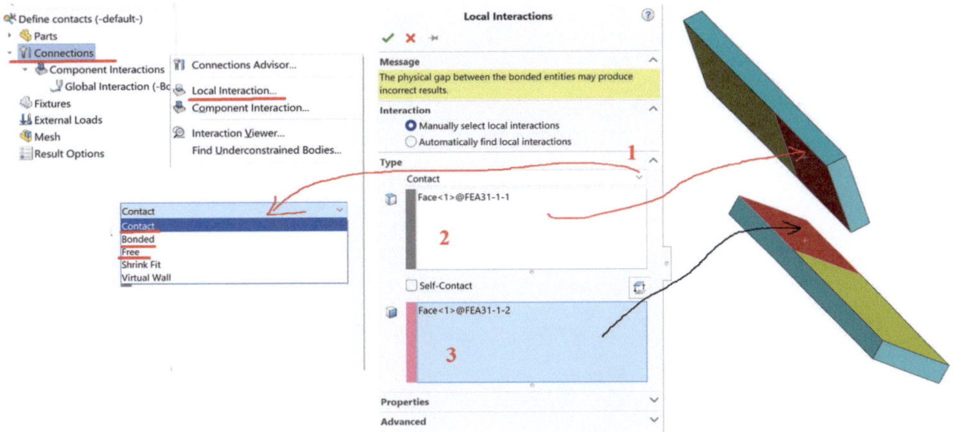

Fig. 4.6   Schematic of "Local interactions"

component. (3) Select the second entity for the interface, which can be faces, edges, and vertices of the other component.

The "Local Interaction" is a very important approach to manually define the type of contact on selected sets of entities. Through "Local Interactions", touching entities at an interface can be defined as "contact interaction". None touching entities at an interface can also be defined as "Bonded Interactions". One example of this is the welded parts on the welding interface.

"**Component Interactions**" is used to specify the type of contact on selected components. It can be used to define any one type of contact: free interaction, bonded interaction, and contact interaction. It is not recommended to use this method. When actual contact conditions are not sure or unpredicted, this approach can be used. It is a time-consuming approach.

As shown in Fig. 4.7, in the simulation study tree, right-click the tab "Connections" tab, and select "Component Interaction". The component Property Manager will appear. Follow these steps to define a type of contact through "Component interactions". (1) Select the type of contact: free, bonded, and contact. And (2) select corresponding components. For the choice of "free", there will not be any interaction between any entities of the selected components. For the choice of "Bonded", when the gap between any entities of the selected component is zero or less than 0.01% of the characteristic length of the model, the entities will be defined as bonded interaction. All other entities will be treated as free interactions. These two scenarios have been covered in global interactions. "Component interaction" is typically not used to define free interaction or bonded interaction between components. For the choice of "contact", it will assume that the type of contact for all entities of the selected components will be contact interaction. This will be

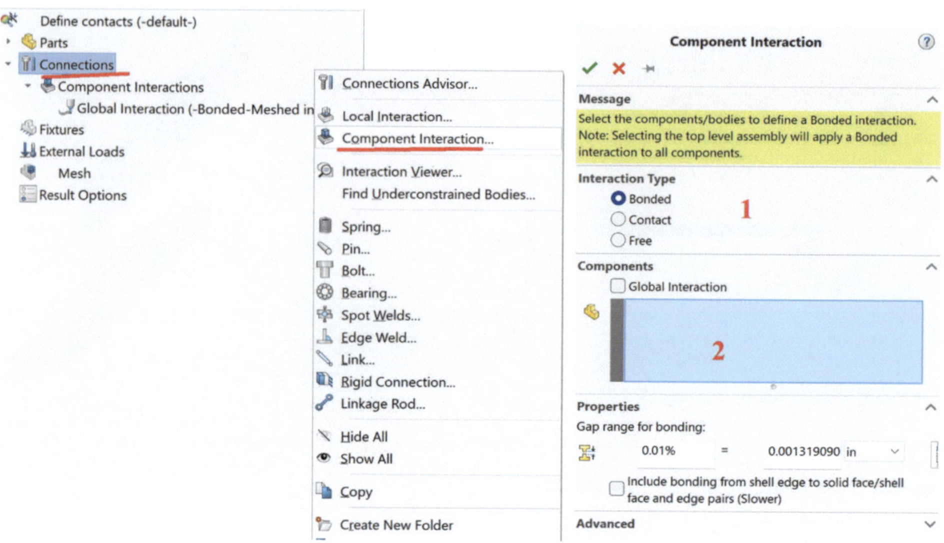

Fig. 4.7 Schematics of "Component Interactions"

time-consuming because, during the FEA simulation, the program will constantly check interaction conditions between all entities of the selected components.

Three types of contacts and three approaches for defining a type of contact have been discussed and explained. With the knowledge and skills from Chap. 3, we can run FEA simulations on assemblies. Now, two examples will be presented to demonstrate how to define types of contacts and to run an FEA simulation on an assembly.

Example 1: Different contacts on an interface (FEA31-asm) Download and open the assembly FEA31-ASM. A device consists of two flat plates as shown in Fig. 4.1a. There is an interface between the top plate and the bottom plate as shown in Fig. 4.1b, which is an exploded view to show the interface. Materials for both parts are alloy steel. Run FEA simulation with a global standard element size of 0.1″ under the following three types of contact cases: (1) The interface is a "contact interaction"; (2) The interface is a "free interaction"; and (3) The interface is a "bonded interaction".

Solution:

(1) **FEA simulation**

Step 1: **Pre-processing**. To define the types of contacts at an interface, it is recommended that an exploded view be created to display each possible entity at the interface. In

Fig. 4.8 Assigning material to each component

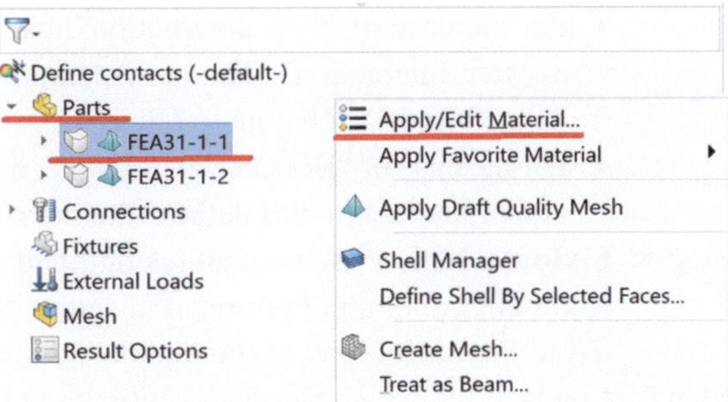

this example, the top plate touches the bottom plate when no load is applied. The exploded view will not affect the FEA simulation. However, after types of contacts have been defined, the exploded view can be collapsed for better post-processing. Since the calculation process for contacting is very complicated and time-consuming, it will be better to specify the actual areas for an interface when possible. As shown in Fig. 4.1b, two split lines have been created to identify the actual areas of the interface. In such a way, it can reduce the calculation time.

Step 2:   **Set up a project**. This is a static stress analysis.

Step 3:   **Select the type of element and assign a material to each component**. In the FEA simulation study tree, expand the "parts" tab and the list of components in this assembly will appear as shown in Fig. 4.8. Follow the procedure explained in Chap. 3.6 to define the type of material and element for each component. Since each component can have one type of element, the assembly with multiple components can have a mixture of different types of elements, which will be discussed later.

Since an assembly may have multiple components, it is recommended that material be assigned to each component at the part level. This way, the material of a few components can be modified when necessary.

In this example, the type of material for both plates is Alloy Steel which is included in the SolidWorks material library. The type of element for both plates in this example will be solid elements with high-quality mesh. No changes to the type of element are needed and Alloy steel can be directly assigned to components.

Step 4:   **Connections**. The automatically generated "global interaction" will be kept without any modifications. In this example, there is no gap in the interface. So, by the default global interaction, the interface will be defined as a bonded interaction. All other entities of both parts don't touch each other. Therefore, the default global interaction will define them as free interactions.

In this example, three FEA study projects will be created for three different types of contacts: contact interaction, free interaction, and bonded interaction.

- For the case of "Contact interaction", use local interaction as described in Fig. 4.6 to define the interface as "contact interaction". This new definition will overwrite the bonded interaction defined in the global interaction.

- For the case of "Free interaction", use local interaction to define the interface as "free interaction". This new definition will overwrite the bonded interaction defined in the global interaction.
- For the case of "Bonded interaction", no additional action is needed because the global interaction has defined the interface as "bonded interaction".

Step 5: **Fixtures**. Follow the procedures outlined in Chap. 3.7: Fixtures to define fixtures. Right-click the tab "Fixture" and select "Fixture geometry" to apply the "Fixed" fixture on the left end of the top plate and the right end of the bottom plate.

Step 6: **Loads**. Follow the procedures outlined in Chap. 3.8: Loads to define the loading of the components. Right-click the tab "External loads" and select "pressure" to define a 1000 psi on the top surface of the top plate.

Step 7: **Meshing**. In this example, there is no stress concentration area. The shortest dimension is the thickness of 0.5 inch of the plates, the global element size of 0.1″ will be used.

Step 8: **Run Simulation**. Right-click the project name and select "Run". This will start the FEA simulation.

Step 9: **Post-processing**

Case of "contact interaction"

The procedures for post-processing of FEA simulation on assemblies are the same as those for FEA simulation on components except for the following two aspects.

- When there is a contact interaction, there will be contact pressure on this interface. It is a type of stress as shown in Fig. 4.9a. In this example, for the case of contact interaction, the contact pressure distribution plot is shown in Fig. 4.9b.
- When a plot is created, it can be the plot for the whole assembly as shown in Fig. 4.10. However, it will be more convenient for mechanical component design when the plot for

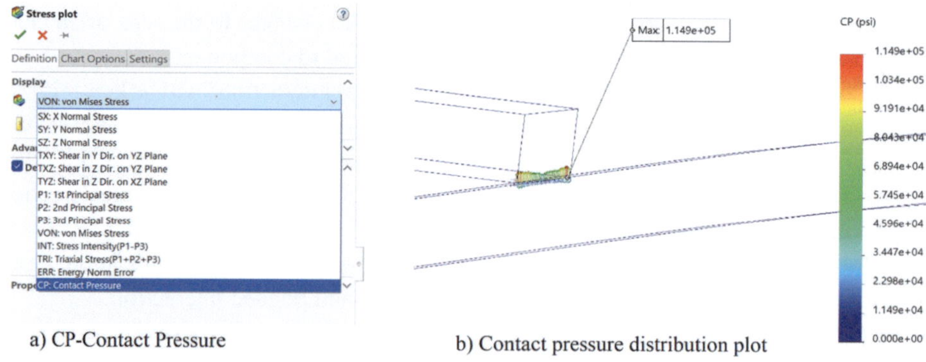

a) CP-Contact Pressure   b) Contact pressure distribution plot

Fig. 4.9 Contact pressure distribution plot of the case of "contact interaction"

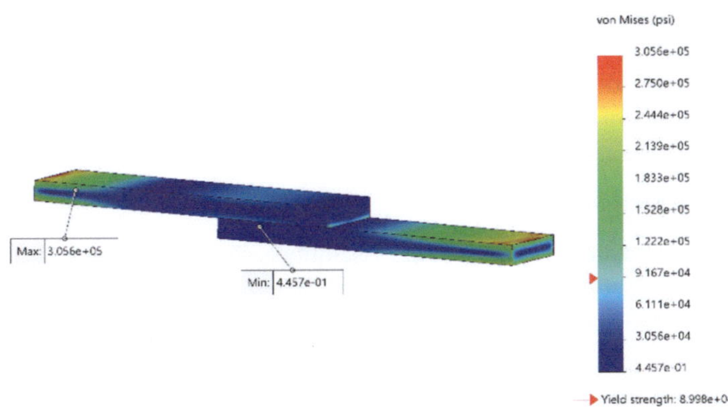

Fig. 4.10   Von Mises stress distribution plot of the assembly of the case of "contact interaction"

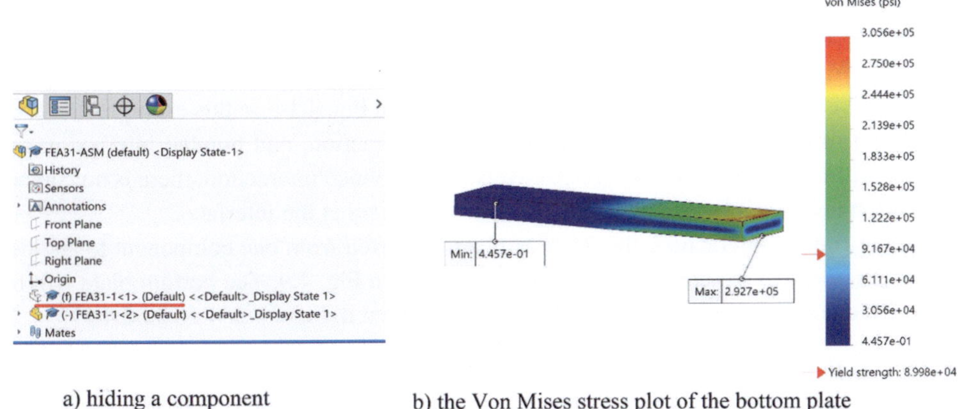

a) hiding a component   b) the Von Mises stress plot of the bottom plate

Fig. 4.11   Von Mises stress plot of the bottom plate of Case of "contact interaction"

a single component is created as shown in Fig. 4.11b. The approach to display the plot of a specified component is (a) Create a plot for the assembly, (b) In the model feature tree as shown in Fig. 4.11a, all other components should be hidden, (3) return to the simulation study tree, and the created plot will be the plot of the specified component. Please note that components will be only hidden and not suppressed. Suppressing a component will change the models of the assembly under the FEA simulation and cause the loss of all simulation data.

The resultant displacement plot of the Case "contact interaction" is shown in Fig. 4.2.

Case of "Bonded Interaction"

The Von Mises stress distribution plot of the assembly is shown in Fig. 4.12a. The von Mises stress distribution plot of the bottom plate is shown in Fig. 4.12b. The resultant displacement distribution plot of the assembly is shown in Fig. 4.3.

Case of "free interaction"

The Von Mises stress distribution plot of the assembly is shown in Fig. 4.13a. The von Mises stress distribution plot of the bottom plate is shown in Fig. 4.13b. The resultant displacement distribution plot of the assembly is shown in Fig. 4.4.

(2) **Interpretations and Discussions**

Among three typical types of contacts, the contact interaction will have the longest computing time and the bonded interaction will have the least computing time. In this example, the global element size, loading, and fixtures of the three simulations are the same except for the type of contacts. The total computing time of the simulation can be found through this approach: (1) Right-click the tab "Results" in the simulation study tree, (2) Select "Solver Messages", and the solver message window will appear as shown in Fig. 4.14. In this example, the total computing times for the contact interaction, free interaction, and bonded interaction are 5:31 min, 2:39 min, and 1:45 min, respectively. For a bonded interaction, there is no contact calculation because it is treated as one welded-component at the interface.

For the contact interaction, the force can be transferred from one component to another component through the contacting pressure as shown in Fig. 4.9. The bottom plate will be deformed due to the interaction at the interface as shown in Figs. 4.2, 4.10, and 4.11b.

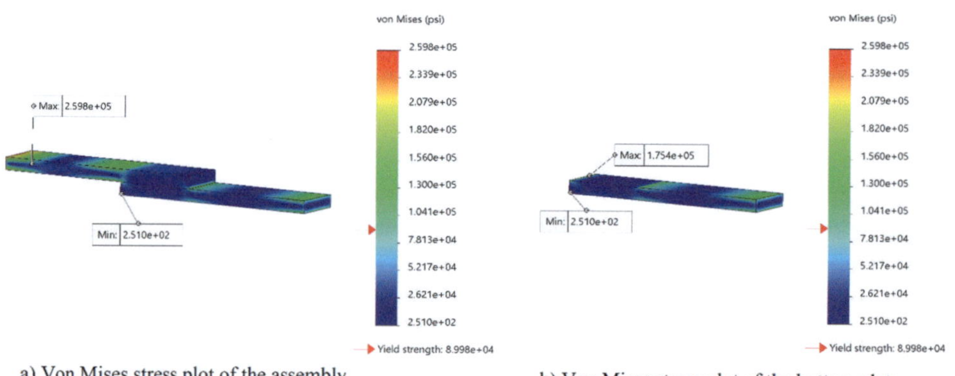

a) Von Mises stress plot of the assembly   b) Von Mises stress plot of the bottom plate

Fig. 4.12   Von Mises stress distribution plots of the case of "bonded interaction"

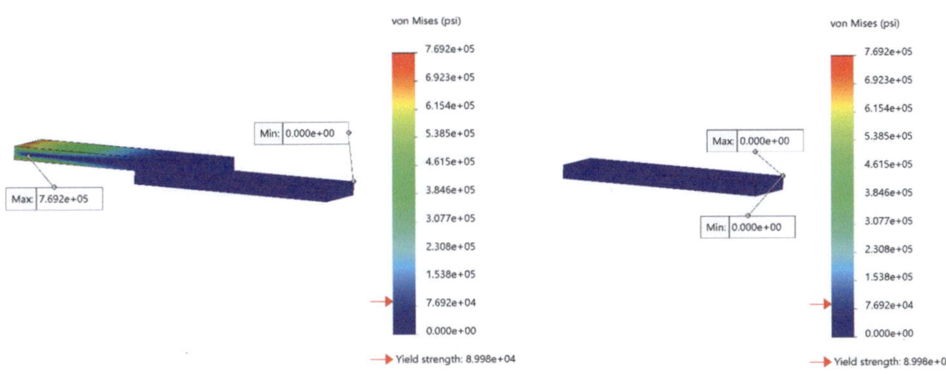

<table>
<tr><td>a) Von Mises stress plot of the assembly</td><td>b) Von Mises stress plot of the bottom plate</td></tr>
</table>

Fig. 4.13 Von Mises stress distribution plots of the case of "free interaction"

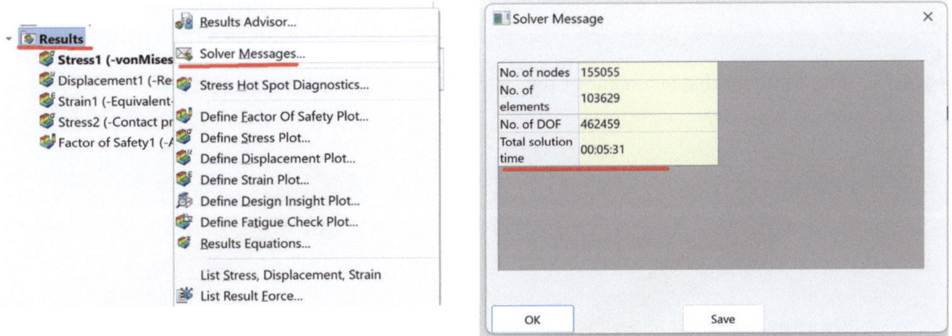

Fig. 4.14 The total computing time

For the bonded interaction, the components will be treated as one welded component through the interface. Through this bonded interface, the bottom plate will be deformed as shown in Figs. 4.3, and 4.12. By using the tool "list-selected", resultant displacements of the outer edges of the top & bottom plates on the bonded interface can be created as shown in Fig. 4.15. The shared points on the bonded interface have the same resultant displacement because they are bonded together.

For free interaction, there is no interaction on the interface. The top plate will be deformed due to the 1000 psi pressure on its top surface as shown in Figs. 4.4 and 4.13a. However, due to no interaction on the interface, no force will be generated at the interface. Therefore, the bottom plate will not be deformed and will have zero stress as shown in Fig. 4.13b.

Example 2: Different assumptions of the same interface (FEA32-ASM) Download and open FEA32-ASM, which consists of three components: a beam, a pin, and a supporter as

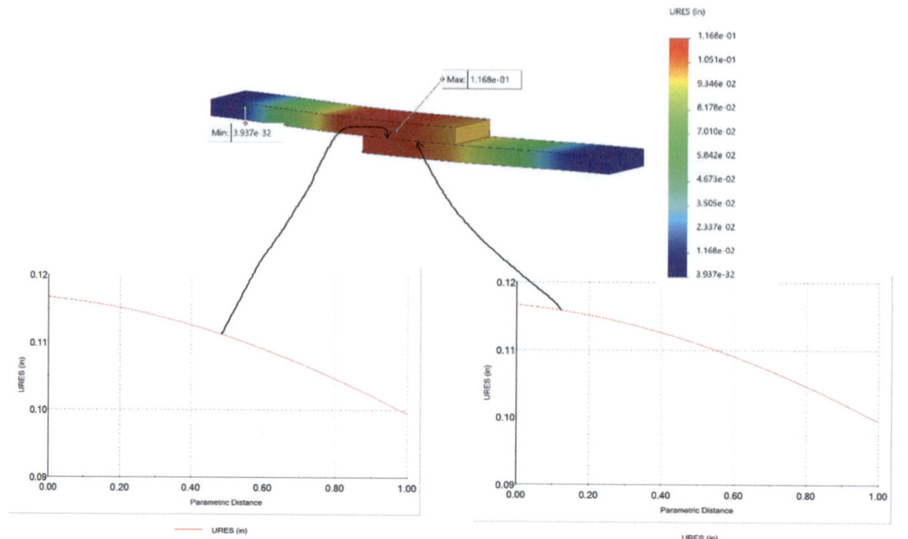

a) The resultant displacement of the outer edge on the top plate   b) The resultant displacement of the outer edge on the bottom plate

Fig. 4.15 The resultant displacement of the outer edges of top and bottom plates

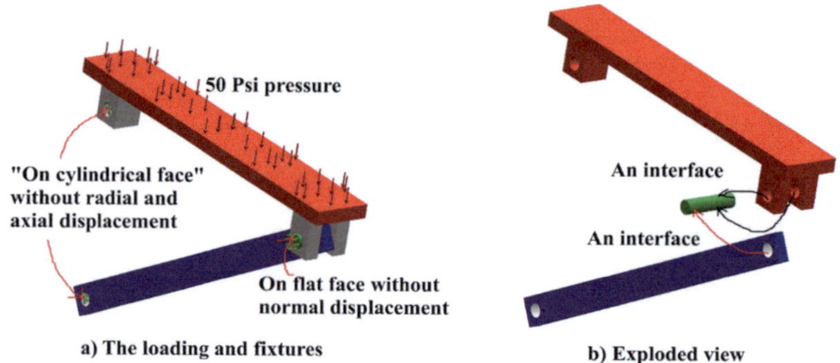

a) The loading and fixtures   b) Exploded view

Fig. 4.16 Loading & fixtures and an exploded view

shown in Fig. 4.16a. There is 50 psi pressure on the top surface of the beam. The material of the pin is Alloy steel. The material of the beam and supporter is plain carbon steel. The pin is used to connect the beam and the supporter. There are two interfaces in this example as shown in Fig. 4.16b. One interface is between two holes of the beam and the pin. Another interface is between the pin and the hole of the supporter. The factor of safety of this device is required to be 1.5. Run FEA simulation under the following two cases: (a) Both interfaces are contact interactions and (b) Both interfaces are bonded interactions.

Solution

(1) FEA simulation

Step 1: **Pre-processing**. The model FEA32-ASM is ready for FEA simulation.

Step 2: **Set up a project**. This is a static stress analysis.

Step 3: **Select the type of element and assign a material to each component**. Assign Alloy steel to the pin and plain carbon steel to the beam and the supporter. In this example, the 3D solid elements with high-quality mesh will be used, we don't make any changes to the type of elements.

Step 4: **Connections**.

The automatically generated "global interaction" will be kept without any modifications. One simulation project will be created for each case.

- In the case of "contact interaction": Two contact interactions will be defined. The first one is between two holes of the beam and the cylindrical surface of the pin. The second one is between the hole of the supporter and the cylindrical surface of the pin as shown in Fig. 4.16b.
- In the case of "Bonded interaction": The first bonded interaction is between two holes of the beam and the cylindrical surface of the pin. The second bonded interaction is between the hole of the supporter and the cylindrical surface of the pin as shown in Fig. 4.16b.

Step 5: **Fixtures**. The left ends of the beam and the supporter are pivoted to the ground or other frame as shown in Fig. 4.16a. Typically, there are some restraints to prevent the beam or the supporter from sliding along the pin. Define the fixture as "On cylindrical faces" without radial and axial displacement on these two pivoted holes. The pin in this example connects the beam to the supporter. There is no axial load on this pin. Typically, there are retaining rings or fastener pins to prevent pins from sliding out. In this simulation, we can specify a fixture on one end of the pin, which will be "On flat faces" without normal displacement. Information about fixtures is provided in Chap. 3.7: Fixtures.

Step 6: **Loads**. Right-click the tab "External loads" and select "pressure" to define 50-psi pressure on the top surface of the beam.

Step 7: **Meshing**. For this simulation, there is no need for mesh control. The smallest dimension is the thickness of $3/8''$ in this example. Use "standard meshing" with a global element of $0.075''$, which is one-fifth of the thickness dimension.

Step 8: **Run Simulation**. Right-click the project name and select "Run". This will start the FEA simulation.

Step 9: **Post-processing**

The factor of safety plots with red areas under the value of 1.5 for the case of "contact interaction" and "bonded interaction" are shown in Fig. 4.17a and b, respectively. For an assembly, it is better to show the factor of safety with red areas under the value of 1.5 for each component.

Figure 4.18a–c are the factors of safety with red areas under the value of 1.5 for the beam, the pin, and the supporter of the case of "contact interaction", respectively.

Figure 4.19a–c are the factors of safety with red areas under the value of 1.5 for the beam, the pin, and the supporter of the case of "bonded interaction", respectively.

(2) Interpretations and Discussions

Different assumptions on the type of interaction at the interfaces will significantly affect the simulation results. The factors of safety of the assembly for the cases of "contact interaction" and "bonded interaction" are shown in Fig. 4.17a and b, respectively. For the case of "contact interaction", failure happens in the beam with a minimum factor of safety of 0.9 and in the supporter with a minimum factor of safety of 1.12. For the case of "bonded interaction", three components fail with a minimum factor of safety of 0.56 in the supporter.

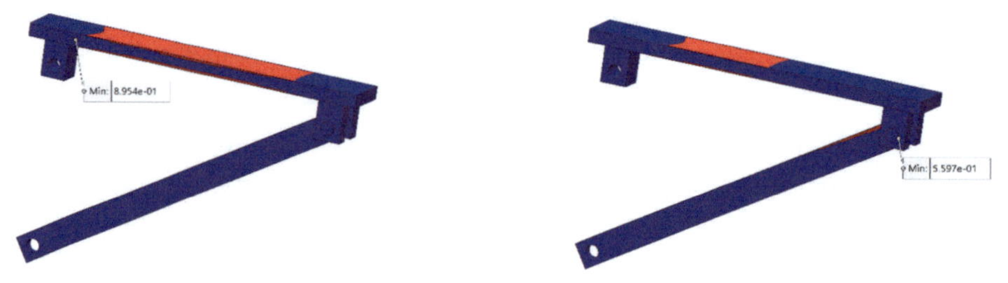

a) Factor of safety for the case of "contact interaction"   b) Factor of safety for the case of "bonded interaction"

Fig. 4.17   The factor of safety with the red area under the factor of safety 1.5

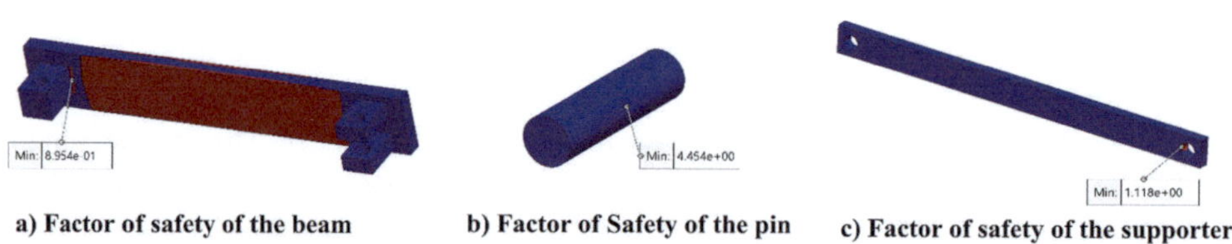

a) Factor of safety of the beam   b) Factor of Safety of the pin   c) Factor of safety of the supporter

Fig. 4.18   The factor of safety of each component for the case of "contact interaction"

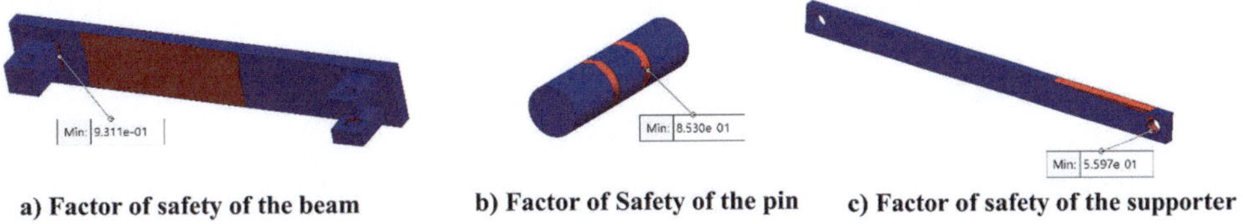

a) Factor of safety of the beam   b) Factor of Safety of the pin   c) Factor of safety of the supporter

Fig. 4.19 The factor of safety of each component for the case of "bonded interaction"

For the case "contact interaction", the following results can be observed about failures per Fig. 4.18.

- For the beam, there are failures in the connections between the top plate and the support columns. This failure is simply due to stress concentration. There are failures on the top and bottom surfaces of the beam plate. This is because of bending stress.
- For the pin, there is no failure in the pin because the minimum factor of safety is 4.45.
- For the supporter, there is a failure at the hole of the supporter that connects with the pin. The minimum factor of safety is 1.12. This is due to the contacting pressure.

For the case "bonded interaction", the following results can be observed about failures per Fig. 4.19.

- For the beam, there are three types of failures. There are failures in connections between the top plate and the support columns. The minimum factor of safety for the beam is 0.93. This failure is simply due to stress concentration. There are failures on the top and bottom surfaces of the beam plate. This is because of bending stress. There is also a failure in the hole connected with the pin. This is because of the assumption of "bonded interaction".
- For the pin, it also fails. The minimum factor of safety of the pin is 0.85. This is because of the assumption of "bonded interaction".
- For the supporter, there is a failure on the hole that connects with the pin. The minimum factor of safety is 0.56. The top and bottom layers of the supporter near the pin are failures too. This is due to bending because of the assumption of "bonded interaction".

In this example, two different simulations performed on the same interfaces result in quite different failure results. For the contact interaction, the pin will not fail. But for the bonded interaction, the beam, the pin, and the supporter fail. Therefore, proper assumption for interfaces in an assembly is very important. The proper assumption for the type of interaction between the two interfaces in this example should be "contact interaction".

4.4   Several Typical Connectors

4.4.1   Introduction

The main purpose of FEA simulation is to explore and determine the strain/stress of components for component design. Each component of a mechanical device during FEA simulation will be meshed into small elements. Therefore, the FEA simulation of a mechanical device or system is typically a time-consuming process. When the same quality of simulation results is maintained, it will be better to simplify the simulation system to save computing time.

Mechanical devices typically contain many components. These components can be grouped into three different types: design-company manufactured components, purchased special components, or standard components. Most of these components are designed or selected through the calculation of maximum stress or stain. However, for some of the purchased or standard components such as pivot pins and bolts & nuts, their selections are not based on stress and strain. For example, pivot pins are usually purchased parts with a limited list of commercially available dimensions. Because pivot pins typically have higher hardness on the outer surface and should be still ductile inside, manufacturing of these pins usually involves specific heat treatment. Another example is hexagon bolts and nuts which are widely used in mechanical devices and are standardized parts. There is a specific design procedure for the selection of bolts because there are limited available standard nominal bolt dimensions. For such specific or standardized types of components, SolidWorks Simulation provides a useful tool: connectors to simplify the simulation models and reduce the computing time without reducing the quality of simulation results.

A connector in SolidWorks Simulation is a mechanism that defines how an entity is connected to another entity or the ground. This connection is completely simulated by mathematical models or equations. So, there is no need to have this SolidWorks component model when it is simplified as a connector.

After a simulation project on an assembly has been created, right-click the tab "connections" in the simulation study tree and the drop-down list appears as shown in Fig. 4.20. The red box in Fig. 4.20 is a list of connectors such as:

- A "Pin" connector is used to connect the cylindrical faces of two components.
- "Bolt" connector is used to define bolt joints between two components or between a component and the ground.

In this section, two frequently and widely used connectors will be demonstrated and discussed: bolt connector and pin connector. In SolidWorks simulation, the interference fit is treated as a type of contact and is not a connector. However, the interference fit is modeled by mathematical equations and is included in this section.

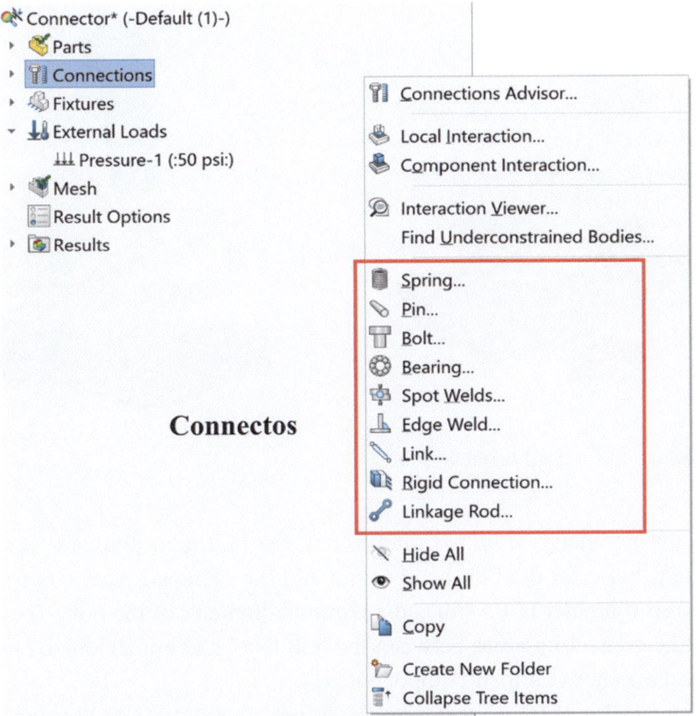

Fig. 4.20 The list of connectors in SolidWorks Simulation

4.4.2 Bolt-Connector

A bolted joint is one of the most common structures in mechanical devices. Typically, several pairs of bolts and nuts with specified tightening torques or pre-axial loads are used to clamp other components together. One schematic of a typical bolted joint structure is shown in Fig. 4.21. In this example, there are four interfaces. The first interface is between two plates. The second interface is between the bolt head and one clamped plate. The third interface is between a nut and another clamped plate. The last interface is between the bolt's male threads and the nut's female threads. All of these are contact interfaces. However, since blots and nuts are standard parts and there is a bolted joint design and analysis procedure for selecting the number of bolts and bolt's nominal dimension, detailed information about bolt stress/stain is not required for a bolted joint design. Therefore, it is recommended that a bolted connector be used to replace bolts and nuts. This will significantly simplify the models and reduce computing time. To define a bolt connector, three groups of information about a bolted joint are required. They are:

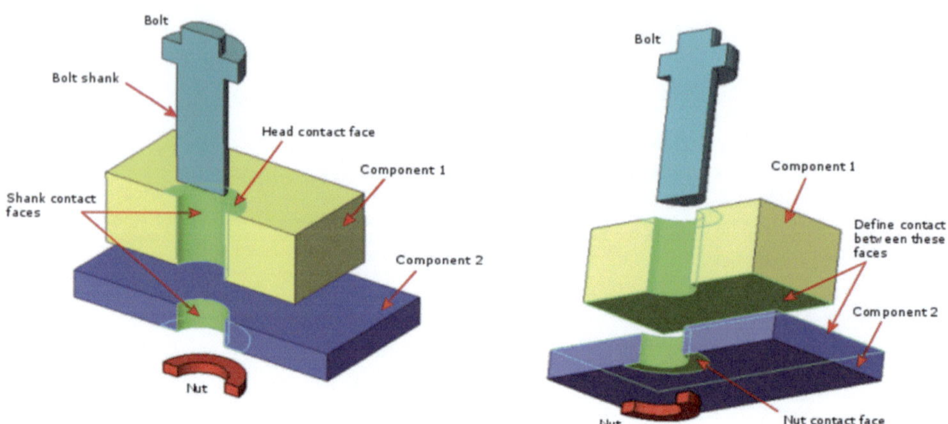

Fig. 4.21 Schematic of a typical bolted joint [1]

(1) Bolt and nut geometry: nominal diameter of the bolt, and diameter of effective contacting areas between the bolt head or nut and the clamped plates. For standard bolts and nuts, this diameter is 1.5 times the nominal diameter of the bolt. They will be used to define the contacting areas between the bolt head and one clamped component, and the nut and another clamped component.
(2) Material properties of bolts: at minimum, Young's modulus and Poisson's ratio should be specified. Typically, the type of material should be assigned to the bolt.
(3) Pre-axial load or tightening torque, which are the forces tightening parts together.

Now, one example is used to demonstrate how to define and run bolt connectors for FEA simulation.

Example 3: Bolt connectors (FEA33-ASM) Download and open FEA33-ASM. A bolted joint as shown in Fig. 4.22a consists of two plates and four pairs of bolts & nuts. The right end of the top plate is fixed, and the left end of the bottom plate is subjected to 1000 lb normal force. Two plates are clamped by four pairs of bolts and nuts which have a nominal dimension ½". Figure 4.22b is the FEA simulation model in which four bolt connectors replace bolts and nuts. The materials for the two plates are alloy steel. The necessary information for defining bolt connectors is displayed in Fig. 4.22a. Run FEA simulation with bolt connectors.

Solution

(1) FEA simulation

Step 1: **Pre-processing**. This assembly model contains two configurations: original and simplified assembly. The configuration "original" is the bolt joint with 4 pairs of

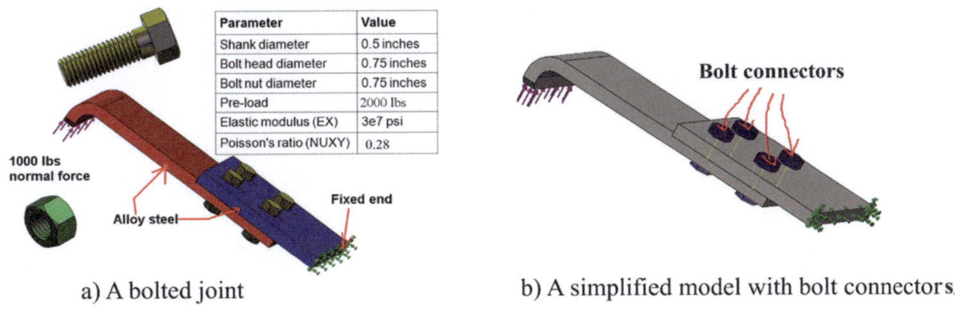

a) A bolted joint   b) A simplified model with bolt connectors.

Fig. 4.22   A bolted joint

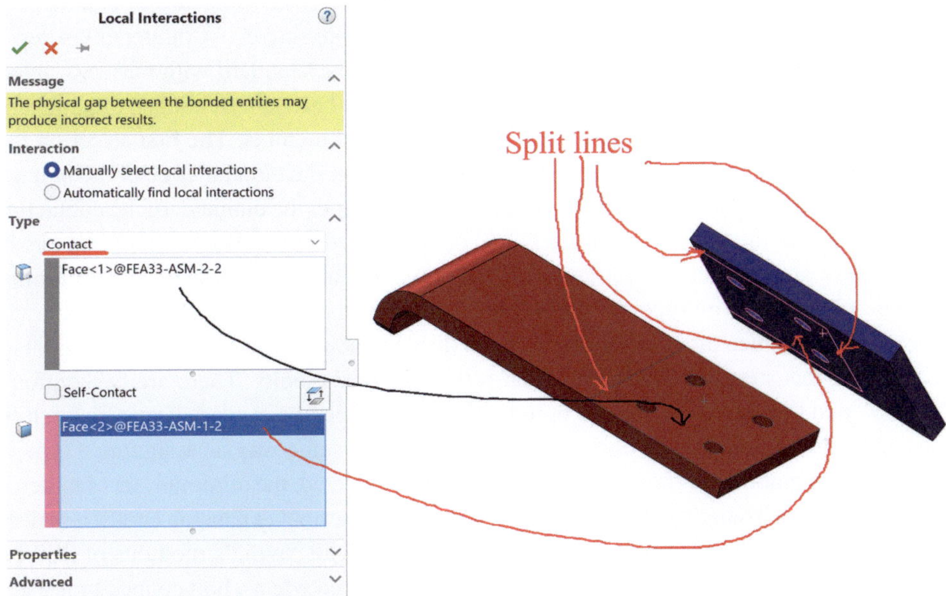

Fig. 4.23   Defining a contact interaction

bolts and nuts. The configuration "simplified" is the simplified assembly where the bolts and nuts are removed (suppressed) and bolt connectors must be defined and added to conduct FEA simulation. Select the configuration "Simplified" for this FEA simulation. To reduce the computing time for contact interaction, some split lines as shown in Fig. 4.23 are added to define actual contact areas on the interface between two plates.

Step 2: **Set up a project**. This is a static stress analysis.

Step 3: **Select the type of element and assign a material to each component**. 3D solid elements with high-quality mesh will be used. Alloy steel is included in the Solid-Works material library. No changes to the type of elements are needed and Alloy steel can be directly assigned to components.

Step 4: **Connections**. In this FEA simulation, the interface between two plates is a contact interaction. Activate the exploded view, right-click the tab 'Connections" in the simulation study tree, and then select "Local interaction". For this simulation, the interface between two plates is "contact" as shown in Fig. 4.23.

For defining a bolt connector, right-click the tab "Connection" in the simulation study tree, then select "Bolt" connector and this will prompt the bolt connector Property Manager as shown in Fig. 4.24. Follow these six steps to define a bolt connector as shown in Fig. 4.24. (1) There are 5 types of bolts. From the left to the right, they are "Standard or counterbore with nut", "Countersink with nut", "Standard or counterbore screw", "Countersink screw", and "Foundation bolt". In this example, the type of bolts is "standard with nut". So, select the first one. (2) Select the contact surface between the bolt head and a clamped member and the contact surface between the nut head and one clamped member. The first selection is a circular edge of the bolt head hole, and the second selection is a circular edge of the bolt nut hole. These two circular edges must be concentric. Then, input the diameter of the contacting areas and the nominal diameter of the bolt. For a standard bolt and nut, the diameters of the bolt head and the nut contact areas are the same and are 1.5 times the bolt's nominal diameter. (3) Connection type of bolted joint. Typically, the "Distributed" connection will be selected to produce more realistic stress and displacement on the bolt's and nut's contact areas. (4) In this step, specify the material properties of the bolts. There are two options: "Library" and "Custom". When the option "library" is selected, a type of material can be selected from the SolidWorks library and assigned in the same way as assigning a type of material to a component. When the option "Custom" is selected, the minimum set of material properties such as Young's modulus, Poisson's ratio, and thermal expansion coefficient need to be specified. In this example, use "Custom" material with Young's modulus of 3E7 psi and Poisson's ratio of 0.28. (5) "Strength data" is not checked when a bolt connector is used, because the stresses of the bolts are not relevant. (6) There are two options for pre-load. The first option is pre-axial tension force. The second option is pre-tightening torque. In this example, the pre-load is 2000 lb axial load.

After four bolt connectors have been defined, the mathematical bolts will appear as shown in Fig. 4.22b.

Step 5: **Fixtures**. Right-click the tab "Fixture" and select "Fixture geometry" to apply the fixed end on the right end of the top plate as shown in Fig. 4.22a.

Step 6: **Loads**. Right-click the tab "External loads" and select "Force" to define a 1000 lb normal force on the left end of the bottom plate as shown in Fig. 4.22a.

Fig. 4.24 Schematic of defining a bolt connector

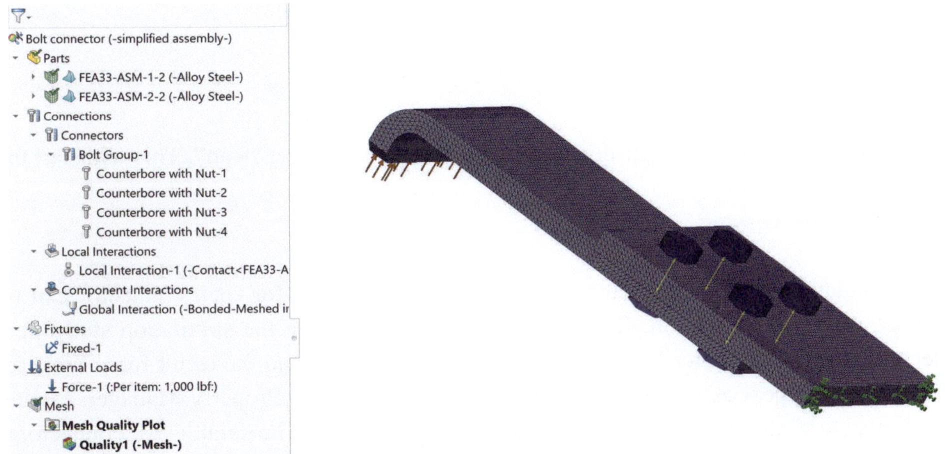

Fig. 4.25 The FEA simulation settings of this example

Step 7: **Meshing**. For this example, use standard mesh with a global element size of $0.1''$, which is one-fifth of the plate thickness of $0.5''$.

    The FEA simulation settings of this example are displayed in Fig. 4.25. The left image of this Figure is the simulation study tree, which displays the types of materials for each component, four bolt connectors, one contact interaction, one global interaction (default), one fixed fixture, and one external force. From the meshing image, it is clearly shown that there is no meshing on the bolt connectors because it is a mathematical model.

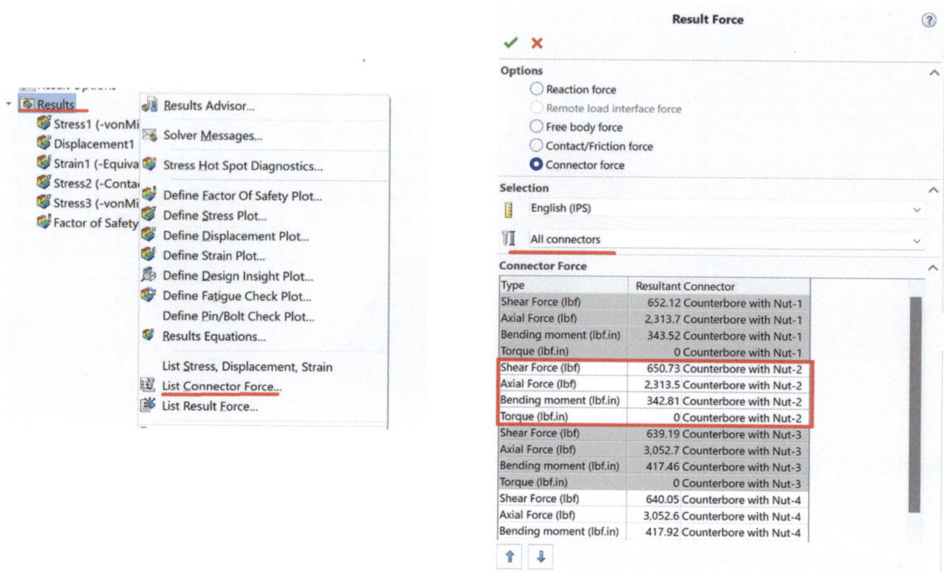

Fig. 4.26 The list of connector forces of bolts in an bolted joint

Step 8: **Run Simulation**. Right-click the project name and select "Run". This will start the
         FEA simulation.

Step 9: **Post-processing**

For a simulation with a connector, a list of connector forces for each connector can be
obtained after the simulation. Right-click the tab "Results" in the simulation study tree,
select "List connector force" in the prompt drop-list window, and the result force Property
Manager will appear as shown in Fig. 4.26. In the result force Property Manager, expand
"All connectors" to select a connector for displaying its forces. The default "All connectors"
can be used to display all connector forces. For each bolt connector, the four types of forces
are shear force, axial force, bending moment, and torque. The shear force and bending
moment of a connector are the forces on the cross-section of the bolt on the contact interface
between two clamped plates. The axial load of a connector is the resultant axial load, which
is different from the pre-load of a bolt.

Follow the procedures described in Ch3.2: General procedure for FEA simulation and
post-processing to create a contact pressure distribution plot as shown in Fig. 4.27, resultant

displacement distribution plot as shown in Fig. 4.28a, von Mises stress distribution plot as shown in Fig. 4.28b, the factor of safety plot of the bottom plate as shown in Fig. 4.29a, and the factor of safety plot of the top plate as shown in Fig. 4.29b.

(2) **Interpretations and Discussions**

Connector forces in Fig. 4.26 are different for each connector even though the pre-load is the same. These differences are due to the applied external force of 1000 lb on the left end of the bottom plate. This external force pushes two plates together on the left edge of the

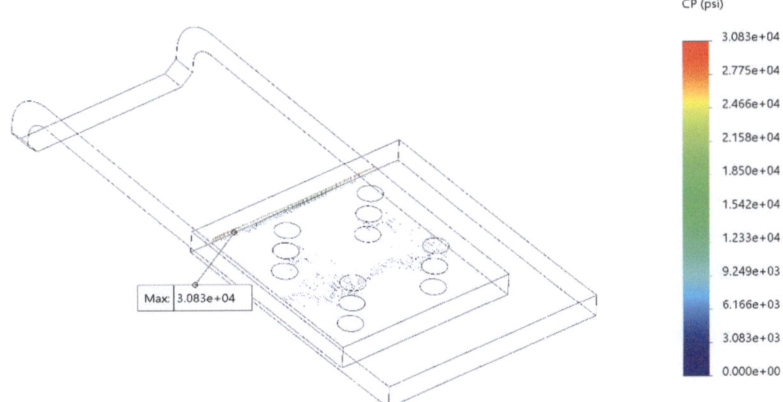

Fig. 4.27 Contact pressure distribution plot

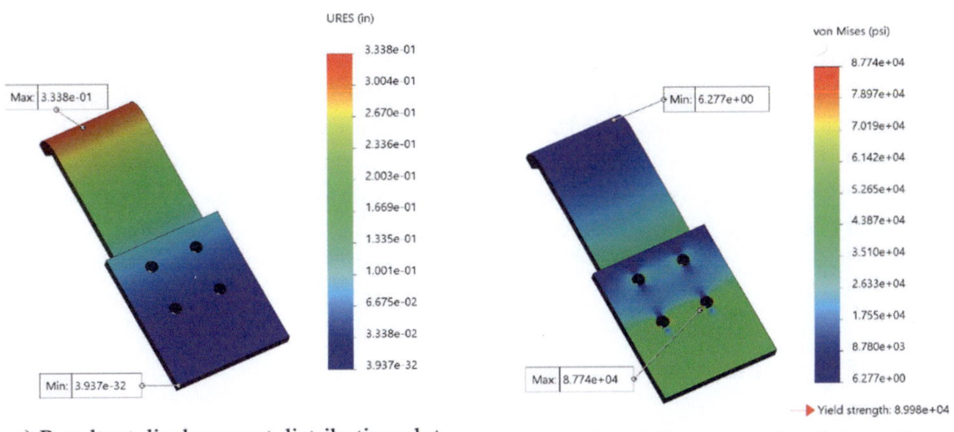

a) Resultant displacement distribution plot   b) Von Mises stress distribution plot

Fig. 4.28 Resultant displacement and von Mises distribution plots

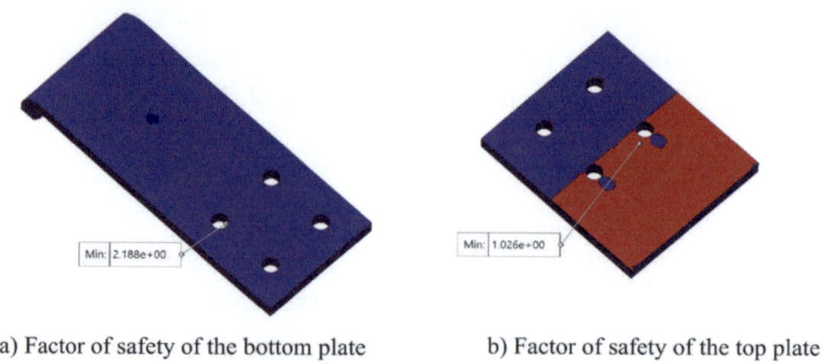

a) Factor of safety of the bottom plate   b) Factor of safety of the top plate

Fig. 4.29 The factor of safety plots with red areas under the value of 2.5

contact area (interface) and tends to separate the right edge of the interface. Therefore, the axial load of bolt connectors near the left edge of the interface is less than the axial load of bolt connectors near the right edge of the interface as shown in Fig. 4.26.

The contact pressure distribution as shown in Fig. 4.27 shows that the contact pressure is not uniform on the contact area. The maximum contact pressure with a value of 3.083×10^4 (*psi*) happens on the starting line of the contact area. This is expected because the external force on the left end of the bottom plate pushes the bottom plate against the top plate.

The resultant displacement distribution plot is shown in Fig. 4.28a. The maximum displacement happens at the left end of the bottom plate. This is expected because the assembly is fixed on the right end of the top plate and behaves like a cantilever beam.

The von Mises stress distribution plot is shown in Fig. 4.28b. The maximum von Mises stress happens inside the contact areas and near a bolt hole and does not happen at the fixed end. One explanation is that there are contact stresses in the contact area.

If the required factor of safety of this device is 2.5, Fig. 4.29a and b are the factors of safety with red areas under the value osf 2.5 for the bottom plate and top plate, respectively. There is a small red area for the bottom plate near the bolt-hole edges. The minimum factor of safety for the bottom plate is 2.188. There are large red areas on the top and bottom surfaces of the top plate as shown in Fig. 4.29b. This is mainly due to the bending stress. Therefore, these results agree with the logical reasoning of this simulation example per theoretical knowledge from mechanical design.

4.4.3 Pin Connectors

Mechanical devices typically have moving components to perform a variety of functions. Pins are widely used as pivot shafts in mechanical design to connect rotating components to frames or a moving component to another moving component. It is expected that there is relative rotation between connected components and pins. Due to this feature, pins usually have higher hardness to resist wear or abrasion. Manufacturing of pins usually requires special heat treatment. Therefore, pins are typically semi-standard purchased components for most design companies. There are some specific procedures to select purchased pins for mechanical design so pin connectors can be used in this situation.

A Pin connector in SolidWorks Simulation replaces the pin model with a mathematic model to connect coaxial cylindrical surfaces of two or more components. They can simulate the behaviors of physical pins.

Now, one example is used to demonstrate how to define a pin connector and run this simulation.

Example 4: A pin connector (FEA32-ASM) Download and open FEA32-ASM, which has been used in Example 2 in Ch4.3. This assembly consists of three components: a top beam, a pin, and a supporter as shown in Fig. 4.16. A 50-psi pressure is applied on the top surface of the beam. A pin as a pivot shaft connects the beam with the supporter. The materials of the beam and the supporter are plain carbon steel. The material of the pin is alloy steel. Use a pin connector to replace the pin model as shown in Fig. 4.30 to run the FEA simulation and then compare simulation results with the results obtained from the assembly with the pin model.

Solution

(1) FEA simulation

Step 1: **Pre-processing**. There are two configurations: "Default" and "Pin connector" for the FEA32 assembly model. The configuration "Default" is the original assembly with a beam, a pin, and a supporter. The configuration "Pin connector" is a simplified assembly with the beam and the supporter only. In this configuration, the pin model is suppressed (or removed). A pin connector to simulate the behaviors of the pin model needs to be defined.

Fig. 4.30 Schematic of a device with a pin connector

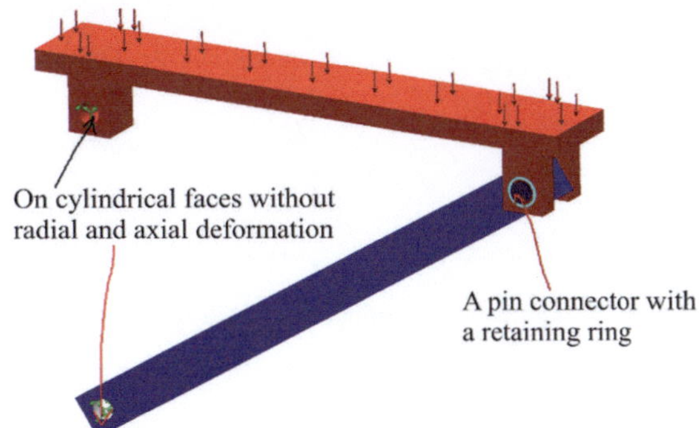

Step 2: **Set up a project**. This is a static stress analysis.

Step 3: **Select the type of element and assign a material to each component**. The 3D solid elements with high-quality mesh will be used. Plain carbon steel is included in the SolidWorks material library. No changes to the type of elements are required and plain carbon steel can be assigned to components.

Step 4: **Connections**. Since the pin model has been removed, a pin connector to connect the beam with the supporter through cylindrical surfaces as shown in Fig. 4.31 is needed. Follow these steps to define a pin connector as shown in Fig. 4.31: (1) Right-click the tab "connections" in the simulation study tree; (2) select "Pin" on the prompt window; (3) In the prompt connector Property Manager, select coaxial cylindrical surfaces of two or more components. More than two cylindrical surfaces for one pin definition can be selected. In this example, one pin can be used to connect three cylindrical surfaces. We can also define two separate pin connectors. Each pin connector will only connect two coaxial cylindrical surfaces; (4) There are two options of possible restraints for the pin: with a retaining ring (No axial translation); and with a key (No rotation). Since pin connectors are typically used to simulate the behaviors of the pivot shaft, select "With retaining ring"; (5) select "Distributed" which provides more realistic behaviors of pivot pin; (6) and (7) Since the pin's stress/strain is not a concern, there is no need to specify the axial stiffness and rotational stiffness. We also don't check "strength data".

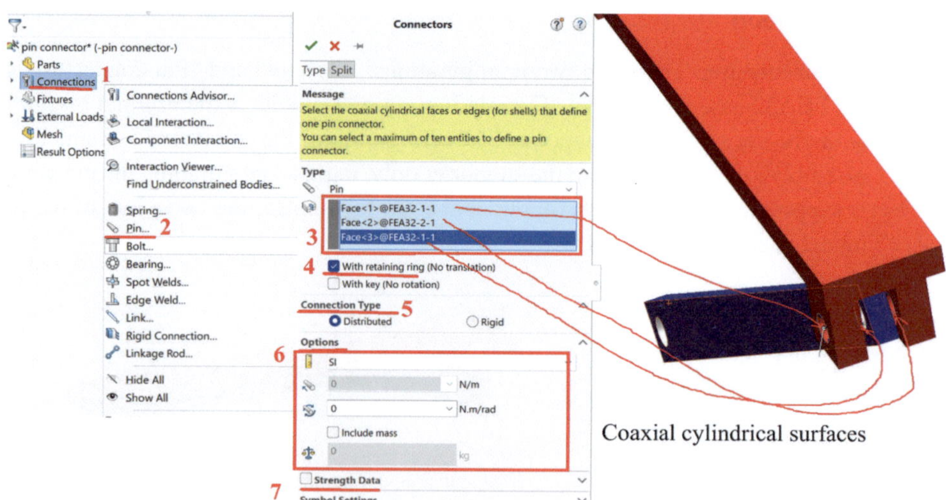

Fig. 4.31 Define a pin connector with a retaining ring

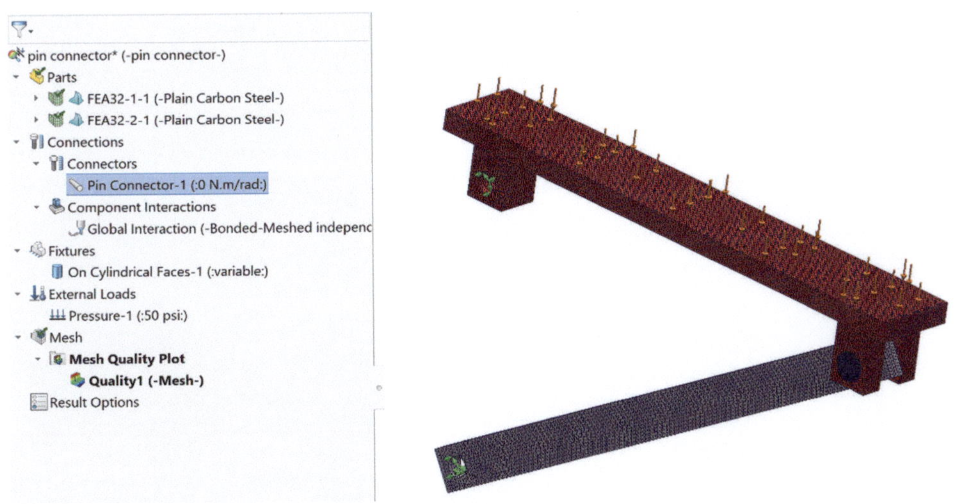

Fig. 4.32 The FEA simulation settings and meshing

After the pin connector is defined, an image of the pin connector will appear in the assembly model as shown in Fig. 4.30. This pin connector is only a mathematical model and will not be meshed.

Step 5: **Fixtures**. Right-click the tab "Fixture" in the simulation study tree, select "Advanced fixture", and then select "On cylindrical faces" to apply "on cylindrical faces" without radial and axial displacement on two holes as shown in Fig. 4.32.

Step 6: **Loads**. Right-click the tab "External loads" in the simulation study tree and select "Pressure" to define a 50-psi pressure on the top surface of the beam.

Step 7: **Meshing**. The smallest thickness in this example is $3/8''$. The standard mesh with a global element size of $0.075''$ will be used, which is one-fifth of the smallest thickness. The FEA simulation setting and meshing are shown in Fig. 4.32. The left side image is the FEA simulation setting which shows types of materials, pin connector, Global Interaction, fixture, and loading. The image of meshing shows that the pin connector is not meshed because it is a mathematical model or relationship.

Step 8: **Run Simulation**. Right-click the project name and select "Run". This will start the FEA simulation.

Step 9: **Post-processing.** Pin connector forces can be listed by the following procedure. Right-click the tab "Results" in the simulation study tree, select "List Connector Force" on the prompt window, and then the result force Property Manager for pin connectors will appear as shown in Fig. 4.33. In this example, there is only one pin connector. However, it connects three coaxial cylindrical surfaces. At each cylindrical surface, there will be a joint. This is the reason that the pin connector

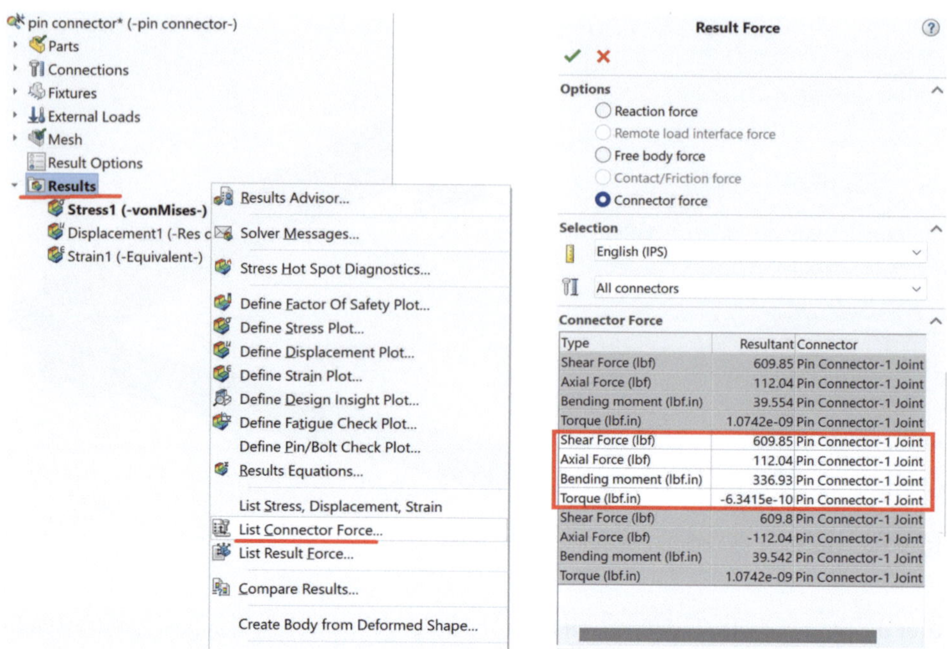

Fig. 4.33 The list of connector forces of pins in an assembly

forces in this example provide three groups of pin connector forces. Forces for each coaxial cylindrical surface are shear force, axial force, bending moment, and torque on a joint.

We can use and compare the simulation results from the case "contact interaction" of Example 2 in Chap. 4.3 and the results from this example with a pin connector. In the following figures, "pin connector" refers to the simulation results in which a pin connector is used to replace the physical pin, and "Physical pin" refers to the simulation results in which contact interaction is the type of interface between physical pins and other components. We create the following plots: the resultant displacement distribution plots of the beam (Fig. 4.34), the factor of safety plots with red areas under the value 1.5 of the beam (Fig. 4.35), the resultant displacement distribution plots of the supporter (Fig. 4.36) and the factor of safety plots of the supporter with red areas under the value of 1.5 (Fig. 4.37).

(2) **Interpretations and Discussions**

Figures 3.34a and b show the resultant displacement distribution plots of the beam obtained from the case "pin connector" and the case "physical pin". The patterns of the resultant displacement, their maximum value, and their location are almost the same.

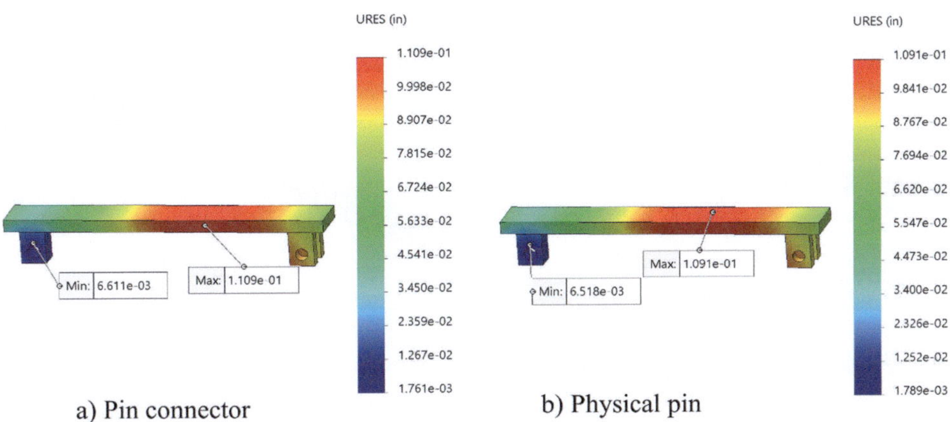

a) Pin connector   b) Physical pin

Fig. 4.34 Resultant displacement distribution plot of the beam

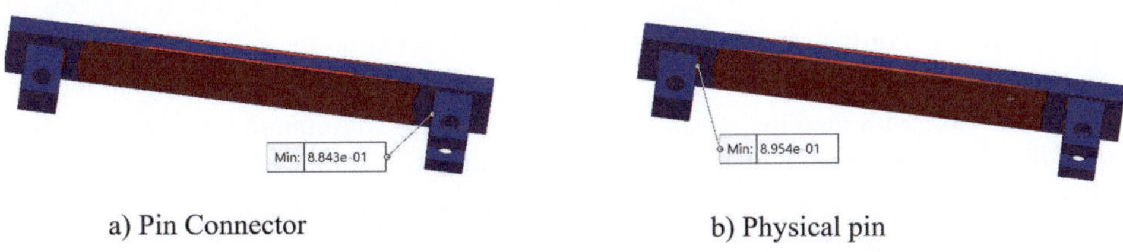

a) Pin Connector   b) Physical pin

Fig. 4.35 The factor of safety of the beam with red areas under the value of 1.5

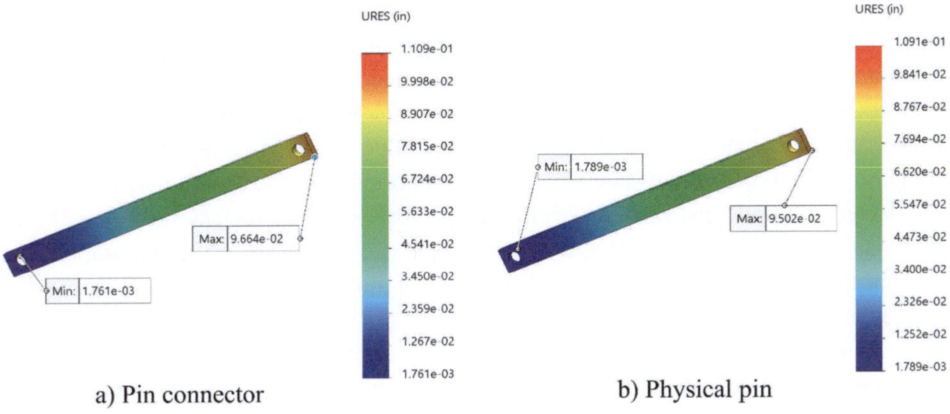

a) Pin connector   b) Physical pin

Fig. 4.36 Resultant displacement distribution plot of the supporter

Fig. 4.37 The factor of safety of the supporter with red areas under the value of 1.5

In Fig. 4.35, the red areas are the areas where the factor of safety of the beam is less than 1.5. The minimum factors of safety are almost the same for the case "pin connector" and the case "physical pin". However, the locations of the minimum factors of safety are slightly different. For the case with a pin connector, the location for the minimum factor of safety is on connections between the top plate (beam) and the support columns near the right end. For the case with the physical pin, the location for the minimum factor of safety is on connections between the top plate (beam) and the support columns near the left end. From Fig. 4.35a and b, the patterns of red areas are almost the same as obtained from using a pin connector and using a physical pin. The failure areas (red areas) on the connections between the top plate and the support columns are due to stress concentration (sharp edges). The failure areas on the top and bottom surfaces of the plate are due to bending stress. Therefore, it can be concluded that using a pin connector can obtain almost the same factor of safety plots obtained from the original assembly with a physical pin.

Figure 4.36a and b show the resultant displacement distribution plots of the supporter obtained from the case of the pin connector and the case of the physical pin. The patterns of the resultant displacement, and their maximum value and location are almost the same.

Figure 4.37a is the factor of safety plot of the supporter under the case of the pin connector with the red area under the value of 1.5. From this plot, the minimum factor of safety is 2.82, which is located at the fixture location of the supporter. However, the minimum factor of safety of the supporter under the case of the physical pin is 1.118 on the cylindrical surface of the hole where the interface between the physical pin and the hole is contact interaction. These results are different. In Fig. 4.37b, use the processing tool "probe" to find that the minimum factor of safety on the fixture is 2.75. Therefore, it can be concluded that the pin connector might not provide more realistic stress on the cylindrical surfaces of the pin connector, but stresses on other areas are almost the same.

4.4.4 Interference Fit (Shrink Fit) Analysis

An interference fit, also called a **shrink fit**, or **press fit** is a fit between two objects in which the external dimension of an inner object slightly exceeds the internal dimension of a cavity of an outer object. Due to the contact pressure developed on the interface, the inner object shrinks while the outer object expands.

In SolidWorks Simulation, an interference fit is treated as one type of contact: shrink fit, in which the interference on the interface is controlled by a mathematical equation. Figure 4.38 displays how to define a shrink fit. The image of the two components is an exploded view of an interface with an interference fit. Right-click the tab "Connections" in the simulation study tree, select "Local interaction" on the prompt window, and the local interaction Property Manager will appear. Expand the type of contacts and select "Shrink fit". Finally, select surfaces on the interface which are treated as an interference fit (shrink fit).

Now, we will show how to run an interference fit analysis in SolidWorks Simulation.

Example 5: An interference fit (shrink fit) (FEA34-ASM) Download, unzip, and open FEA34-ASM. The copper ring (FEA34-1) is clad with a 6061-alloy aluminum bar (FEA34-2) through a special procedure as shown in Fig. 4.39a. The outer diameter of the aluminum bar is Ø3.005″ and the inner diameter of the copper ring is Ø3.000″. Treat this interface (interference fit) as "shrink fit" and run the FEA simulation.

Solution

(1) FEA simulation

Step 1: **Pre-processing**. The model FEA34-ASM is ready for FEA simulation.
Step 2: **Set up a project**. This is a static stress analysis.

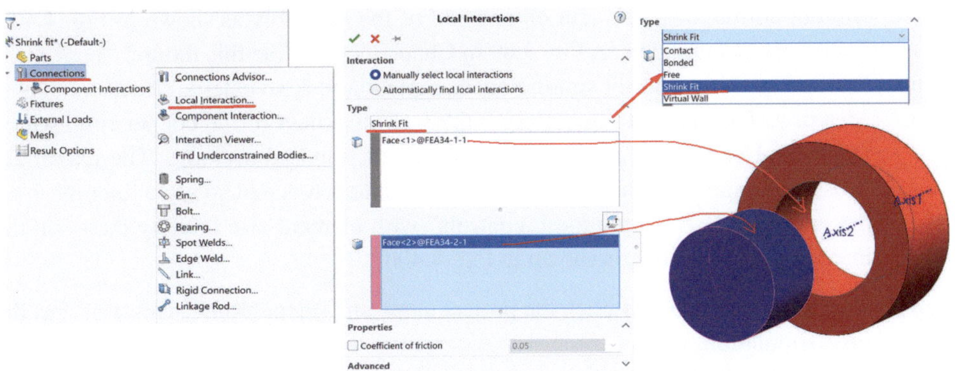

Fig. 4.38 Defining an interference fit (shrink fit)

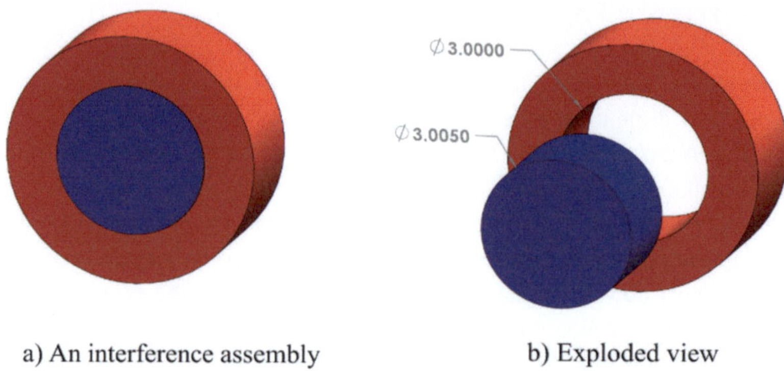

a) An interference assembly   b) Exploded view

Fig. 4.39 An Assembly with an interference fit

Step 3: **Select the type of element and assign a material to each component**. The 3D solid elements with high-quality mesh will be used. Both Copper and Aluminum 6061-T6 are included in the SolidWorks material library. No changes to the type of elements are needed and Copper and Aluminum 6061-T6 can be directly assigned to components.

Step 4: **Connections**. In this example, the interference fit on the interface will be treated as a type of contact "shrink fit". Right-click the tab "Connections" in the simulation study tree and select "local interaction". The local interaction Property Manager will appear as shown in Fig. 4.38. Select "Shrink fit" as a type of contact and then select interface surfaces to define a shrink fit.

Step 5: **Fixtures**. There is no fixture for this example because they are in equilibrium due to the internal contacting forces. To run this FEA simulation, some fixtures can be assumed, and it will not affect the shrink fit analysis. The interference fit (shrink fit) will expand the outer ring and compress the inner column. The assembly also should be able to expand along its cylindrical axis so "On flat faces" can be used without normal displacement on one end of the assembly as shown in Fig. 4.40a. Follow the notes shown in Fig. 3.69 in Chap. 3.7 to define this fixture.

Step 6: **Loads**. For this shrink-fit simulation example, there is no external load.

Step 7: **Meshing**. The shortest dimension is the wall thickness of the copper ring, which is 1 inch. Use $0.2''$ as a global element size of standard meshing. The maximum stress will happen on the interface due to the interference fit so use a finer mesh on the interface. Use "Apply mesh controls" with element size $0.2/5 = 0.04''$ on the interface. The meshing is shown in Fig. 4.40b.

Step 8: **Run Simulation**. Right-click the project name and select "Run". This will start the FEA simulation.

Step 9: **Post-processing**.

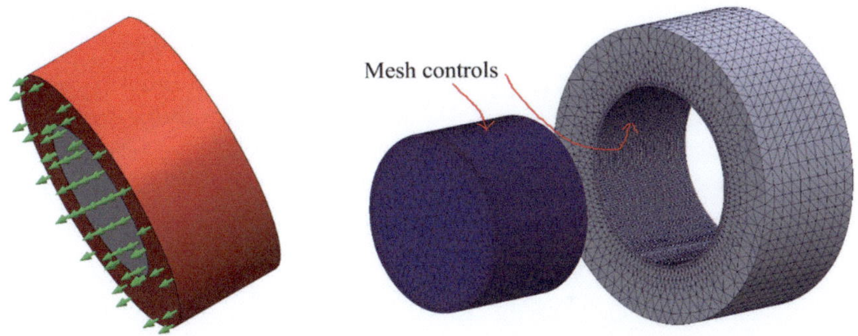

a) On flat faces without normal displacement   b) The global mesh 0.2" and mesh control 0.04"

Fig. 4.40 The fixture and mesh controls

The normal stress σ_x and σ_y distribution plots of this shrink-fit assembly are displayed in Fig. 4.41a and b. These stress plots are in a Cartesian coordinate system.

In this example, stress is axial symmetrical about its axis due to the shrink fit (interference fit) on the interface. Using a cylindrical coordinate system as shown in Fig. 4.42 makes it easier to analyze. In a cylindrical coordinate system, three normal stresses are σ_θ (circumferential stress), σ_r (radial stress) and σ_z (z-direction normal stress) as shown in Fig. 4.42. σ_θ is a circumferential stress, or hoop stress, which is a normal stress on the θ surface. σ_r is a radial stress, that is, normal stress along the radial direction, which is normal to the r-surface. σ_z is a normal stress along the z-axis on the z-surface, which is the same type of stress as it is in the Cartesian coordinate system.

The procedure for defining a circumferential stress (hoop stress) σ_θ as shown in Fig. 4.43a is: (1) create a σ_y (y-normal stress). (2) Then, select the cylindrical axis of the cylindrical

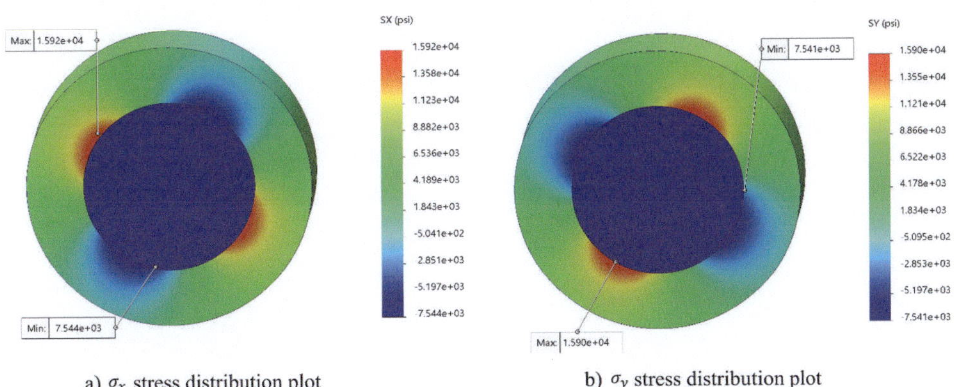

a) σ_x stress distribution plot   b) σ_y stress distribution plot

Fig. 4.41 The normal stresses σ_x and σ_y distribution plots in the Cartesian coordinate system

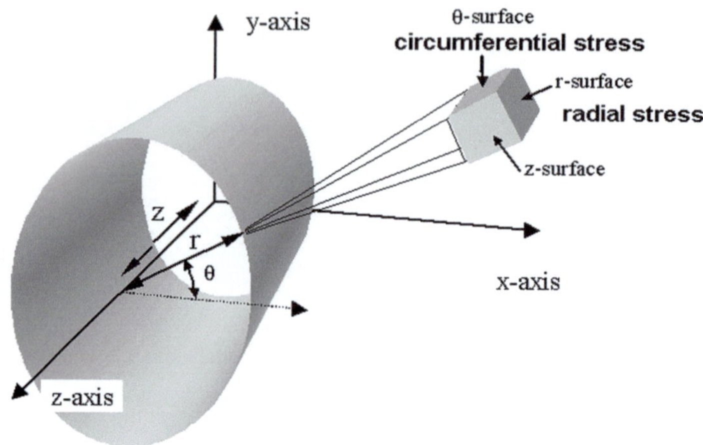

Fig. 4.42 Schematic of Cylindrical coordinate system [1]

coordinate system. In this example, the central axial of the aluminum column or the copper ring can be selected. In this way, the circumferential stress or hoop stress σ_θ distribution plot is created as shown in Fig. 4.43b. To differ from the σ_y stress, there is a cylindrical coordinate system symbol that will be automatically added on the right bottom corner as shown in Fig. 4.43 b).

The procedure for defining a radial stress σ_r as shown in Fig. 4.44a is: (1) Create a σ_x (x-normal stress). (2) Then, select the cylindrical axis of the cylindrical coordinate system.

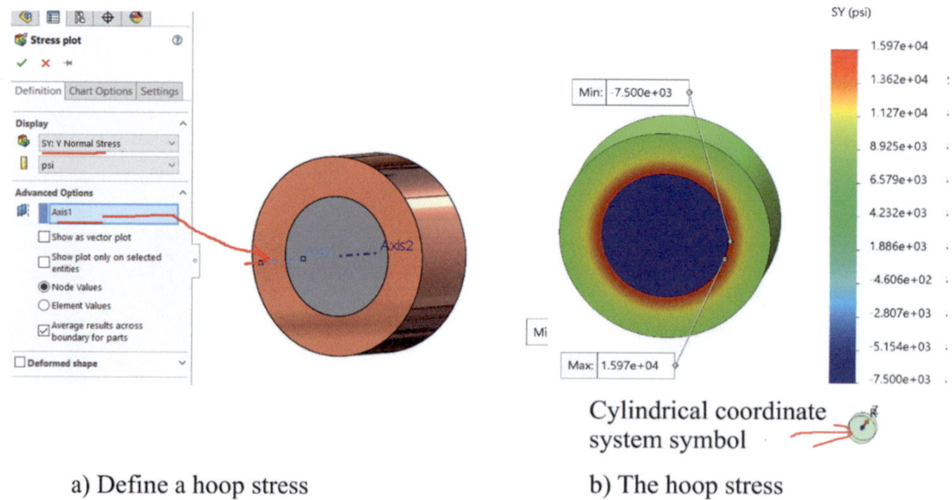

a) Define a hoop stress   b) The hoop stress

Fig. 4.43 The circumferential stress distribution plot

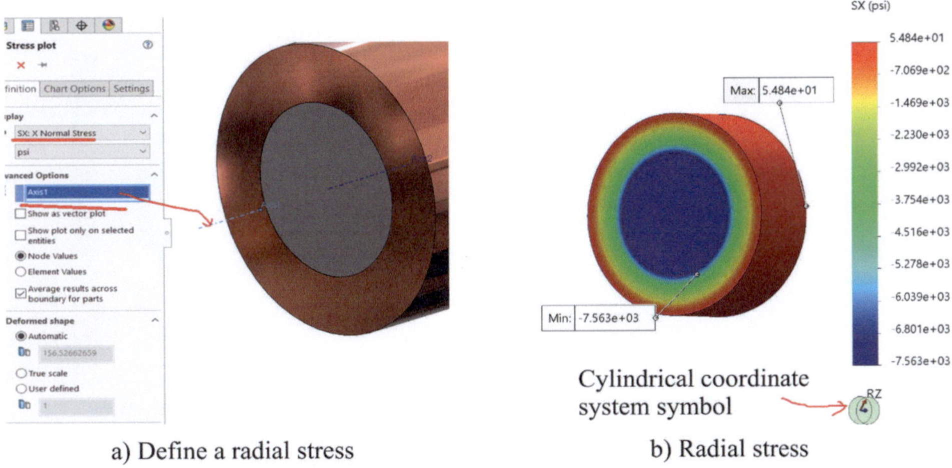

a) Define a radial stress   b) Radial stress

Fig. 4.44 The radial stress distribution plot

In this example, the central axial of the aluminum column or the copper ring can be selected. In this way, the radial stress σ_r distribution plot is created as shown in Fig. 4.44b. To differ from the σ_x stress, there is a cylindrical coordinate system symbol that will be automatically added on the right bottom corner as shown in Fig. 4.44b.

The von Mises stress and the factor of safety distribution plots are displayed in Fig. 4.44a and b, respectively.

(2) Interpretations and Discussions

When Fig. 4.41 is compared with Figs. 4.43b and Fig. 4.44b, the expression of stress distribution plots in the cylindrical coordinate system is better than that in the Cartesian coordinate system when stresses are axisymmetric.

From Figs. 4.43b and 4.44b, the copper ring is under expansion and the aluminum column is under compression. Those results are what is expected.

In this example, the interference is only $0.005''$. But it still causes significant stress. The maximum von Mises stress is $2.068 \times 10^4 \, psi$ and the corresponding minimum factor of safety is 1.814 as shown in Fig. 4.45a and b. The maximum von Mises stress happens on the shrink-fit interface.

The hoop and radial stresses for an interference fit can be calculated through a theoretical formula [2]. The terms and schematics of an interference fit are shown in Fig. 4.46. The maximum stress will happen on the interface. The contact pressure (stress) on the interface is calculated by Eq. (4.1).

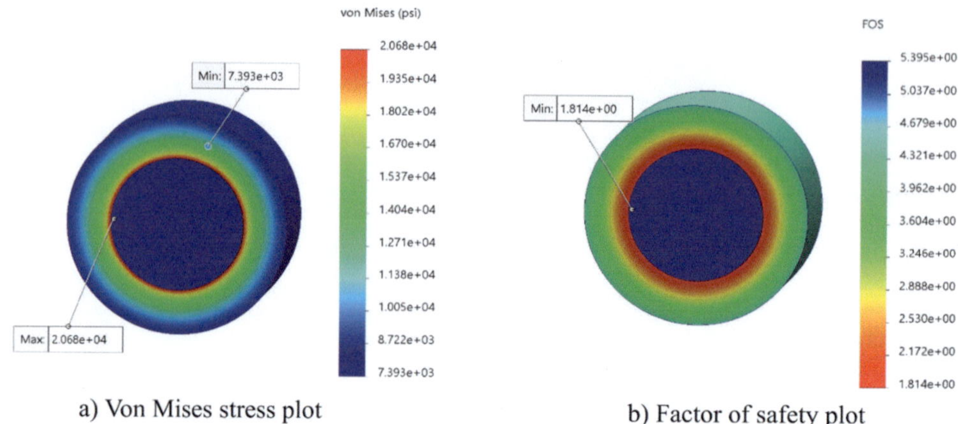

a) Von Mises stress plot   b) Factor of safety plot

Fig. 4.45   The Von Mises stress and the factor of safety plots of the assembly

$$\sigma_p = \frac{\delta}{R\left[\frac{1}{E_o}\left(\frac{r_0^2+R^2}{r_0^2-R^2}+v_o\right)+\frac{1}{E_i}\left(\frac{R^2+r_i^2}{R^2-r_i^2}+v_i\right)\right]} \tag{4.1}$$

The maximum radial stress σ_r happening at the interface is a compression stress and can be calculated by Eq. (4.2).

$$\sigma_r = -\sigma_p = -\frac{\delta}{R\left[\frac{1}{E_o}\left(\frac{r_0^2+R^2}{r_0^2-R^2}+v_o\right)+\frac{1}{E_i}\left(\frac{R^2+r_i^2}{R^2-r_i^2}+v_i\right)\right]} \tag{4.2}$$

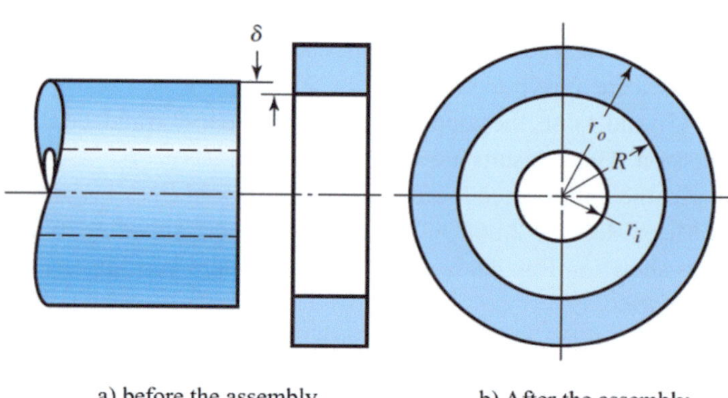

a) before the assembly   b) After the assembly

Fig. 4.46   Terms and schematics of an interference fit [2]

The maximum extension circumferential or hoop stress σ_{ho} for the outer ring happens at the interface and can be calculated by Eq. (4.3).

$$\sigma_{ho} = p\left(\frac{r_0^2 + R^2}{r_0^2 - R^2}\right) \tag{4.3}$$

The maximum compression circumferential or hoop stress σ_{hi} on the inner column happens at the interference and can be calculated by Eq. (4.4)

$$\sigma_{hi} = -p\left(\frac{R^2 + r_i^2}{R^2 - r_i^2}\right) \tag{4.4}$$

where the subscripts o and i refer to the outer ring and the inner column; E_o and E_i are Young's modulus of the outer ring's and inner column's materials; v_o and v_i are Poisson's ratio of the outer ring's and inner column's materials. The geometrical dimensions δ, R, r_o and r_i are displayed in Fig. 4.46.

For this example: $E_o = 15954151$ (*psi*); $v_o = 0.37$; $E_i = 10007604$ (*psi*); $v_i = 0.33$; $\delta = 0.0025''$; $R = 3.000''$; $r_o = 5.000''$; and $r_i = 0$.

The theoretical compression radial stress on the interface is -7463 psi by Eqs. (4.1) and (4.2). The radial stress on the interface obtained from FEA simulation as shown in Fig. 4.44b is -7563 psi. The relative difference is 1.3%. The theoretical circumferential stress of the outer ring on the interface per Eq. (4.3) is 15858 psi. The circumferential stress of the outer ring on the interface obtained from FEA simulation as shown in Fig. 4.43b is 15970 psi. The relative difference is 0.7%. The theoretical circumferential stress of the inner column on the interface per Eq. (4.4) is -7463 psi. The circumferential stress of the inner column on the interface obtained from FEA simulation as shown in Fig. 4.43b is -7500 psi. The relative difference is 0.6%. Therefore, the FEA simulation results for an interference fit can provide acceptable results with a very small relative difference.

4.5   Truss Analysis

This section will use one example to show how to run a truss analysis in SolidWorks Simulation.

A truss consists of straight members connected at pivot joints, the length of which is much bigger than the largest dimension of its cross-section. The geometrical feature of a member is a straight slender member with constant cross-section. Trusses are commonly used in architectural and structural applications such as bridges, roofs, power towers, and others.

A schematic of a truss is shown in Fig. 4.47. Each end of a truss member is a pivot joint. When the truss member is in equilibrium, it can only have axial loads: tension or

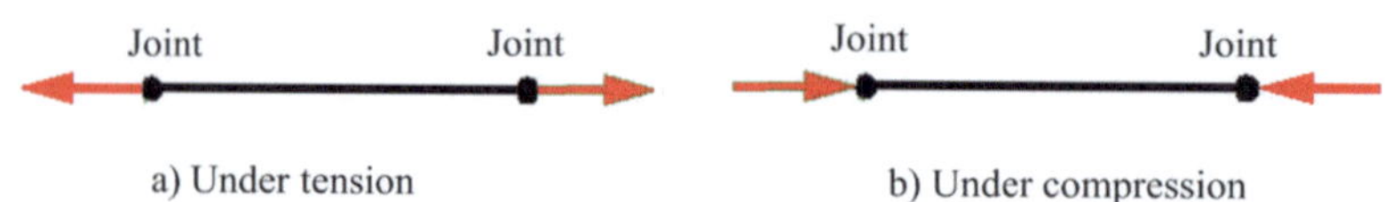

Fig. 4.47 Schematics of a truss member

compression. In a truss analysis, all external loads will be only concentrated forces and applied on joints.

Each joint of a truss member has only three translational degrees of freedom. When fixtures are applied on joints, there are only three translational restraints (zero or a specific value). When a joint is treated as a fixed, it is immovable (no translations). Defining fixtures on joints has been explained in Figs. 3.64 and 3.66 of Chap. 3.6: Fixtures and will be demonstrated in the following example.

A truss member in SolidWorks will be meshed by a truss element, which is one of four types of elements (truss, beam, shell, and solid elements). When each member of a truss is defined as a truss, the joints of all truss members will be automatically determined by the program and are at the interactions of member centers. The shared joints in a truss analysis will be automatically treated as a pivoted point. So, there is no need to define interfaces between truss members. A schematic of the joints of a truss will be displayed in the following example.

Since a truss member is a simple constant-cross-section straight component under axial load, there is no need to specify more elements in a truss member. The program will use the default number of truss elements to mesh a truss. However, the mesh parameters can be changed on selected truss members by applying mesh controls.

Example 6: A truss analysis (FEA38-ASM) Download, unzip, and open FEA38-ASM, which is a planar truss. The restraints, loadings, and their list of joints are shown in Fig. 4.48a. The materials of all truss members are plain carbon steel and have been assigned at the component level. The supporter on point B is a fixed and the supporter on point A is a roller. There are 4000 lb concentrated forces applied on points C and D, which are the points for joints. The cross-section of the truss member is 3″ by 2″. The geometric dimensions of the truss are shown in Fig. 4.48a. When all members are defined as "truss" members, the list of joints is automatically calculated and determined by the program as shown in Fig. 4.48b. Run FEA simulation on this truss by tress elements. Interpret and discuss simulation results.

Solution

(1) FEA simulation

Step 1: **Pre-processing**. The model FEA38-ASM is an assembly. There are some over-
  lapped geometries on interfaces between truss members, which is normally not

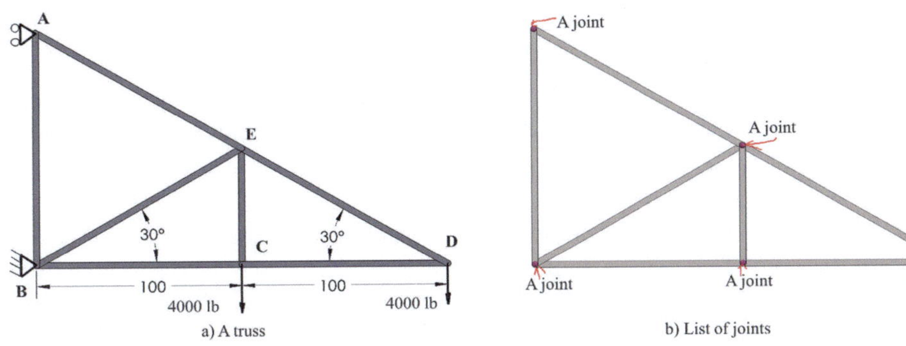

a) A truss

b) List of joints

Fig. 4.48   Schematics of a truss and its joints

allowed. However, every component is treated as truss members, and FEA simulation is run by truss elements, these overlapping geometries will be ignored and replaced by joints. So, the model is ready for FEA simulation.

Step 2: **Set up a project**. This is a static stress analysis.

Step 3: **Select the type of element and assign a material to each component**.

In this example, a material (plain carbon steel) has been assigned to each component at the part level. Follow this procedure to convert solid elements into truss elements as shown in Fig. 4.49. (1) In the simulation study tree, expand "Parts" to show each part, which indicates that each component will be meshed by solid elements. Right-click a part and select "Treat as beams". (2) After step 1 is completed, the component has been converted as a beam and will be meshed by beam elements. (3) Right-click the beam component, and (4) select "Edit definition". The Apply /Edit beam Property Manager will appear. (5)Select "Truss". This will convert a beam member into a truss member.

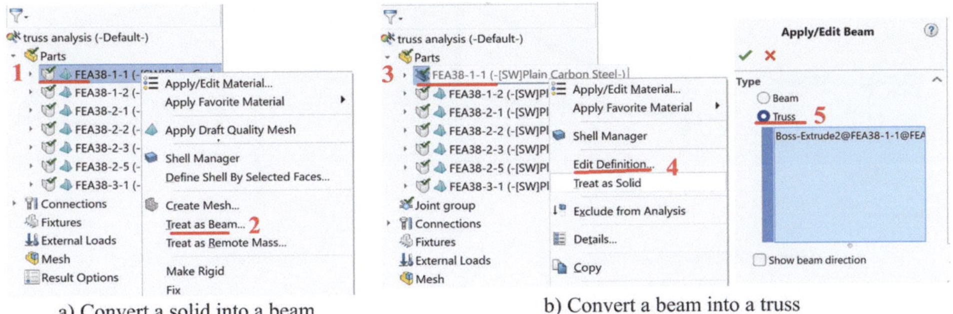

a) Convert a solid into a beam

b) Convert a beam into a truss

Fig. 4.49   Schematics of defining truss members

Repeat this for every component until all components have been converted into truss members that will be meshed by truss elements.

Step 4: **Connections**. There is no need to specify interfaces between truss members because joints will be automatically calculated and determined by the program and the shared joints will be automatically treated as a pivot point. The list of joints in this example is displayed in Fig. 4.48b.

Step 5: **Fixtures**. A joint for a truss element will only have three translational degrees of freedom. There are mainly two approaches for defining a fixture for a truss: "Immovable" for a fixed joint and "Use Reference Geometry" for all other fixtures.

For "immovable" at point B, right-click the tab "Fixtures" in the simulation study tree and select "Fixed Geometry". The fixture Property Manager will appear. Then, select "Immovable" and select the joint B as shown in Fig. 4.50a.

For "Use Reference Geometry" at point A, right-click the tab "Fixtures" in the simulation study tree and select "Fixed Geometry". The fixture Property Manager will appear. Then, select "User Reference Geometry", select joint A, select a flat surface as a reference, and then select two directions for zero translation displacement as shown in Fig. 4.50b.

This example is a planar truss. Joints at points C, D, and E as shown in Fig. 4.48 will not have any displacement normal to the plane of the truss. Zero displacements on these joints along the normal direction to the plane of the truss need to be specified. Use "Use reference Geometry" to do this. The image of fixtures of all joints of this truss analysis is shown in Fig. 4.51.

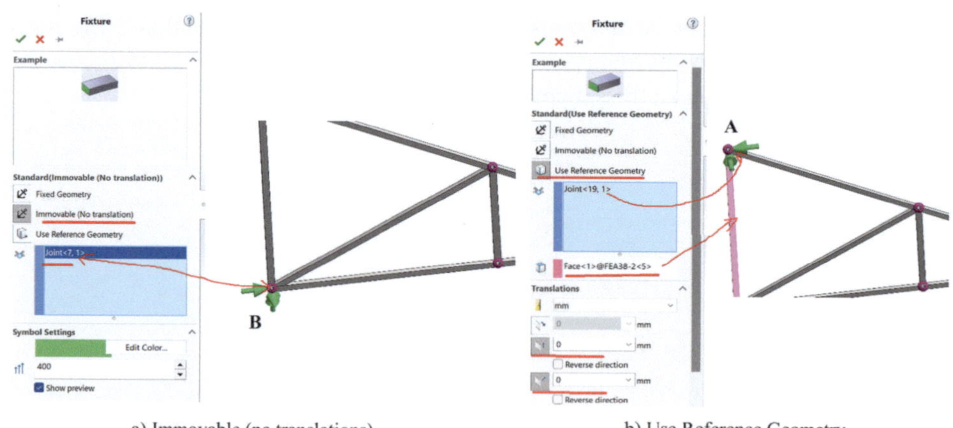

a) Immovable (no translations)   b) Use Reference Geometry

Fig. 4.50 Defining a fixture for joints of truss elements

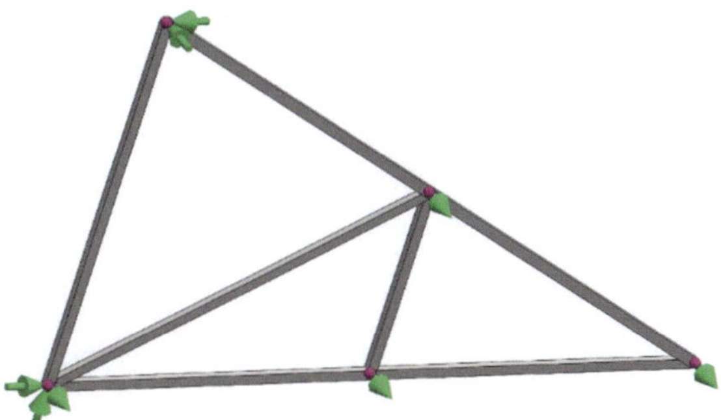

Fig. 4.51 Fixtures of this truss analysis

Step 6: **Loads**. Right-click the tab "External loads" and select "Force" to define a 4000 lb force on points C and D as shown in Fig. 4.52.

Step 7: **Meshing**. The default number of truss elements for truss analysis is typically used. Right-click "Mesh" in the simulation study tree and select "Create mesh" to complete meshing for all truss members. Now the FEA setting for this truss analysis has been completed. The project is ready to run.

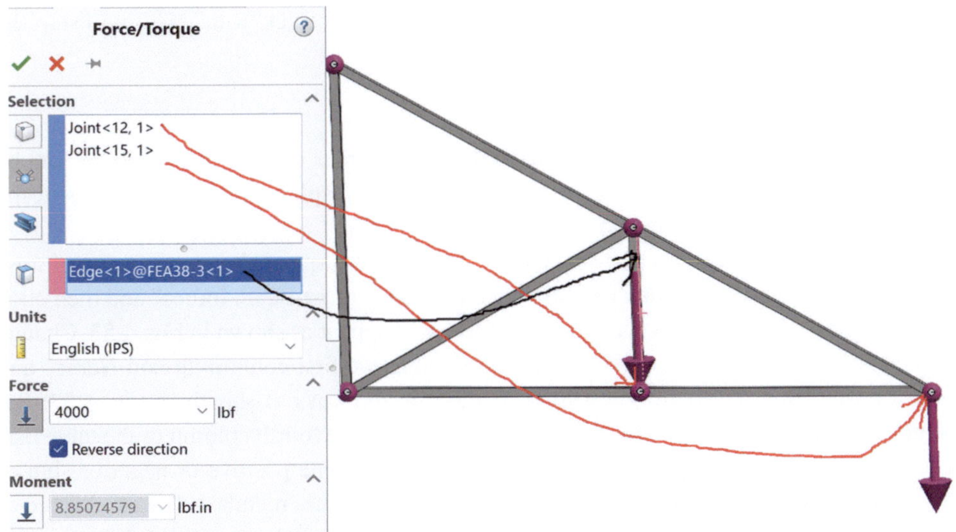

Fig. 4.52 Loads of this truss analysis

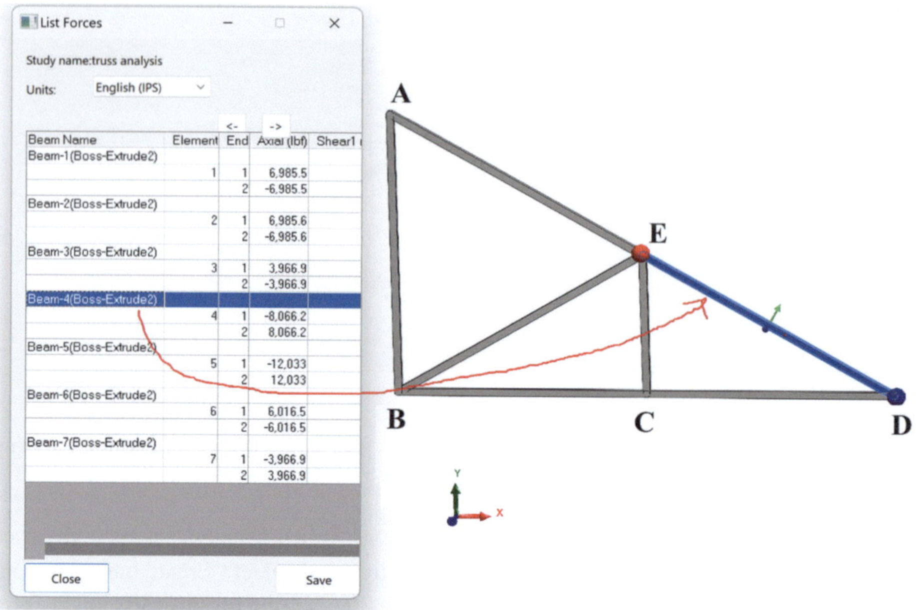

Fig. 4.53 List of truss member axial forces

Step 8: **Run Simulation**. Right-click the project name and select "Run". This will start the
FEA simulation.

Step 9: **Post-processing**.

After the truss analysis has been completed, the axial forces of each truss member can be
listed and the axial stress of each truss member can be displayed. Reaction forces at the
restraint locations can be obtained.

In SolidWorks simulation, a truss member is treated as a special beam. For a list of truss
member axial forces, right-click the tab "Results", select "List Beam forces" and then click
"Ok". A table of lists of every truss member force will appear as shown in Fig. 4.53. Change
the unit for the force and then move the cursor along the first column "Beam Name" and
click a beam. This will highlight a truss member in the truss as shown in Fig. 4.53. For
example, "Beam 4" in the table is the truss member ED. The fourth column of the table lists
the axial forces of two ends (joints) of the truss member. The positive or negative sign of
the axial forces is the sign for the forces on two ends of a truss member. The axial force of
truss member ED per simulation result in Fig. 4.53 is a tension force of 8066.2 lb.

For axial stress of truss members, right-click the tab "Results" in the simulation study
tree and select "Define Stress Plot". The stress plot Property Manager will appear. Then,
change the unit and select "Axial" stress. This will generate an axial stress plot of the truss.
After this axial stress plot has been created, right-click the tab of the plot and select "Probe".

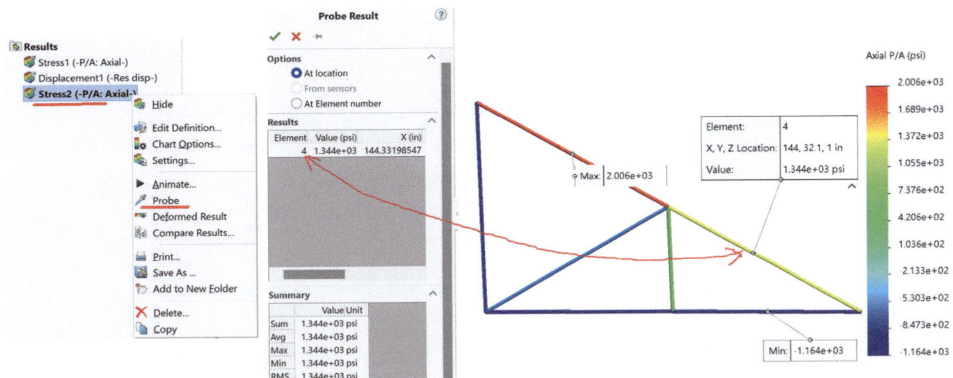

Fig. 4.54 The axial stress plot of the truss

The probe Property Manager will appear as shown in Fig. 4.54. Now, when a truss member is selected, the axial stress of this truss member will be displayed as shown in Fig. 4.54.

The post-processing tools can be used to create the resultant displacement plot as shown in Fig. 4.55, and the reaction forces on the supports as shown in Fig. 4.56.

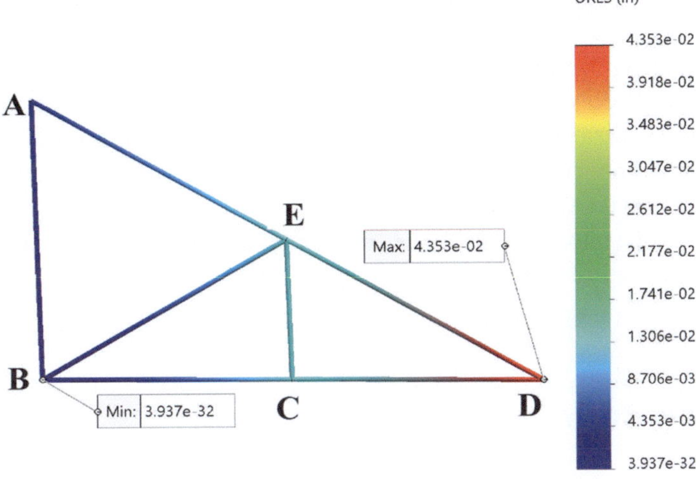

Fig. 4.55 The resultant displacement plots of the truss

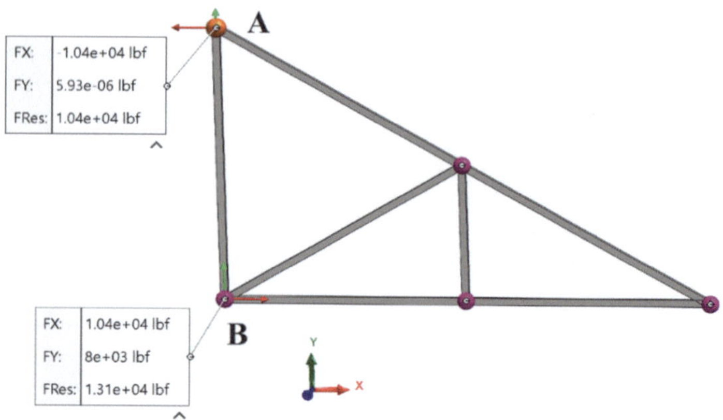

Fig. 4.56 The reaction forces at Points A and B

(2) **Interpretations and Discussions**

By using the method of joints [4], the theoretical value of the axial force of the truss member ED is 8000 lb. The simulation results of the axial force of the truss member ED per Fig. 4.53 is 8066.2. The relative difference is 0.83%. Both results are very close.

By using theoretical calculation of engineering statics [4], the reaction forces at the supporting point B are the reaction force along the x-direction $R_{Bx} = 8000\,lb$ and the reaction force along the y-direction $R_{By} = 10392\,lb$. The corresponding FEA simulation results of reaction forces at supporting point B per Fig. 4.56 are: the reaction force along the x-direction $R_{Bx} = 8000\,lb$ and the reaction force along the y-direction $R_{By} = 1.04E04\,lb$. The relative differences between the theoretical calculation results and the FEA simulation results are less than 0.08%.

Per Fig. 4.55, the maximum displacement happens at Point D. This is expected from the point view of theoretical analysis.

4.6   Beam Analysis

This section will use one example to show how to run a beam analysis in SolidWorks Simulation.

A beam is a structural member with a constant cross-section that primarily resists loads applied laterally to its axis. The geometric feature is that its length is much bigger than the longest dimension of its cross-section. In SolidWorks Simulation, a beam member can be straight or curved. A slender straight beam with a constant cross-section will be demonstrated only in this book. A typical application of beams is a machine frame.

A joint is automatically identified at the free ends of a beam member and the intersection of two or more beam members. A beam member is defined by a straight line connecting two joints at its ends. Shared joints between beam members are automatically treated as bonded joints. A joint of a beam member is treated as a point but has six degrees of freedom including three translations and three rotations. There are two typical approaches for defining fixtures on joints of beam members, which have been discussed in Figs. 3.63 and 3.66 of Chap. 3.6: Fixtures. The first approach is to use a fixed fixture when a joint is fixed. No three translational deformations and no three rotational deformations are defined. The second approach is to "Use reference Geometry" for defining restraints in specified directions. These will be demonstrated in the following example.

Three types of loads can be applied on a beam member, which are concentrated forces, concentrated bending moments, and distributed lateral load. Concentrated forces and bending moments can be applied to any point of a beam. This point must be on the surfaces of a beam and created by inserting a reference point as a reference geometry. The distributed force, typically a force per unit length, can be applied only on a beam member. This means that if a distributed force is applied to a specified partial section of a beam, the beam needs to be split into several beams to make the specified partial section into a beam.

In SolidWorks Simulation, a beam member will be meshed by beam elements. Typically, the default meshing for beams is used. But "Apply mesh controls" on selected beams can be used to change the number of beam elements.

After the FEA simulation is completed, the stresses of each beam element are presented in their local directions of a local coordinate system. On a beam cross-section, a local coordinate system is automatically created. Direction 1 is parallel to the longest length direction of the cross-section and direction 2 is normal to direction 1 as shown in Fig. 4.57a. Display the local coordinate system of a beam. Expand the parts, right-click a beam member in the simulation study tree and select "Edit Definition". The Apply/Edit Beam Property Manager will appear. Then, selecting "Show beam direction" will result in displaying beam direction as shown in Fig. 4.57b.

Now, an example is used to show how to run beam analysis and some post-processing techniques will be demonstrated.

Example 7: Beam analysis (FEA39-ASM) Download, unzip, and open FEA39-ASM. A simple supporter beam with a span of $60''$ is a rectangular tubing with a height of $3''$, width of $2''$, and thickness of $0.1875''$. Its material is plain carbon steel. The loading conditions and restraints are displayed in Fig. 4.58. Run FEA simulation by using beam elements, then interpret & discuss results.

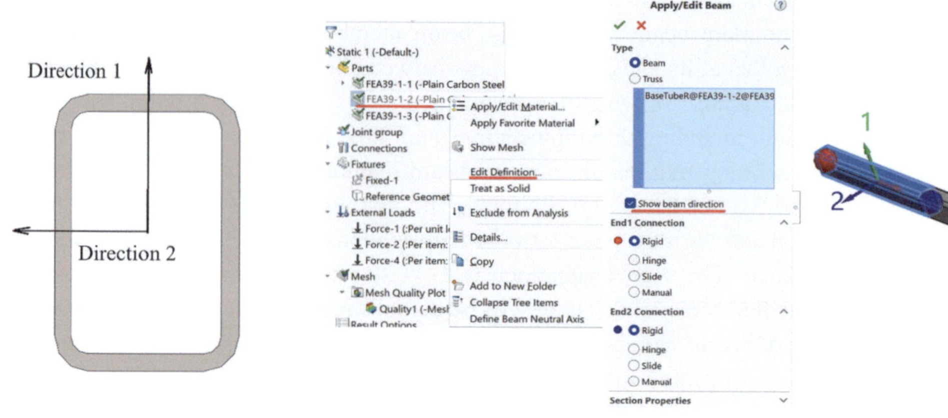

Fig. 4.57 Schematics of a local coordinate system of a beam cross-section

Fig. 4.58 A simple supported beam

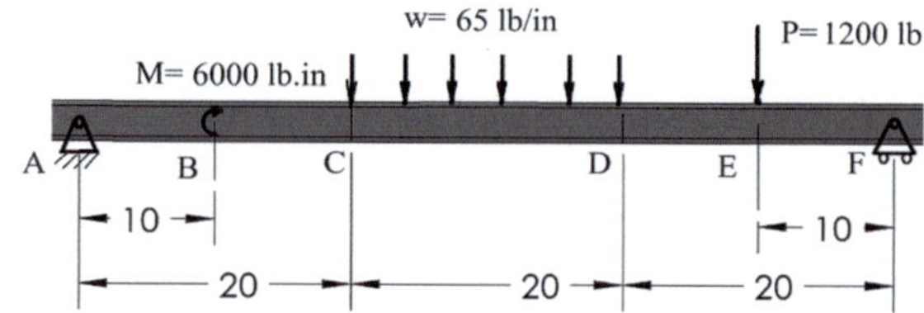

Solution

(1) FEA simulation

Step 1: **Pre-processing**. This is one simple supported beam. However, a distributed force of 65 lb/in is applied on the segment CD. This beam needs to be split into three beam members: beam AC, beam CD, and beam DF. To apply a concentrated force of 1200 lb at point E, a reference point 1 is created on the top surface of beam DF by inserting the "Reference Geometry" as shown in Fig. 4.59a. To apply a concentrated moment of 6,000 lb-in at point B, reference point 2 is created on the side surface of the beam AC as shown in Fig. 4.59a. The model is now ready for FEA simulation.

Step 2: **Set up a project**. This is a static stress analysis.

Step 3: **Select the type of element and assign a material to each component**.

Materials of plain carbon steel have been assigned in part level. Follow the procedure shown in Fig. 4.49a to convert each solid component into beams. Four joints of all beam members will be automatically calculated and generated as shown in Fig. 4.59b.

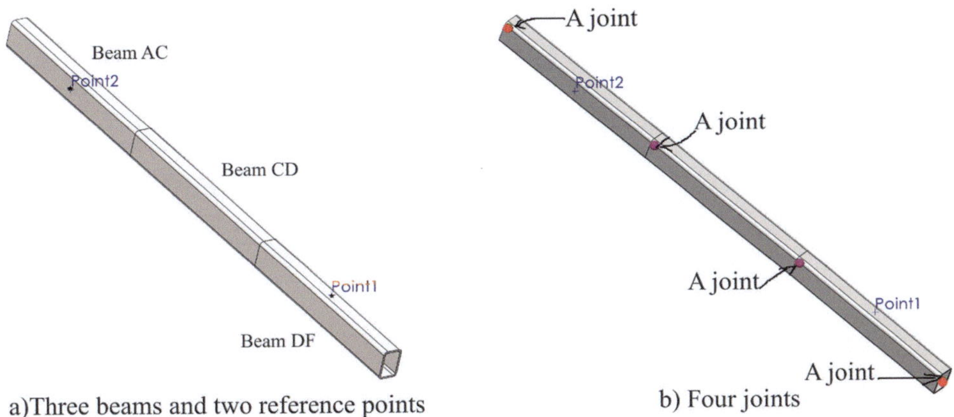

a)Three beams and two reference points   b) Four joints

Fig. 4.59 Schematics of beams and the list of four joints

Step 4: **Connections**. Each beam member will have two joints at its ends. Shared joints between beam members will be automatically bonded.

Step 5: **Fixtures**. Since the restraint at point A for this beam analysis is a fixed pivot pin, the joint at point A can freely rotate around the pivot pin and all other five deformations including three translations and two rotations will be zero. Right-click the tab "Fixtures" in the simulation study tree and select "Use Reference Geometry". The Fixture Property Manager will appear as shown in Fig. 4.60a. Select the corresponding joint and the proper reference surface, and then set zero deformation for five degrees of freedom except the rotation about the pivot pin as shown in Fig. 4.60a. The restraint at point F for this beam analysis is a roller. The joint at point F will have zero deformation along the vertical direction. Right-click the tab "Fixtures" in the simulation study tree and select "Use Reference Geometry". The Fixture Property Manager will appear. Select the corresponding joint and the proper reference geometry such as an edge, and then set zero translational deformation as shown in Fig. 4.60b. The schematics of fixtures of this example are displayed in Fig. 4.60.

Step 6: **Loads**. For defining a distributed force of 65 lb/in on the middle beam CD as shown in Fig. 4.58, right-click the tab "External loads" and select "Force". The force/Torque Property Manager will appear as shown in Fig. 4.61a. Select "Beams" from three possible choices: "vertices, points", "Joints" and "beams" in the window. Select the middle beam and a reference edge that is parallel to the distributed force as shown in Fig. 4.61a. Then select "per unit length" and finally define the value of 65 lb/in.

To define a concentrated bending moment at point B as shown in Fig. 4.58: Right-click the tab "External loads" and select "Torque". The force/Torque Property Manager will appear as shown in Fig. 4.61b. Select "Vertices, Points" from three possible choices: "vertices,

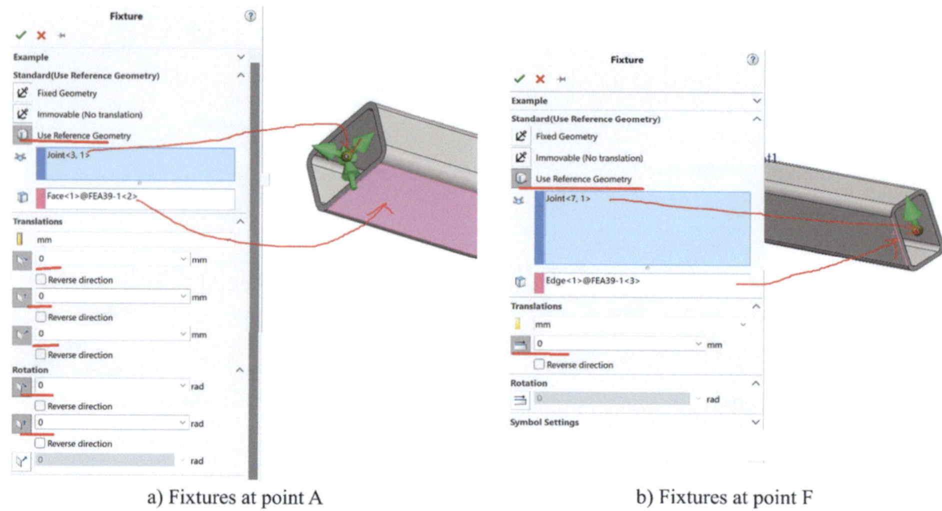

a) Fixtures at point A   b) Fixtures at point F

Fig. 4.60 Fixtures of the beams

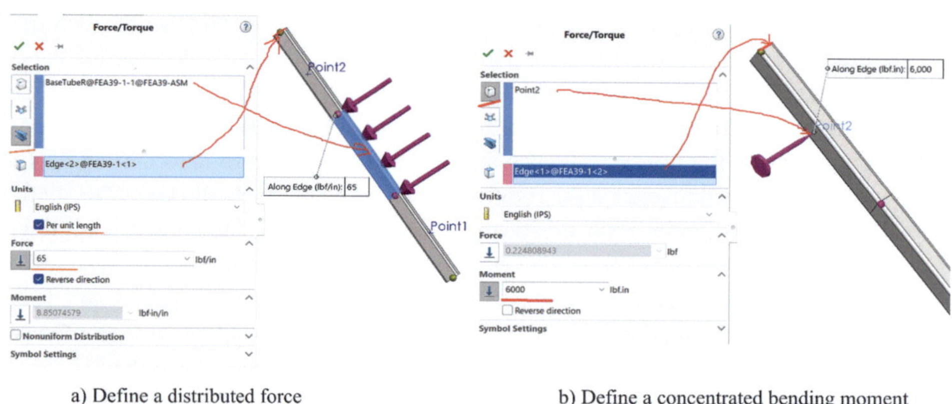

a) Define a distributed force   b) Define a concentrated bending moment

Fig. 4.61 Define a distributed force and a concentrated bending moment

points", "Joints" and "beams" in the window. Select point 2 and a reference edge that is parallel to the bending moment as shown in Fig. 4.61b. Then define the bending moment of 6000 lb.in.

For defining a concentrated force at point E as shown in Fig. 4.58, right-click the tab "External loads" and select "force". The force/Torque Property Manager will appear as shown in Fig. 4.62a. Select "Vertices, Points" from three possible choices: "vertices, points", "Joints" and "beams" in the window. Select point 1 and a reference edge that is parallel to the

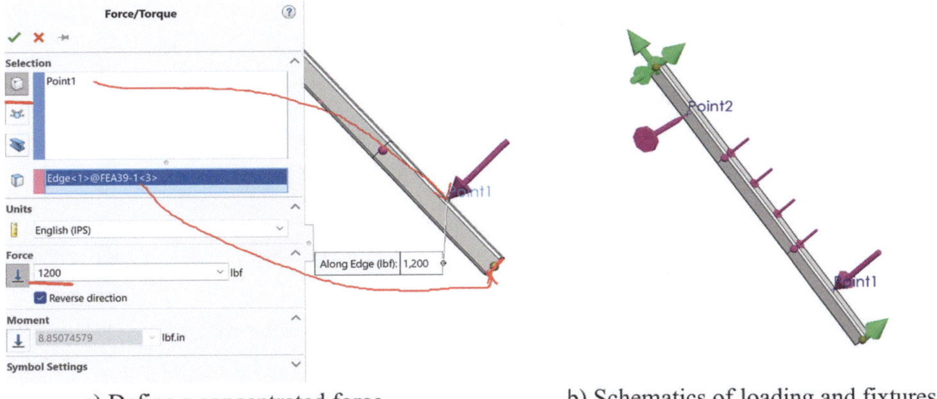

a) Define a concentrated force   b) Schematics of loading and fixtures

Fig. 4.62 Define a concentrated force and image of loads and fixtures

force as shown in Fig. 4.62a. Then define the concentrated force of 1200 lb. The schematics of fixtures and loads of this beam analysis are displayed in Fig. 4.62b.

Step 7: **Meshing**. In this example, all members are beams and will be meshed by beam elements. The meshing will be the default meshing. However, the number of beam elements in each beam can be changed by using "Apply mesh controls".

Step 8: **Run Simulation**. Right-click the project name and select "Run". This will start the FEA simulation.

Step 9: **Post-processing**. After the FEA simulation on this beam analysis is completed, use post-processing techniques to obtain the required information.

In this example, the shear force diagram will be the diagram along the direction 1. Right-click the tab "Results" in the simulation study tree and select "Define beam diagrams". In the Beam Diagrams Property Manager, select "Shear force diagram in Dir 1" to generate the shear force diagram as shown in Fig. 4.63a.

In this example, the bending moment diagram will be the diagram along the direction 2. Right-click the tab "Results" in the simulation study tree and select "Define beam diagrams". In the Beam Diagrams Property Manager, select "Moment about Dir 2" to generate the bending moment diagram as shown in Fig. 4.63b).

The stress for beam analysis is based on beam theory. The typical choice for a stress plot in beam analysis will be the upper bound axial and bending stress. Right-click the tab "Results" and select "Define stress plot". The stress plot Property Manager will appear. Select "Upper bound axial and bending" to generate the stress plot for the maximum stress

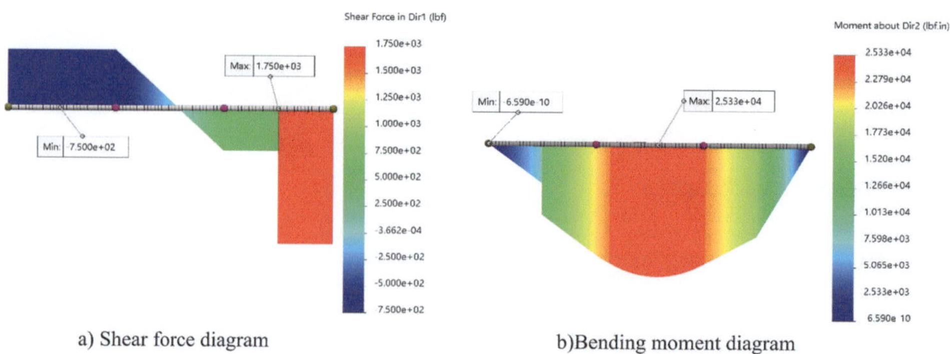

a) Shear force diagram   b)Bending moment diagram

Fig. 4.63   Shear force and bending moment diagrams

as shown in Fig. 4.64a. The resultant displacement plot is shown in Fig. 4.64b. The reaction forces at points A and F (Fixtures) are displayed in Fig. 4.65.

(2) **Interpretations and Discussions**

At supporter point A as shown in Fig. 4.48, the significant reaction force is in the vertical direction. The values from the theoretical calculation and FEA simulation per Fig. 4.65 are the same, 750 lb. At the supporter point F as shown in Fig. 4.48, the significant reaction force is in the vertical direction. The values from the theoretical calculation and the FEA simulation per Fig. 4.65 are the same, 1750 lb.

The maximum shear force from the theoretical calculation is 1750 lb, which is the same value from the FEA simulation per Fig. 4.63a.

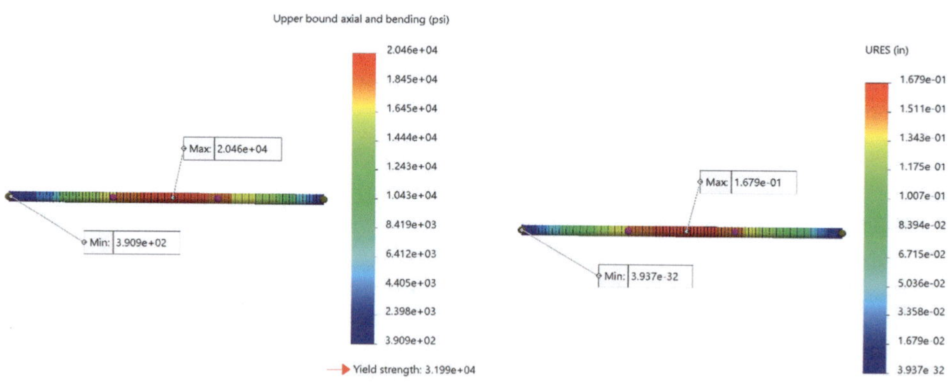

a) Upper bound axial and bending stress plot   b) Resultant displacement plot

Fig. 4.64   Stress and displacement plots

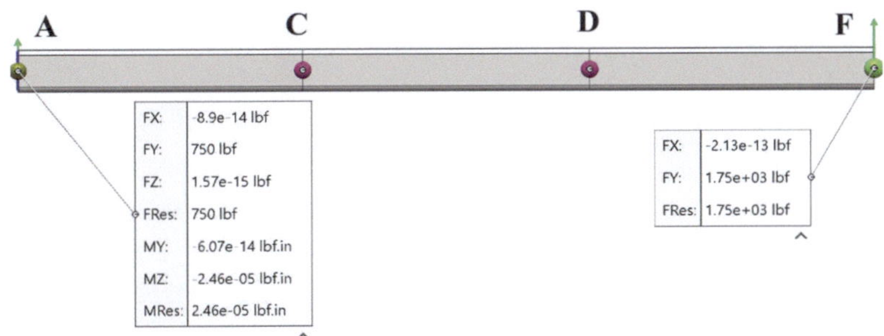

Fig. 4.65 Reaction forces on fixtures

The maximum bending moment from the theoretical calculation is 25250 (lb.in). The maximum bending moment from the FEA simulation per Fig. 4.63b is 25330 (lb.in). The relative difference is only 0.3%.

The maximum stress and maximum deformation happen in the middle of the beam per Fig. 4.64a and b. These results are expected for this beam theoretical analysis.

4.7   A Mixture of Different Types of Elements

In SolidWorks Simulation, each component can be meshed by one type of element, which includes solid elements, shell elements, truss elements, and beam elements. Therefore, FEA simulation on an assembly can have a mixture of different types of elements.

Every mechanical component is a 3D part. All mechanical components can be meshed by solid elements. However, for practical applications and the purpose of saving computing time, some components can be simplified and meshed by truss elements, beam elements, and shell elements due to their geometries and loading conditions. The main features of each type of element in SolidWorks simulation have been discussed and explored in Chap. 3.5: The type of elements and assigning materials. For a slender component under only axial loads, it can be simplified as a truss. For a lender component under axial loading, shearing, and bending, it can be simplified as a beam. For a thin plate or sheet metal component, it can be simplified as a shell.

FEA simulation on an assembly is very complicated due to meshing and contacts on interfaces. FEA simulations on an assembly with a mixture of different types of elements are more complicated mainly due to contacts on interfaces.

Typical types of mixture meshing are solid with beam, solid with shell, shell with beam, and solid with beam & shell. When a component is simplified as a beam or a truss, it is simplified as a group of joints. When a thin plate or sheet metal is simplified as a shell, it becomes a surface. There is no general guideline for properly defining contacts on

Fig. 4.66 Picture of an assembly consisting of a base, tubing, and a bar

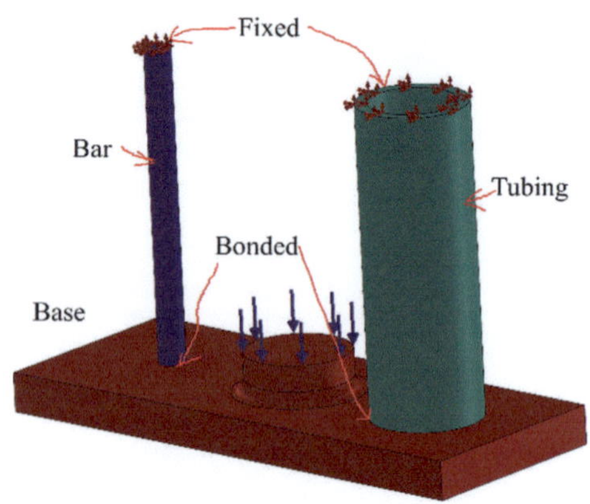

interfaces for mixture meshing. However, when contacts on interfaces for mixture meshing such as bonded or contact between solid and shell or beam are defined, the information for beam or shell should be placed in the first window in the contact Property Manager. When the contact is on the interface between the shell and beam, the information for the beam should be placed in the first window.

When a mixture meshing is used on a FEA simulation on an assembly, simulation results will not be the same. In Example 8 which is shown next, there are two cases of FEA simulation on the same assembly. Case one uses all solid elements. Case two will use solid elements, beam elements, or shell elements. The simulation results on a component with the same type of element in both cases will be very similar and close. However, the simulation results on a component with different types of elements in the two cases are different. Now, one example will be used to demonstrate a mixture meshing.

Example 8: A mixture of solid, shell, and beam elements (FEA35-ASM) Download, unzip, and open FEA35-ASM, which consists of three components: a solid round bar with a diameter of 0.25″, tubing with an OD 1.05″ and a wall thickness of 0.113″ and a base as shown in Fig. 4.66. The material of the tubing and the base are plain carbon steel. The bar is made of alloy steel. The top ends of the bar and tubing are fixed. The bottom ends of the bar and the tubing are welded to the base. Run the following FEA simulation of two cases. In Case One, all components are meshed by solid elements. In Case Two, the base is meshed with solid elements, the tubing is meshed with shell elements, and the bar is meshed with beam elements.

Solution

(1) **FEA simulation**

Step 1: **Pre-processing**. Since the tubing will be meshed by shell elements in Case Two, a mid-surface is created in the part model of the tubing. The assembly FEA35-ASM is ready for FEA simulation.

Step 2: **Set up a project**. This is a static stress analysis. Two simulation projects will be created. One is for Case One and another is for Case Two. But before the simulation project Case One is created, the mid-surface in the tubing (FEA35-1) should be suppressed. Before the simulation project Case Two is created, the mid-surface in the tubing (FEA35-1) should be unsuppressed.

Step 3: **Select the type of element and assign a material to each component**. In this example, materials have been assigned at part level.

For Case One:

- Nothing needs to be changed for the models because they will be meshed by solid elements.

For Case Two:

- The tubing (FEA35-1) will be meshed by shell elements. Expand "Parts" in the simulation study tree and then expand "FEA35-1". Right-click "SolidBody" and select "Exclude from Analysis". Then right-click "SurfaceBody" and select "Edit definition" to specify the thickness of the shell in the shell definition Property Manager as shown in Fig. 4.67. In this example, the thickness is $0.113''$. Now, this component will be meshed by shell elements.
- The bar (FEA35-2) will be meshed by beam elements. Right-click "FEA35-2" and select "Treat as beam" as shown in Fig. 4.67.
- Nothing needs to be changed for the base (FEA35-3) model. It will be meshed by solid elements.

Step 4: **Connections**

For Case One: In this example, the bottom ends of the bar and the tubing are touching the base. The default "Global interaction" already specifies the bonded contacts between the bottom end of the bar with the base and between the bottom end of the tubing with the base. Local interaction can still be used to specify the bonded contacts on interfaces.

For Case Two: Use local interaction to define bonded contact between the bottom joint of the beam and the base as shown in Fig. 4.68a. The beam (joint) is selected in the first window.

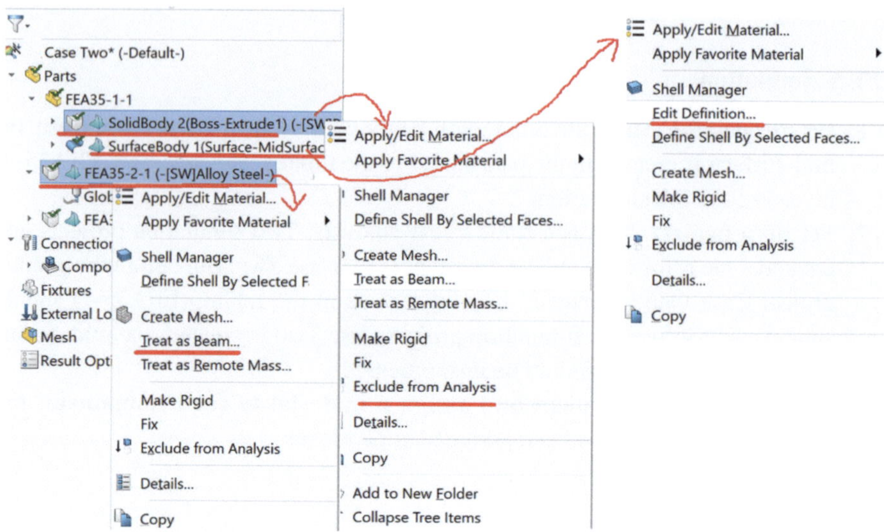

Fig. 4.67 Setting of types of elements

The bonded contact between the bottom edge of the shell and the base is defined through local interaction as shown in Fig. 4.68b. The shell edge is selected in the first window.

Step 5: **Fixtures**

For Case One: Right-click the tab "Fixture" and select "Fixture geometry" to apply the fixed fixtures on the top ends of the bar and the tubing.

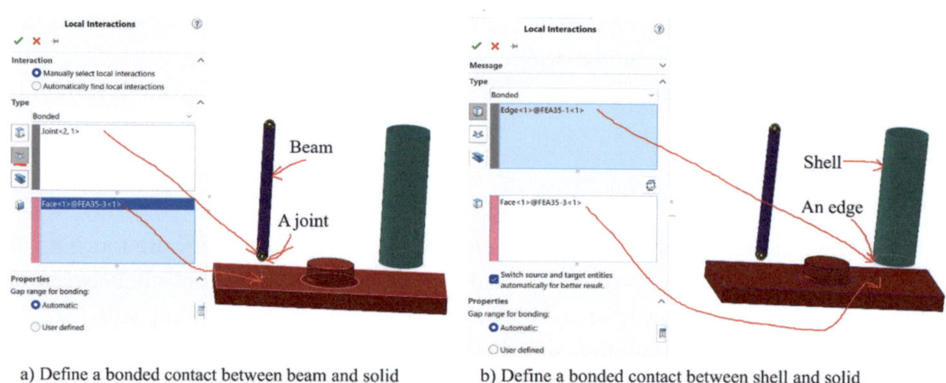

a) Define a bonded contact between beam and solid   b) Define a bonded contact between shell and solid

Fig. 4.68 Define bonded contacts for mixture meshing

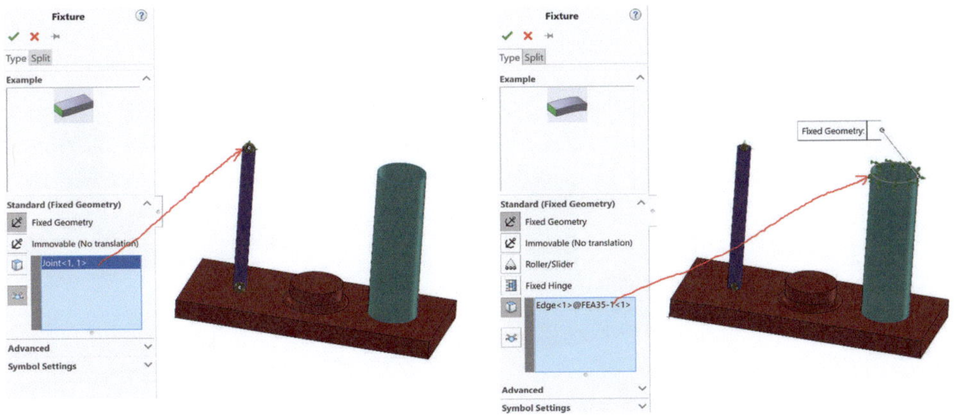

a) Fixed fixture for the beam joint   b) Fixed fixture for the shell edge

Fig. 4.69 Define fixed fixtures

For Case Two: For the bar, the fixed fixture is applied on the top joint as shown in Fig. 4.69a. For the tubing, the fixed fixture is applied on the top edge of the shell as shown in Fig. 4.69b.

Step 6: **Loads**. Right-click the tab "External loads" and select "Force" to define a 1000 lb force on the top surface of the component base as shown in Fig. 4.66.

Step 7: **Meshing**

For Case One: The bar with a diameter of 0.25″ will be meshed by a solid element with the local element size of 0.25″/5 = 0.05″. The tubing with a wall thickness of 0.113″ will be meshed by a solid element with a local element size of 0.113/5 = 0.0226″. The base with the shortest dimension of 0.375″ will be meshed by solid elements with a global element size of 0.375″/5 = 0.075″. In this example, "Apply mesh control" will be used on the bar and the tubing with local element sizes. The tubing will be used as an example to show how to do this. Right-click the tab "Mesh" in the simulation study tree and then select "Apply mesh control". The mesh control Property Manager will appear. First, select "Use Per part size" and then click the tubing as shown in Fig. 4.70a. To define local element size, un-select "Use per part size" and now the mesh parameters will appear as shown in Fig. 4.70b. Enter the local element size as 0.0226″. Repeat the same procedure to define the local element size for the bar (FAE35-2) as 0.05″. Finally, use "Create mesh" to define the global element size of 0.075″ for the base (FEA35-3). The meshing image is shown in Fig. 4.71a.

For Case Two: Meshing for the bar (beam elements) is the default. The globe element size of 0.075 for the base can be used for the tubing (shell elements). "Apply mesh control"

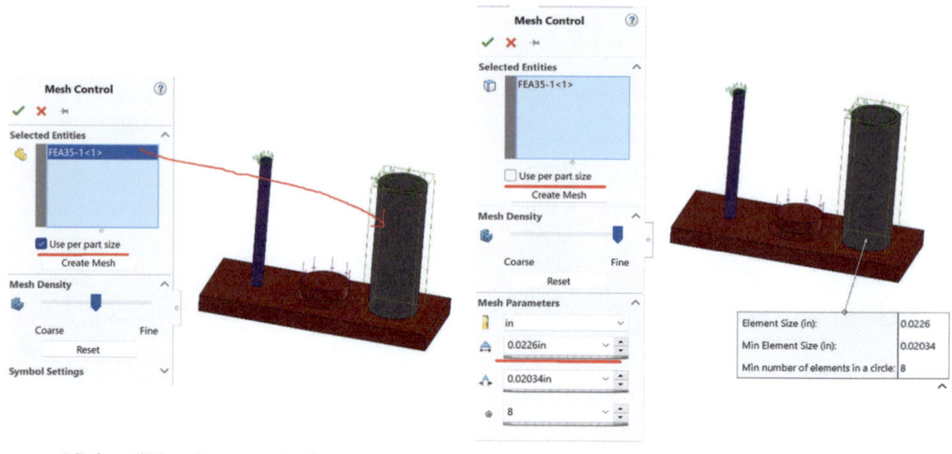

a)Select "User Per part size"   b) Define the local element size

Fig. 4.70 Define a local element size for a component

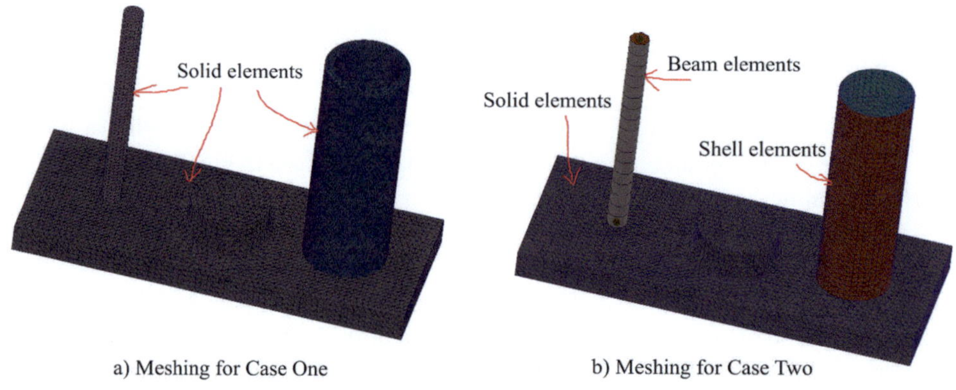

a) Meshing for Case One   b) Meshing for Case Two

Fig. 4.71 Images of meshing

does not need to be used and "Create mesh" with a global standard element size of 0.075″ can be used. The image of the meshing for Case Two is shown in Fig. 4.71b.

Step 8: **Run Simulation**. Right-click the project name and select "Run". This will start the FEA simulation.

Step 9: **Post-processing**

Stress and displacement plots for each component in two cases will be created. The choice of stresses will be Von Mises stress for solid elements and shell elements. The choice of displacement will be the total resultant displacement. For the tubing (FEA35-1), when it is meshed by shell elements, the maximum von Mises stress will be the bigger value from the top and the bottom von Mises stresses. For the bar (FEA35-2), when it is meshed by beam elements, the stress "Upper bound axial and bending" will be used because Von Mises stress is not provided for beam elements. For the bar, the corresponding stress of the bar with solid elements per the orientation of the models will be S_Z normal stress which is normal stress along the z-axis.

The Von Mises stress plots of the base (FEA35-3) in Case One and Case Two are displayed in Fig. 4.72a and b, respectively. The resultant displacement plots of the base (FEA35-3) in Case One and Case Two are displayed in Fig. 4.73a and b, respectively.

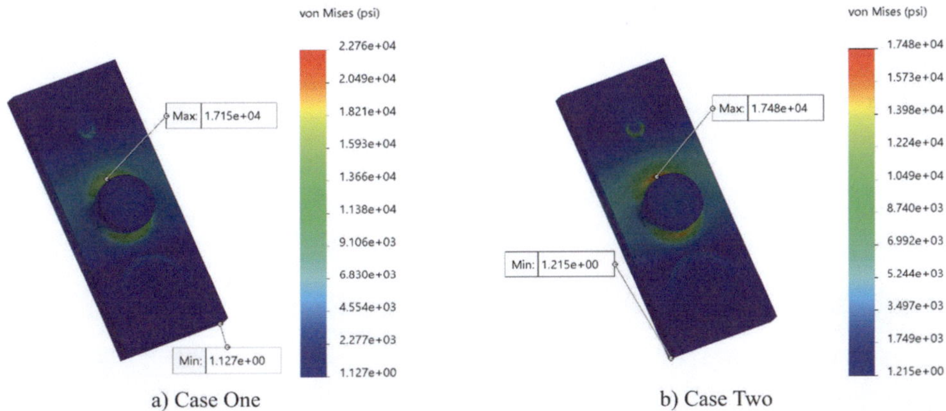

Fig. 4.72 Von Mises stress plots of the base (FEA35-3)

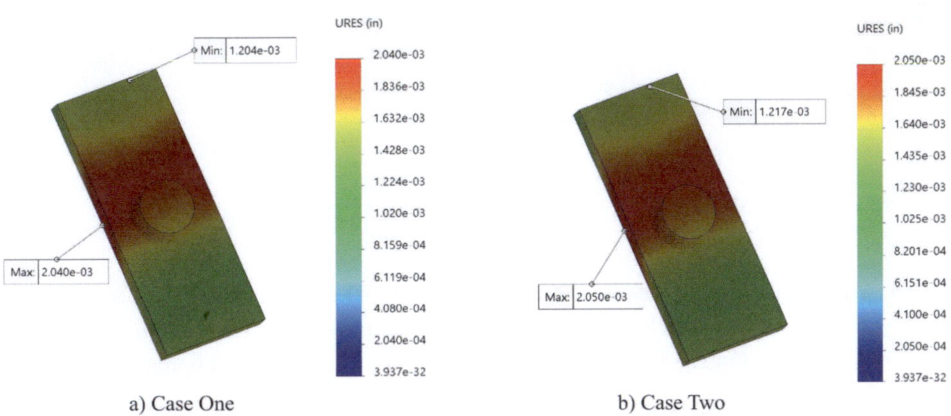

Fig. 4.73 Resultant displacement plots of the base (FEA35-3)

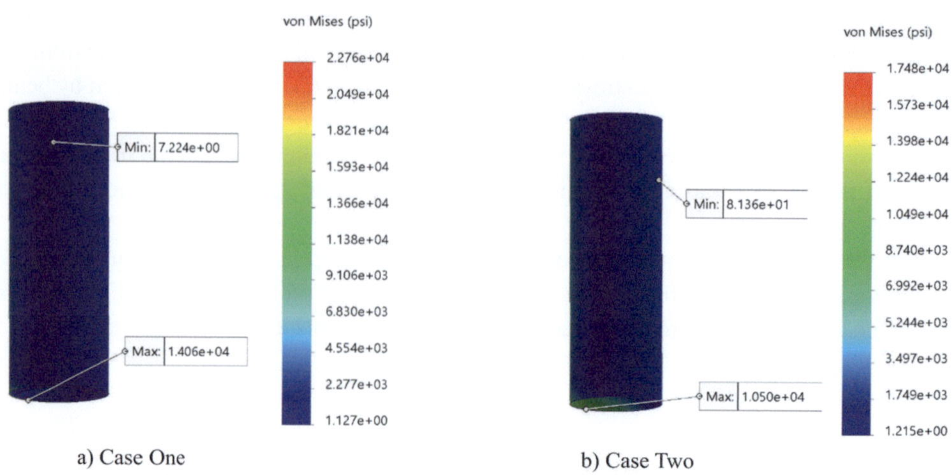

a) Case One   b) Case Two

Fig. 4.74 Von Mises stress plots of the tubing (FEA35-1)

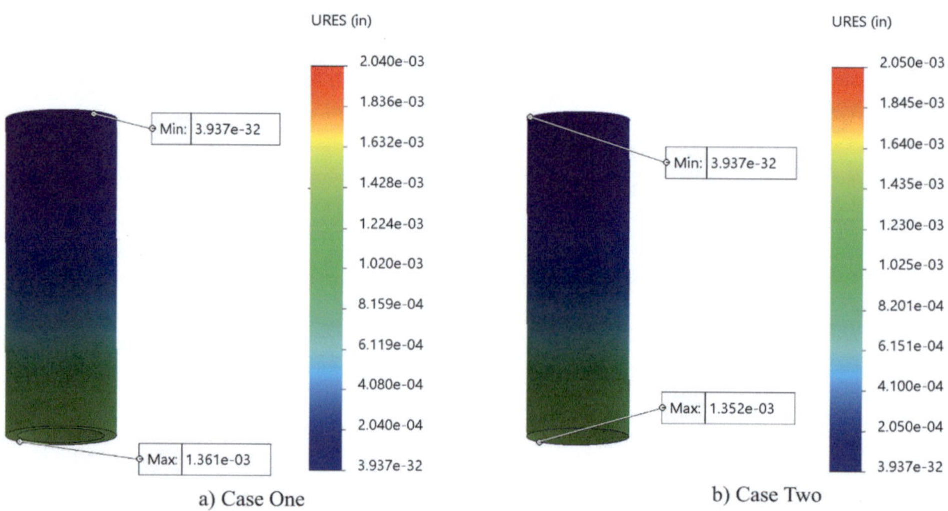

a) Case One   b) Case Two

Fig. 4.75 Resultant displacement plots of the tubing (FEA35-1)

The Von Mises stress plots of the tubing (FEA35-1) in Case One and Case Two are displayed in Fig. 4.74a and b, respectively. The resultant displacement plots of the tubing (FEA35-1) in Case One and Case Two are displayed in Fig. 4.75a and b, respectively.

The z-normal stress of the bar (FEA35-2) in Case One and the "Upper bound axial and bending" stress of the bar (FEA35-2) in Case Two are displayed in Fig. 4.76a and b, respectively. The resultant displacement plots of the bar (FEA35-2) in Case One and Case Two are displayed in Fig. 4.77a and b, respectively.

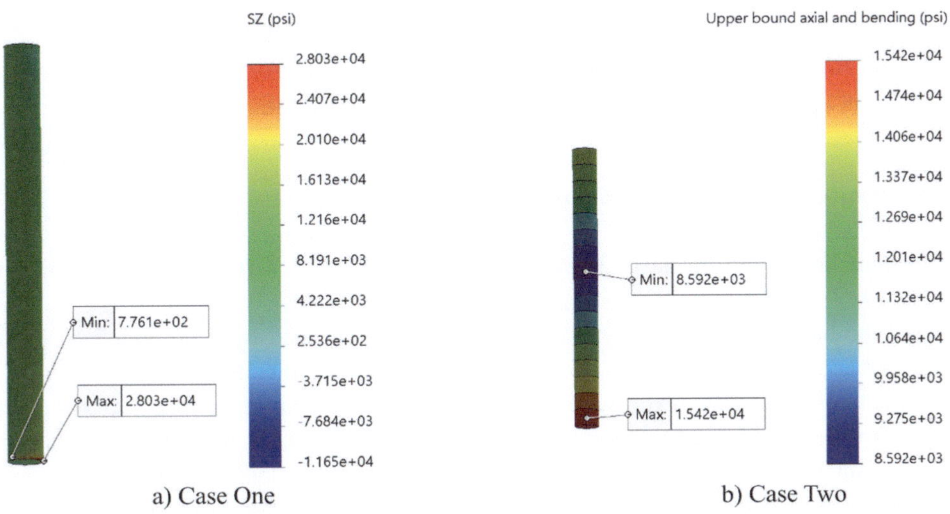

a) Case One b) Case Two

Fig. 4.76 Normal stress plots along the bar central axis of the bar (FEA35-2)

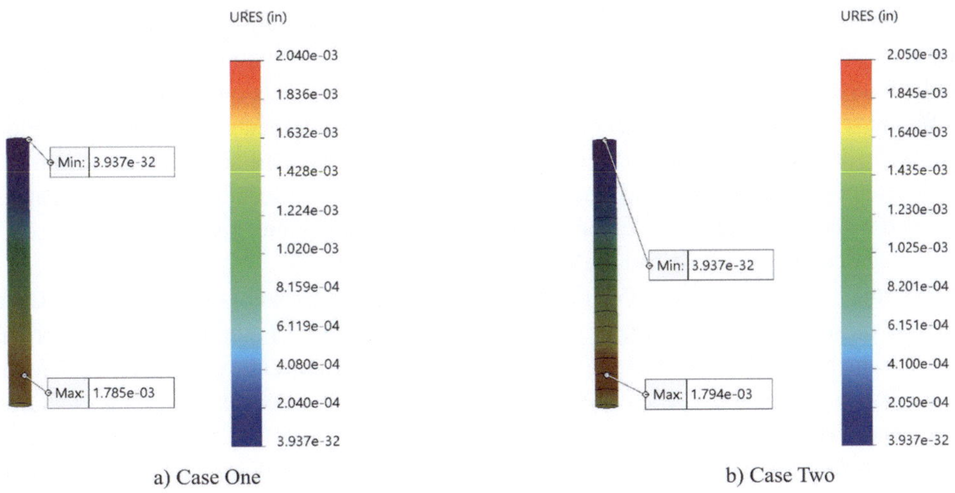

a) Case One b) Case Two

Fig. 4.77 Resultant displacement plots of the bar (FEA35-2)

Table 4.1 Simulation results of stress & resultant displacement for Case One and Case Two

Components and variables	Case One	Case Two
The base (stress)	1.715E4 psi (von Mises)	1.748E4 psi (von Mises)
The base (displacement)	2.040E-3 inch	2.050E−3 inch
The tubing (stress)	1.406E4 psi (von Mises)	1.050E4 psi (von Mises)
The tubing (displacement)	1.361E−3 inch	1.294E−3 inch
The bar (stress)	2.804E4 psi (z-normal stress)	1.524E4 psi (axial and bending)
The bar (displacement)	1.785E−3 inch	1.765E−3 inch

(2) Interpretations and Discussions

The simulation results displayed in Figs. 4.72, 4.73, 4.74, 4.75, 4.76 and 4.77 are shown in Table 4.1.

For the base: The base (FEA35-3) in both cases is meshed by solid elements. From Fig. 4.43 and Table 4.1, the patterns of von Mises stress plots in both cases are very similar. The locations of maximum von Mises stress in both cases are the same. The maximum von Mises stresses in both cases are very close. The relative difference is only 1.9%. The patterns of resultant displacements in both cases are very similar. The locations of the maximum displacement in both cases are the same. The maximum resultant displacements in both cases are very close. The relative difference is only 0.5%. Therefore, it can be concluded that the simulation results on the base in both cases are almost the same.

For the tubing: The tubing (FEA35-1) in Case One is meshed by solid elements and in Case Two is meshed by shell elements. The locations of maximum stress in both cases are the same and happen on the bonded interface per Fig. 4.74. But maximum von Mises stresses in both cases are different. The relative difference is 25.3%. This might be caused by the different settings on the interface. For solid elements to solid elements, the bonded condition is that a surface of solid elements is bonded to another surface of solid elements. But for Solid elements to shell elements, the bonded condition is that a surface of solid elements is bonded with an edge of shell elements. Per Fig. 4.75, the maximum resultant displacements in both cases are very close and the relative difference is 0.66%.

For the bar: The bar (FEA35-2) in Case One is meshed by solid elements and in Case Two is meshed by beam elements. Per Fig. 4.76 and Table 4.1, the maximum z-normal stress, which is normal stress along the bar central axis in Case One, is 2.803E4 psi, and in Case Two is 1.542E4 psi. The difference between these two values is 45%. This is a large difference. This might be caused by different settings on the interface. For solid elements to solid elements, the bonded condition is that the surface of the base is bonded to the bottom surface of the bar. But for Solid elements to beam elements, the bonded condition is that the surface of the base is bonded with the bottom joint of beam elements. However, per

Fig. 4.77 and Table 4.1, the maximum resultant displacements in both cases are very close and the relative error is only 0.50%.

Based on this example, we could draw this conclusion. In FEA simulation with a mixture meshing, the simulation results of components with solid elements will not be affected by the simplification of other components such as beam elements or shell elements. However, stresses obtained on the simplified components might be quite different.

4.8 Major FEA Simulation Design Project

FEA simulation is practical knowledge and skill for mechanical engineering students and engineers. Running more FEA simulations on components and assemblies is the only way to accumulate more experience. A better understanding of FEA simulation, more skills, and knowledge of how to deal with different settings on different simulation problems can be gained with lots of practice.

After this section is completed, major FEA simulation design projects can be performed. A major FEA simulation design project is any FEA simulation on a real mechanical device, product, or sub-assembly. It should not be too complicated and should be manageable. Many different major FEA simulation design projects have been performed in our FEA-related course. These are some recommendations and requirements for a major FEA simulation design project [4].

- It is strongly recommended that a team should work together on a major FEA simulation design project. Team members can learn from each other.
- It should be an assembly with several interfaces where the corresponding types of contacts will be defined.
- Essential fixture conditions or related statements should be specified. Loading conditions should be specified. Otherwise, it might lead to lots of quite different results.
- If possible, the design specifications such as the rated factor of safety and the maximum allowable deformation are specified.
- The redesign of failed components will be too complicated, and a huge amount of work will be required. Therefore, it will be only a design check of a design project.
- A technical report will be required in which interpretation and discussion of FEA simulation results should be included.

Now, one major FEA simulation design project will be listed and explained here.

Major FEA simulation design project: Design check on a lift arm assembly (FEA37-ASM)

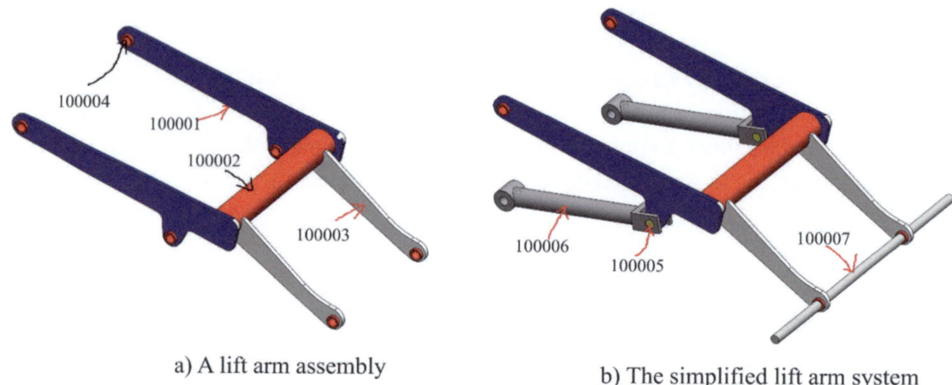

a) A lift arm assembly   b) The simplified lift arm system

Fig. 4.78 The lift arm assembly and the lift arm system (FEA37-ASM)

Download, unzip, and open FEA37-ASM. A lift-arm assembly is a working mechanism for a frond-end loader as shown in Fig. 4.78a. The lift-arm assembly consists of the following parts: (1) two of part 100,001 (plate, lift arm beam); (2) one of part 100,002 (tube, lift arm link bar); (3) two of part 100,003 (plate, lift arm link plate, (4) six of part 100,004 (bushing, purchased part).

A lift-arm assembly is a sub-assembly of a front-end loader. The simplified lift arm system is displayed in Fig. 4.78b, which links the lift arm assembly with a front-end loader for showing loading conditions and some loading configurations. The simplified list arm system includes the lift arm assembly, two of part 100,005 (pin, purchased part), two of part 100,006 (cylinder, purchased part), and one of part 100,007 (bucket, simplified rigid bar).

The Bill of Materials of the list arm system is listed in Table 4.2. The bucket (PN# 100,007) is simplified as a rigid bar. For this simplified rigid bar, a high Young's modulus, and a high strength such as elastic modulus = 3E10 psi, tensile strength 1.3E7 psi, yield strength 9E6 psi, and Poisson's ratio 0.3 could be used.

The components in the lift arm assembly are welded together. One assumption of the fixture conditions of the lift arm system is shown in Fig. 4.79.

It is assumed that the worst loading condition is displayed in Fig. 4.80, which simulates that one end of the bucket is stuck inside the loading material. The cylinder (PN# 100,006) is in its default configuration. The shear force of 4500 lb is applied on any one end of the rigid bar (PN# 100,007) and parallel to the right plane and points down as shown in Fig. 4.80. For the actual design of this project, there are several other important loading configurations. Other loading configurations will be created per understanding of front-end loader working conditions and functions.

The design specifications for the lift arm assembly are: (1) The factor of safety is 2, and (2) The maximum allowable resultant displacement is 0.75″.

Table 4.2 Bill of Materials of the lift arm system (FEA37-ASM)

Part #	Part name	Material	Quantity	Notes
100,001	Plate, list arm beam	Plain carbon steel	2	Design part
100,002	Tube, list arm link bar	Plain carbon steel	1	Design part
100,003	Plate, lift arm link plate	Plain carbon steel	2	Design part
100,004	Bushing, purchase part	Alloy steel	6	Purchase part
100,005	Pin, purchase part	Alloy steel	2	Purchase part
100,006	Cylinder, purchase part	Alloy steel	2	Simplified cylinder with different extension lengths
100,007	Bucket, simplified rigid bar	Material with a high Young's modulus and strength	1	A simplified rigid bar

Fig. 4.79 The fixture conditions of the lift arm system

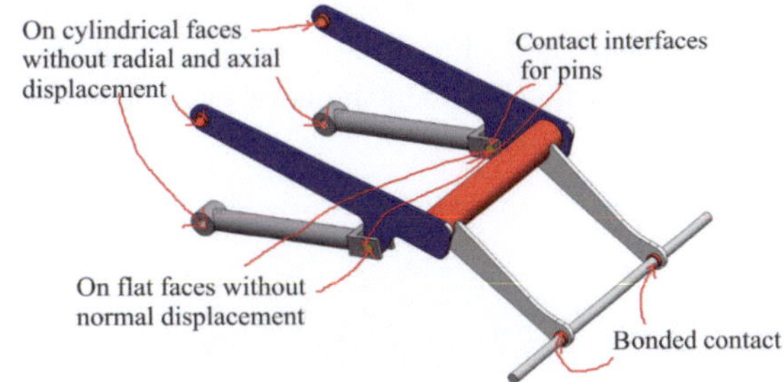

Run an FEA simulation to check whether each component of the lift arm assembly satisfies the design specifications. Write a technical report with proper simulation result plots and results' interpretations.

Fig. 4.80 The worst loading
condition

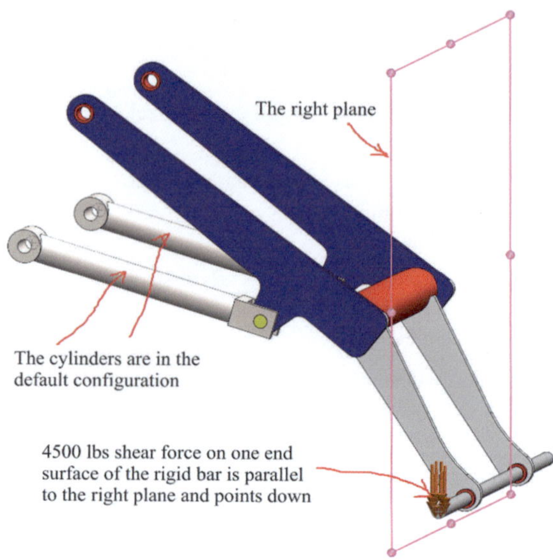

Exercises

4-1 Download, unzip, and open FEA31-ASM. A device consists of two plates. Materials
for both plates are plain carbon steel. The fixture and loading conditions are displayed
in Fig. 4.81. The interface between two plates is treated as three cases: "bonded",
"contact" and "Free". Run FEA simulation to show the von Mises stress plot and
resultant displacement plots of each case. Interpret and discuss the FEA simulation
results.

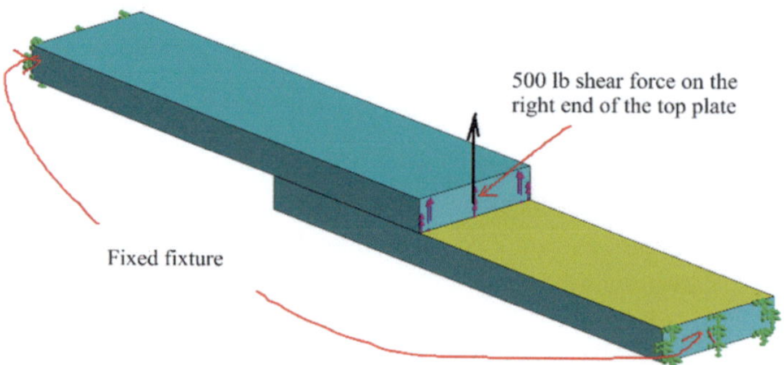

Fig. 4.81 The schematics of loading and fixture conditions (FEA31-ASM)

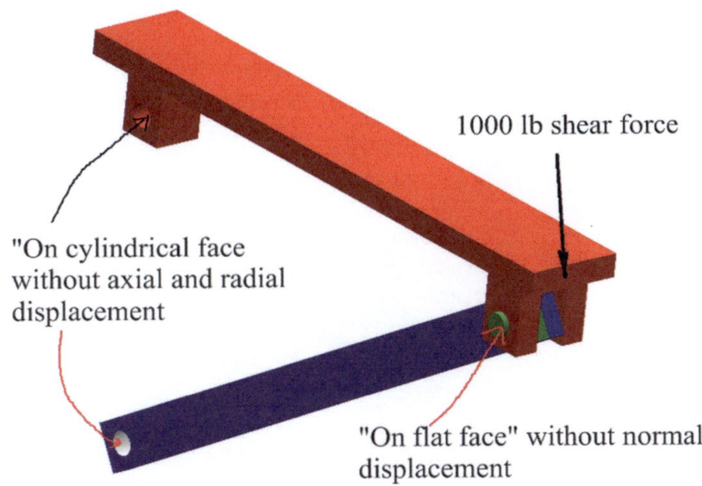

Fig. 4.82 Schematics of loading and fixture conditions (FEA3-ASM)

4-2 Download and open FEA32-ASM, which consists of three components: a beam, a pin, and a supporter. The material of all three parts is alloy steel. The fixture and loading conditions are shown in Fig. 4.82. The interfaces between the beam and the pin, and between the pin and the supporter can be treated as two cases: "bonded" and "contact". Run FEA simulation to show the von Mises stress plot and resultant displacement plots of each case. Interpret and discuss the FEA simulation results.

4-3 Download, unzip, and open FEA33-ASM. It is a simple bolted joint. Materials for both plates are alloy steel. Suppress all bolts and nuts and use bolt connectors to run the FEA simulation. The fixture, loading conditions, and the information for bolts are displayed in Fig. 4.83. The interface between two plates is treated as a "contact" interface. Run the FEA simulation to show the von Mises stress plot and resultant displacement plots of each part. Interpret and discuss the FEA simulation results.

4-4 Run the FEA simulation of Exercise 4.2 but the pin is suppressed, and the pin connector is used to replace the physical pin. Run the FEA simulation to show the von Mises stress plot and resultant displacement plots of each part. Compare the simulation results on this question with those obtained for the "contact" interface in Exercise 4.2. Interpret and discuss the FEA simulation results.

4-5 Download, unzip, and open FEA34-ASM, which is an interference fit (shrink fit). The material of the outer ring (FEA34-1) is alloy steel. The material of the inner column (FEA34-2) is plain carbon steel. Open the inner column (FEA34-2) and change the diameter to Ø3.010″. The fixture condition is displayed in Fig. 4.84a. Treat the interference on the interface as a type of shrink fit and run an FEA simulation. Display von Mises stress, hoop stress, and radial stress of each component. Interpret and discuss the FEA simulation results.

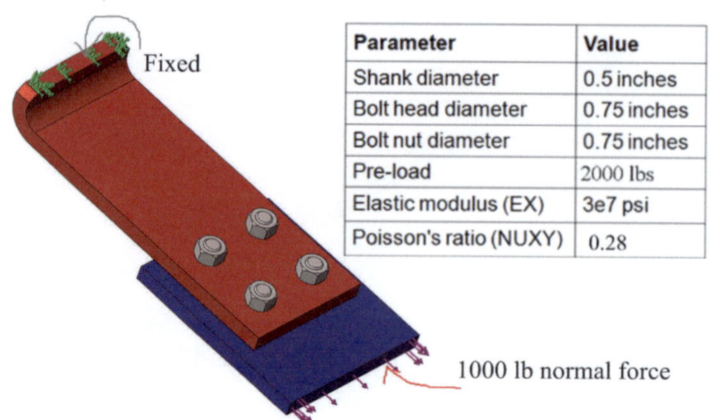

Parameter	Value
Shank diameter	0.5 inches
Bolt head diameter	0.75 inches
Bolt nut diameter	0.75 inches
Pre-load	2000 lbs
Elastic modulus (EX)	3e7 psi
Poisson's ratio (NUXY)	0.28

Fig. 4.83 Schematics of the fixture and loading conditions (FEA33-ASM)

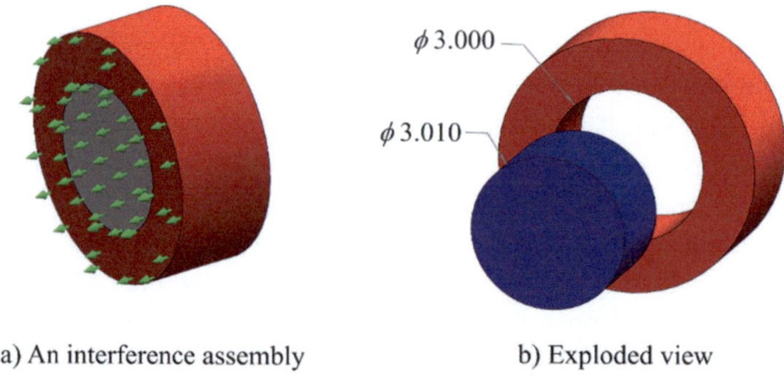

a) An interference assembly   b) Exploded view

Fig. 4.84 Schematics of an interference fit

4-6 Download, unzip, and open FEA38-ASM, which is a planar truss. The restraints and loadings, and a list of joints are shown in Fig. 4.85. The materials of all truss members are alloy steel. The supporter on point B is a fixed and the supporter on point A is a roller. There are 6000 lb concentrated forces applied on point C and a 2000 lb force at point D. The cross-section of the truss member is 3″ by 2″. Run an FEA simulation on this truss by tress elements. Generate axial stress and resultant displacement plots. Show the reaction forces at the support points A and B. Interpret and discuss simulation results.

4-7 Download, unzip, and open FEA39-ASM. A simple supporter beam with a span of 60″ is made from plain carbon rectangular steel tubing with a height of 3″, width of 2″ and a thickness of 0.1875″. The loading conditions and restraints are displayed in Fig. 4.86. Create some reference points when necessary. Run an FEA simulation by using beam elements. Generate one shear force diagram, one bending moment

Fig. 4.85 Schematics of a planar truss

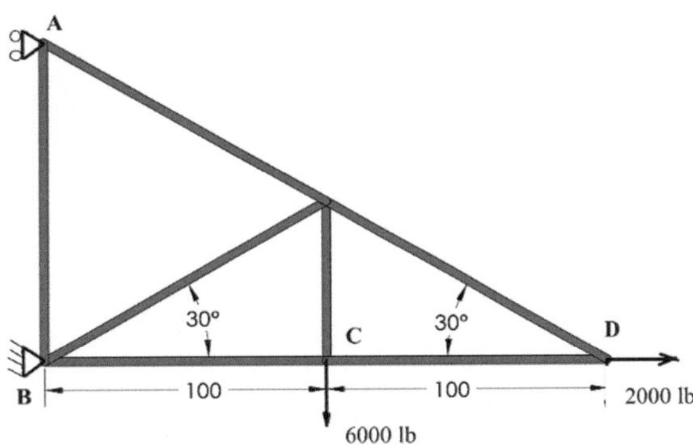

Fig. 4.86 Schematics of a beam

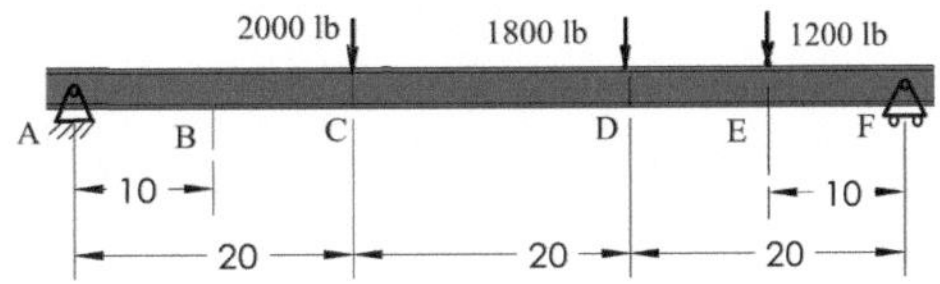

diagram, and the upper bound axial and bending stress plot, the resultant deformation plot, and show the reaction forces at the support points A and F. Then interpret & discuss the results.

4-8 Download, unzip, and open FEAM36-ASM which consists of two parts: the top plate (FEA36-1) and the support column (FEA36-2). The material of the plate is AISI 1020 steel. The material of the column is alloy steel. The fixture and loading conditions are shown in Fig. 4.87. The interface between the two parts is boned. Run FEA simulation on two cases. In Case One, the column is meshed by solid elements. In Case Two, the column is meshed by beam elements. The plate will be meshed by solid elements in both cases. Display the stress and displacement plots of each part of each case. Interpret and discuss the FEA simulation results.

4-9 Download, unzip, and open FEA37-ASM, which is a major FEA simulation design project described in Chap. 4.7: Major FEA simulation design project. Team with other mechanical engineering students or mechanical engineers to complete this project.

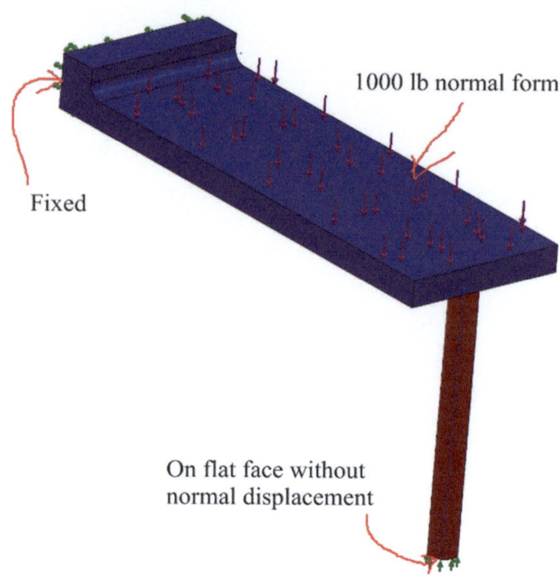

Fig. 4.87 Fixtures and load conditions

References

1. SolidWorks Web Help, https://help.solidworks.com
2. Richard G. Budynas, and J. Keith Nisbett (2015) Shrigley's Mechanical Engineering Design, Tenth Edition, McGraw-Hill Education, New York, NY.
3. Russell C. Hibler (2021) Engineering Mechanics: Statics, 15th Edition, Pearson
4. Xiaobin Le, Ali Moazed, and Anthony Duva (2016) The Design Projects for the Simulation-Based Design Course Paper presented at 2016 ASEE Annual Conference & Exposition, New Orleans, Louisiana. https://doi.org/10.18260/p.26118

Natural Frequency Analysis

5

Abstract

One of the main tasks in mechanical design is to prevent it from experiencing resonance. For this task, we need to know the natural frequencies of the components or mechanical systems. This chapter will show how to use the "Frequency" module in SolidWorks Simulation to estimate the first five natural frequencies. Natural frequency is dependent on the material, dimension, and type of fixture, and is independent of loading. When there are some rigid motions for the component under the simulation, the number of frequencies we need to solve will be the number of rigid motions plus 5, so that the first five natural frequencies can be obtained. For frequency analysis on an assembly, there are only two types of contacts in SolidWorks Simulation. They are "Bonded" and "Free". For the bonded interactions, the components on any connected interface will be treated as if they are welded together. For the free interactions, the analysis separates the assembly on the connected interfaces into different sub-systems or several individual components for frequency analysis. Several examples with step-by-step instructions are presented to demonstrate how to run natural frequency analysis.

5.1 Natural Frequency Analysis

This chapter will show how to use SolidWorks Simulation to estimate the first five natural frequencies of components and an assembly to prevent them from resonance.

Supplementary Information The online version contains supplementary material available at https://doi.org/10.1007/978-3-031-64132-9_5.

Mechanical vibrations refer to the repetitive motions or the oscillations of a mechanical system about its equilibrium position. These vibrations can always occur in various mechanical systems because they always have some type of moving components. The main sources for vibrations include repetitive functions, external forces, sudden impacts, and unbalanced rotating components. Some effects of mechanical vibrations are metal fatigue due to cyclic stresses, rubbing, fretting, or loosening due to repetitive relative motion, and possible resonance which must be prevented.

The natural frequency of a mechanical component or system is the frequency at which the object will vibrate freely after hitting it. Natural frequency is an inherent feature, solely determined by material properties, geometry, and boundary conditions, and is independent of loading conditions. The unit of frequency is Hertz (a cycle per second). For a continuous component, there will be an infinite number of natural frequencies because the component has an infinite number of degrees of freedom. Natural frequencies are arranged according to their values from low to high, so the first natural frequency is the lowest value of frequencies.

Resonance is a phenomenon that occurs when an object or system is subjected to an external force or vibration with a frequency that is equal to one of its natural frequencies. The dynamic characteristics of the SDF (Single Degree of Freedom) system under harmonic excitation force are displayed in Fig. 5.1 [1]. The amplitude ratio M is the ratio of vibration amplitude to the static displacement due to the force amplitude of external excitation force. Frequency ratio r is the ratio of the external excitation frequency to the natural frequency. ζ is the damping ratio of the SDF system. From Fig. 5.1, if a system is resonant, the vibration amplitude approaches infinity when the damping ratio is near zero. This means that the system will be damaged or fail due to resonance. So, one of the main design tasks for mechanical engineers is to prevent resonance for their designs.

Fig. 5.1 Dynamic characteristics of the SDF system under harmonic excitation force

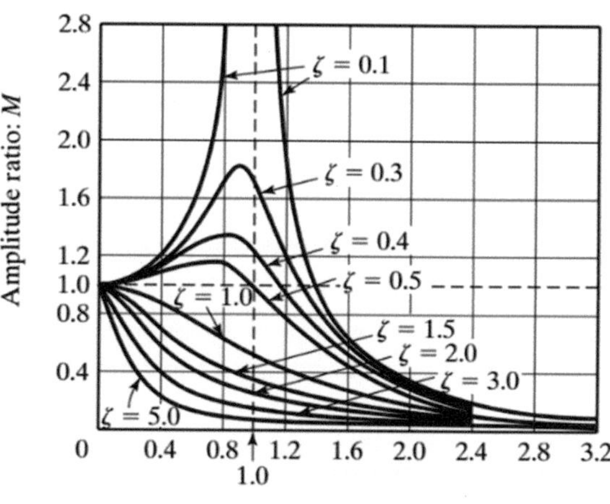

To prevent a resonance, we need to know the natural frequencies of an object. The natural frequency can be theoretically calculated using several methods such as Rayleigh's method, Holzer's method, the Reyleigh-Ritz method, the matrix iteration method, and so on [1]. They can also be determined with experimental methods by frequency analysis, measurement instruments, signal analysis, experimental modal analysis, and so on [1]. Of course, the natural frequencies can also be estimated through FEA simulations.

SolidWorks Simulation has a "Frequency" module, which can be used to estimate the natural frequency of components and assemblies. The general procedure for frequency analysis using the "Frequency" module will be presented and explained in Example 1. It is also worth knowing that for frequency analysis on an assembly, there are only two types of contacts: "Bonded" and "Free". For "Bonded" contact, components on an interface will be welded together. For "Free" contact, there is no interaction between components on an interface. Now we will use two examples to demonstrate how to run frequency analysis.

Example 1: Frequency analysis of a component (FEA40) Download and open FEA40, which is an alloy steel plate. Run FEA simulations to estimate the first five natural frequencies in two cases. In Case One, there is no fixture. In Case Two, one end is fixed as shown in Fig. 5.2.

Solution

(1) FEA simulation

Step one: Create a study by using the "Frequency" module.
Since a very coarse mesh can be used for frequency analysis with reasonable results, it typically isn't required to run pre-processing on the model. The model will be meshed by

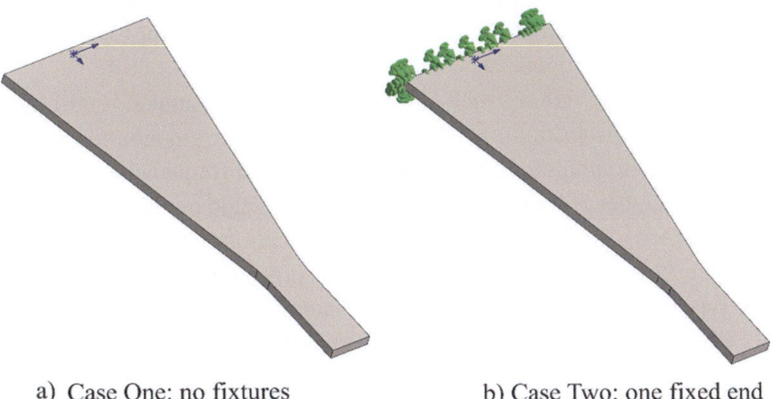

a) Case One: no fixtures   b) Case Two: one fixed end

Fig. 5.2 Two cases of frequency analysis

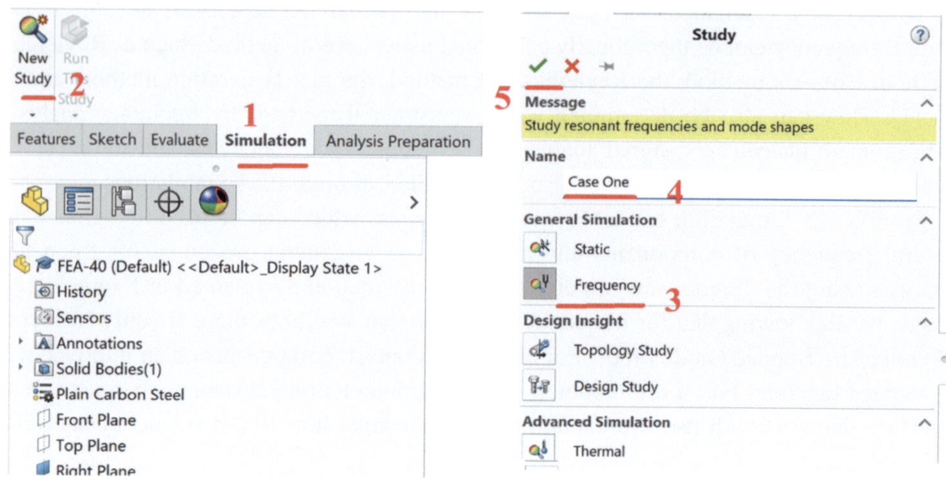

Fig. 5.3 Create a study (simulation project)

solid elements with a default global element size. Follow the procedure shown in Fig. 5.3 to create a frequency analysis study. (1) Click the "Simulation" tab in the toolbars, (2) Click "New Study" and the "Study" Property Manager will appear; (3) Select the "Frequency" module; (4) Enter the name of a study; (5) Finally, click the check mark to complete a creation of a new study.

Step two: Define the first five natural frequencies.

One of the main purposes of natural frequency analysis is to prevent resonance. For mechanical components or mechanical systems, the external excitation frequency is typically not very high. So knowledge of the first five natural frequencies is generally good enough for mechanical design. The default number of natural frequencies in FEA simulation is 5 as shown in Fig. 5.4. In this example, Case Two has a fixed end, so for Case 2 we don't need to change the number of frequencies.

When an object has no fixture, such as Case One in this example, it will have six rigid motions including three transitions and three rotations. These rigid motions will have almost zero frequency in FEA simulation and they are not natural frequencies. However, we will need to define the number of frequencies as the number of rigid motions plus 5 to get the first five natural frequencies. In this case, there are 6 rigid motions and 5 natural frequencies, so we need to define 11 frequencies in total. Follow this procedure to make such modifications as shown in Fig. 5.4. (1) Right-click the study name (Case one) in the simulation study tree; (2) Select "Properties". The "Frequency" Property Manager window will appear; (3) Change the default number of frequencies from 5 to 11; (5) Finally, click the "Ok" button to complete the modification of this simulation's properties.

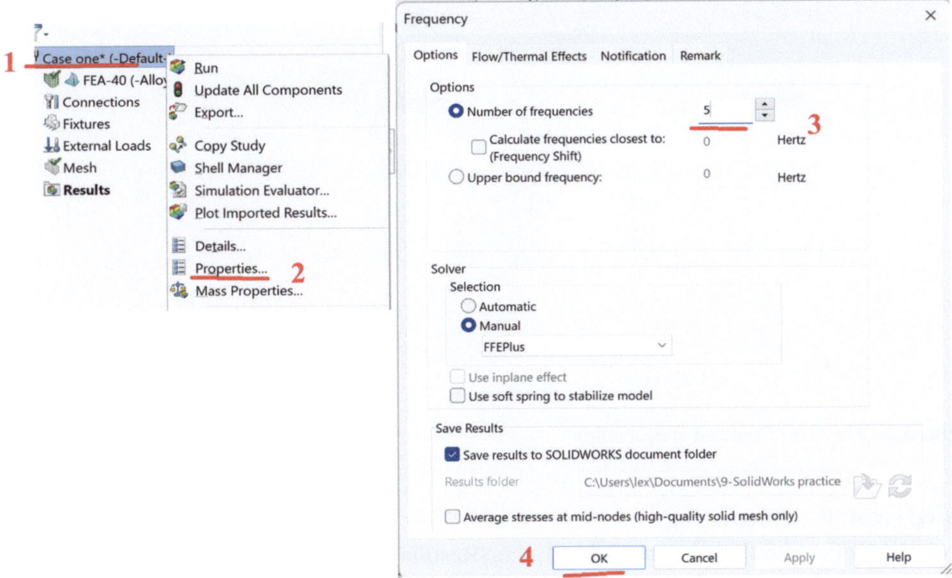

Fig. 5.4 Define the number of frequencies

Step three: Define the type of material and use solid elements for meshing.

We need to assign a type of material to the component. Since a very coarse meshing will be good enough for frequency analysis, components can be always meshed by solid elements. Typically, we just use the default global element size for meshing.

Step four: Define connections.

For frequency analysis, the type of contacts on interfaces in SolidWorks Simulation will be "Bonded" or "Free" only, the concepts of which are the same as those in Chap. 4.

For frequency analysis on a component, there are no interfaces. So, in this example, there is no need to define connections. The connections for an assembly will be discussed in Example 2 of this chapter.

Step five: Define fixtures.

We can use the same technique described in Chap. 3 to define fixtures. Even though there is an "External loads" tab in the simulation study tree, we don't need to specify loads because natural frequencies are independent of loadings.

Step six: Meshing

Since we are only interested in the first 5 natural frequencies, we don't need to have fine meshes or apply mesh control. Typically, we just use the default mesh parameter for meshing.

Step seven: Run simulation

Right-click the tab of the study name and select "run" to run FEA simulations.

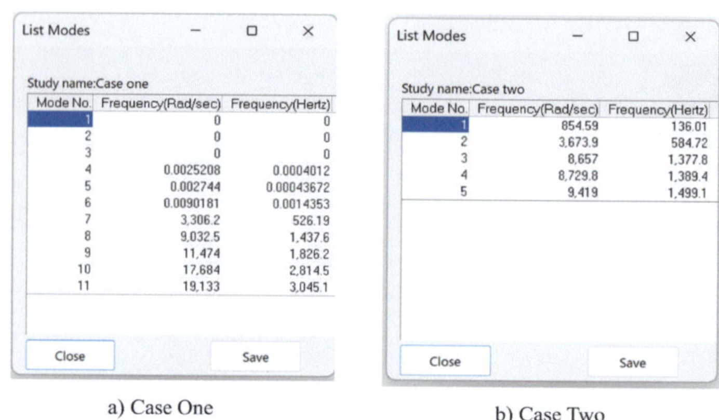

<table>
<tr><td colspan="3">

List Modes   — □ ×

Study name:Case one

Mode No.	Frequency(Rad/sec)	Frequency(Hertz)
1	0	0
2	0	0
3	0	0
4	0.0025208	0.0004012
5	0.002744	0.00043672
6	0.0090181	0.0014353
7	3,306.2	526.19
8	9,032.5	1,437.6
9	11,474	1,826.2
10	17,684	2,814.5
11	19,133	3,045.1

Close   Save

</td><td colspan="3">

List Modes   — □ ×

Study name:Case two

Mode No.	Frequency(Rad/sec)	Frequency(Hertz)
1	854.59	136.01
2	3,673.9	584.72
3	8,657	1,377.8
4	8,729.8	1,389.4
5	9,419	1,499.1

Close   Save

</td></tr>
</table>

a) Case One   b) Case Two

Fig. 5.5 The list of natural frequencies

Step eight: Post-processing

After the FEA simulation is completed, the "Results" tab will appear. Right-click the "Results" tab and select "List Resonant Frequencies". A table of natural frequencies will appear. The lists of natural frequencies for Case One and Case Two are displayed in Fig. 5.5a and b, respectively.

(2) **Interpretations and discussions**

For Case One, where there are no fixtures, the first six frequencies listed in Fig. 5.5a are almost zero because these are rigid motions. Therefore, the first natural frequency is 526.19 (Hz), which is listed in the row for Mode No. 7.

For Case Two where there is one fixed end, the first natural frequency is 136.01 (Hz), which is listed on the row for Mode No. 1.

From Fig. 5.5a and b, the natural frequencies of the same components with the same type of material but different fixture conditions are quite different.

Example 2: Frequency analysis on an assembly (FEA36-ASM) Download, unzip, and open FEA36-ASM, which consists of two components: the top plate (FEA36-1) and the support column (FEA36-2). The material of the top plate is plain carbon steel. The material of the support column is alloy steel. The fixtures of this assembly are shown in Fig. 5.6. The left end of the top plate is fixed. The bottom end of the support column has no normal displacement. The interface between the top plate and the support column can be set according to two cases: Case One—Bonded and Case Two—Free. Run an FEA simulation to estimate the first five natural frequencies.

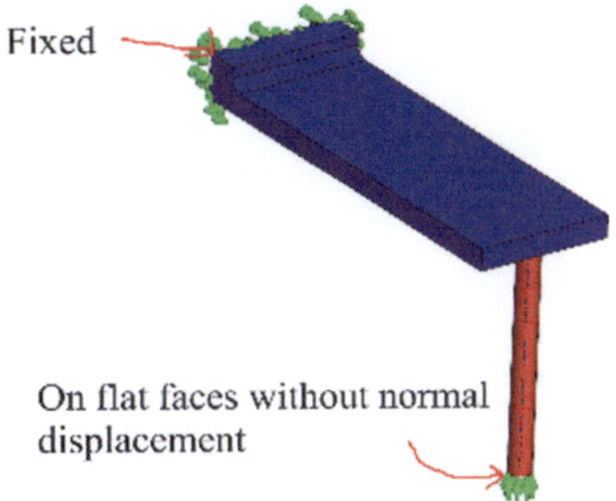

Fig. 5.6 Schematics of fixtures of this assembly

Solution

Follow the procedure described in Example 1 except for some differences in Step Four—Define connections. For frequency analysis on an assembly, the "Frequency" module in SolidWorks Simulation provides two types of contacts: "Bonded" and "Free" as shown in Fig. 5.7.

For Case One—Bonded, we define the interface as a bonded contact by using "Local Interactions" as shown in Fig. 5.7. Alternatively, we could just use the default global interaction, which treats a touching interface as a bonded interface. For this case, the number of frequencies in Step 2 will be 5.

In Case Two—Free, when the interface is defined as a type of free contact, this free contact will separate the assembly model into two different sub-models in this example. Therefore, we will just run frequency analysis on the top plate and the support column,

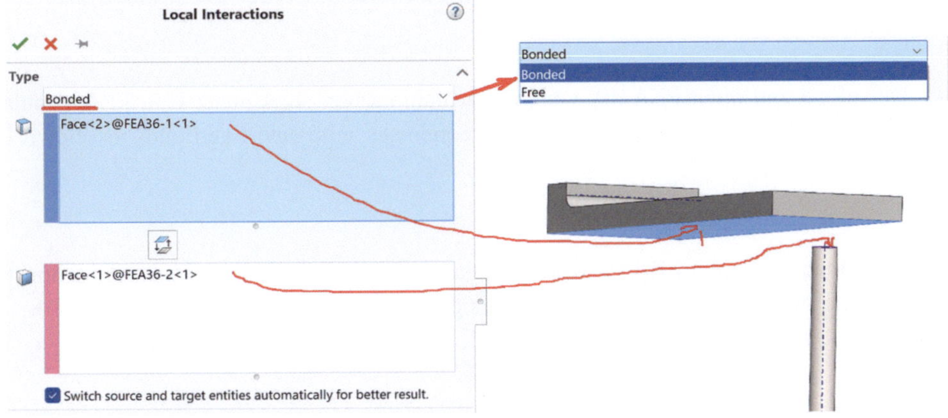

Fig. 5.7 Define a bonded interface for Case One

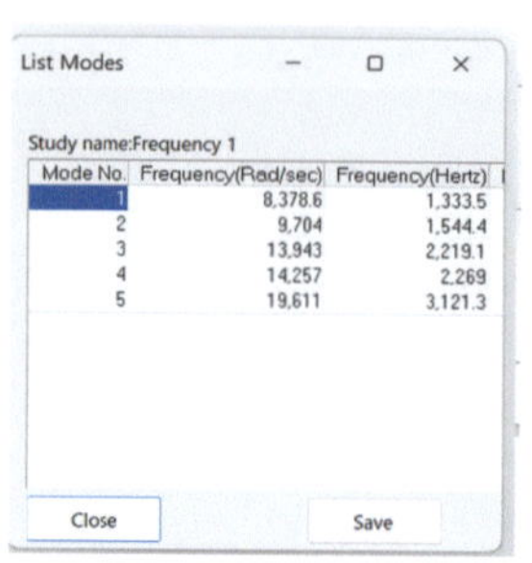

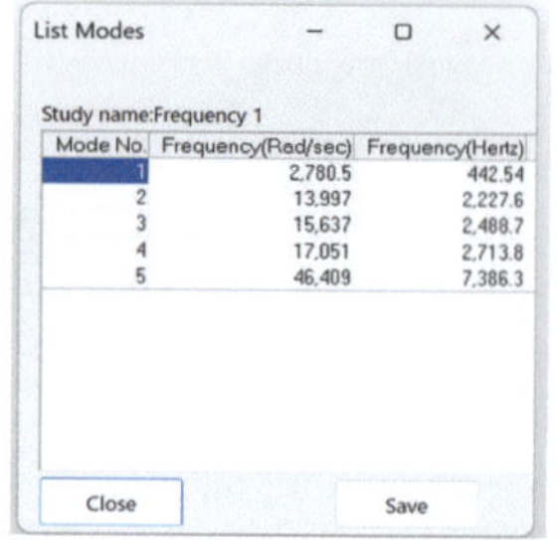

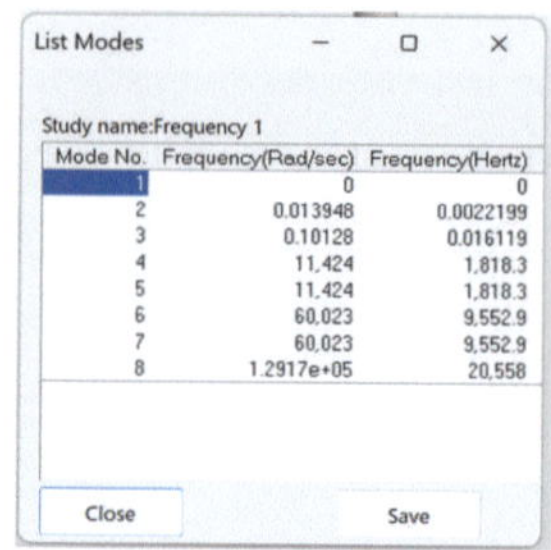

| a) Case One | b) The top plate in Case Two | c) The support column in Case Two |

Fig. 5.8 Lists of natural frequencies

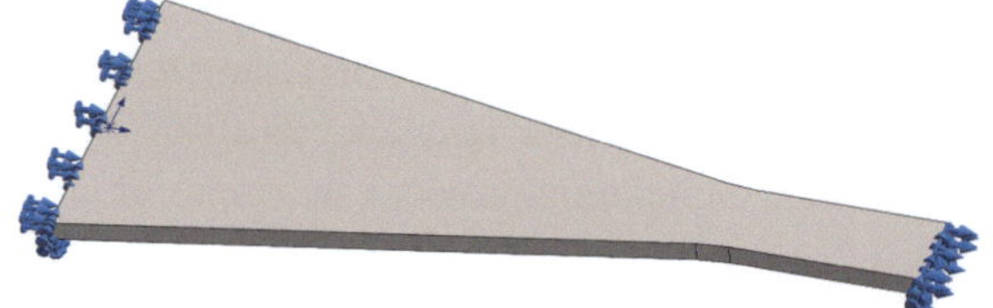

Fig. 5.9 Schematics of the fixtures

separately. For the top plate in Case Two, the number of frequencies in Step Two will be 5. But for the support column in Case Two, the number of frequencies in Step Two will be 8 because there are three rigid motions.

The lists of the first five natural frequencies of this example are displayed in Fig. 5.8.

Exercises

5-1 Download and open FEA-40, which is a plain carbon steel plate. Run FEA simulations to estimate the first five natural frequencies with two fixed ends as shown in Fig. 5.9.

5-2 Download and open FEA41, which is made of Alloy steel. Run FEA simulation to estimate the first five natural frequencies on the two cases as shown in Fig. 5.10.

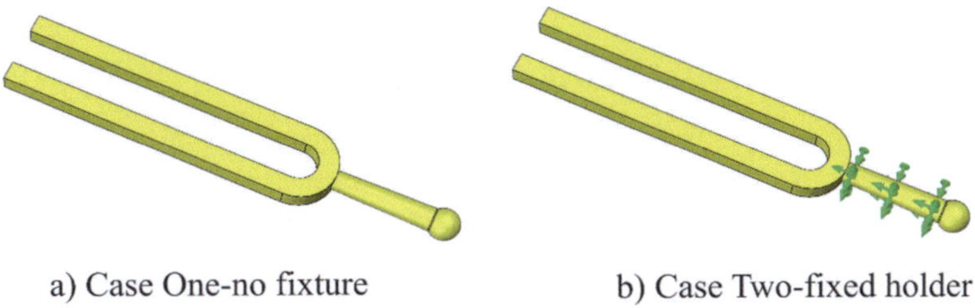

a) Case One-no fixture   b) Case Two-fixed holder

Fig. 5.10 Schematics of fixtures for two cases

References

1. S.S. Raon (2017) Mechanical Vibrations, 6th edition, Prentice Hall
2. SolidWorks Web Help, https://help.solidworks.com

Thermal Analysis and Thermal Stress Analysis

6

Abstract

Stress of mechanical components can be induced by external forces or/and by a change in temperature. Thermal stress of a component is induced when the component cannot be freely expanded or contracted upon a temperature change, or when there is a non-uniform distributed temperature. This chapter discusses and demonstrates how to obtain a temperature distribution and then how to simulate thermal stress by Solid-Works Simulation. For thermal stress analysis, we must use the "Thermal" module in SolidWorks Simulation to obtain the temperature distribution under thermal loads. Then, we need to use the "Static" module to run a thermal stress analysis, where the temperature distribution obtained will be used as a thermal load for thermal stress analysis. In this chapter, we will first discuss the three heat transfer mechanisms: conduction, convection, and radiation. Then, the general procedure for thermal analysis is presented and explained in detail. Examples of steady thermal analysis and transient thermal analysis are presented to demonstrate how to use the "Thermal" module to run thermal analysis. Finally, we use the "Static" module in SolidWorks Simulation to run thermal stress analysis, where the temperature distribution is used as thermal load.

6.1 Thermal Analysis and Thermal Stress Analysis

Stress of mechanical components can be induced by external forces or/and by a change in temperature. Thermal stress of a component is induced when the component cannot freely expand or contract upon a temperature change, or when there is a non-uniform distributed

Supplementary Information The online version contains supplementary material available at https://doi.org/10.1007/978-3-031-64132-9_6.

temperature. This chapter will briefly discuss how to obtain a temperature distribution and how to simulate thermal stress using SolidWorks Simulation.

To determine the thermal stress of a component, we must determine its temperature distribution by using the principles of heat transfer. For mechanical component design, the exchange of thermal energy between physical systems is governed by three fundamental heat exchange mechanisms: conduction, convection, and radiation [1].

Conduction is a heat transfer mechanism on a microscopic scale in which thermal energy transfers from one point to another through the interaction between the atoms or molecules of matter. Conduction occurs in solids, liquids, and gases. Conduction is the most significant means of heat transfer within a solid or between solid objects in thermal contact. The heat transfer by conduction obeys Fourier's law, which states that the rate of heat conduction $q_{conduction}$ is proportional to the heat transfer area (A) and the temperature gradient (dT/dx), that is,

$$q_{conduction} = -kA\frac{dT}{dx} \tag{6.1}$$

where k is thermal conductivity. Its unit is $\frac{W}{m^0 k}$ in SI unit system or $\frac{Btu/s}{in^0 F}$ in the English unit system. Thermal conductivity k can be constant but is typically expressed as a function of temperature T.

Convection is a heat transfer mechanism in which heat transfers between a solid face and its environment through random molecular motion, or by bulk/macroscopic fluid motion. The rate of heat exchange $q_{convection}$ between a fluid of temperature T_f and a face of a solid of area A at a temperature T_s obeys Newton's law of cooling which can be written as:

$$q_{convection} = hA\left(T_s - T_f\right) \tag{6.2}$$

where h is the convection heat transfer coefficient. Its unit is $\frac{W}{m^{2 0}K}$ in SI unit system or $\frac{Btu}{Sin^{2 0}F}$ in the English unit system. The convection heat transfer coefficient h can be constant but is typically expressed as a function of temperature T.

Radiation is a heat transfer mechanism due to the emission of thermal energy in the form of electromagnetic waves. All bodies with temperatures above absolute zero emit thermal energy. The rate of heat exchange $q_{radiation}$ from a hot object with a temperature T_h and an area A_h to a cooler surrounding with a temperature T_c can be expressed with Stefan-Boltzmann Law as:

$$q_{radiation} = \varepsilon \sigma A_h(T_h - T_c) \tag{6.3}$$

where σ is the Stefan-Boltzmann constant. ε is the emissivity coefficient, which is a dimensionless material property in the range of $0 < \varepsilon < 1$ depending on the type of material and temperature of the surface. The emissivity coefficient ε can be treated as a constant but can also be expressed as a function of temperature.

Fig. 6.1 Two models for thermal analysis and thermal stress analysis

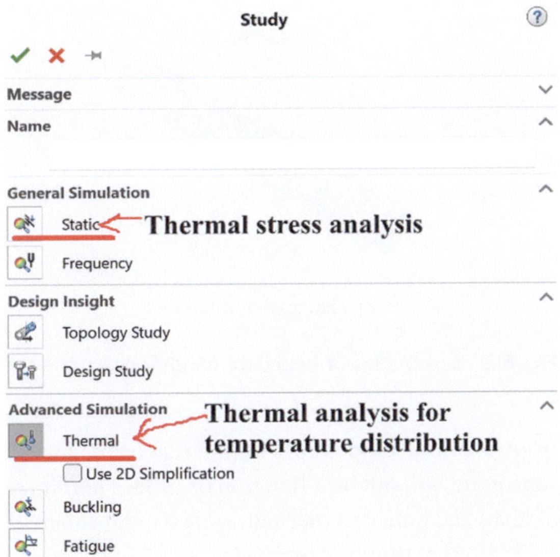

Temperature distribution and the corresponding thermal stress of a component can be theoretically analyzed or solved by FEA simulation. In SolidWorks Simulation, thermal analysis for temperature distribution and thermal stress are completed by using two different modules as shown in Fig. 6.1. First, we will use the "Thermal" module to obtain the temperature distributions, which can be for a steady state or a transient state. Then, we will use the "Static" module to calculate thermal stress, which has been discussed in Chaps. 3 and 4. For thermal stress simulation, the steady or transient temperature distribution will be used as an external load. Now, we will use two examples to demonstrate how to run thermal analysis and thermal stress analysis by using SolidWorks simulation.

Example 1: A bar for steady-state analysis (FAE42) Download and open part FEA42. The material of this part is plain carbon steel. The boundary conditions for thermal analysis and thermal stress analysis are displayed in Fig. 6.2a and b, respectively. The right end has a temperature of 200 °F. The convection coefficient on the cylindrical surface is 1.16E−5 BTU/(s-in^2-F). The ambient temperature is 80 °F. The left end can be treated as an isolated surface. Run a steady-state thermal analysis and thermal stress analysis.

Solution

(1) FEA simulation

(a) Thermal analysis

The main task for thermal analysis is to determine temperature distribution. For a thermal system with heat energy, heat energy flows from a high-temperature area to a low-temperature area through different heat transfer mechanisms. The temperature at any point will change

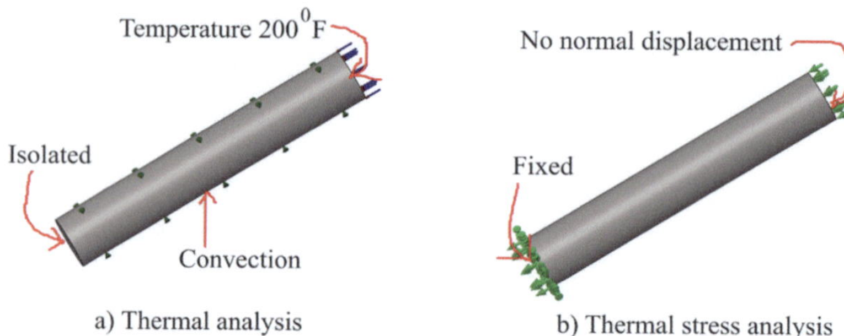

Fig. 6.2 Schematics of boundary conditions

from its initial temperature until it reaches a steady state, that is, the temperature at a steady-state point will not be a function of time. Therefore, for thermal analysis, there are two types of analysis: transient thermal analysis and steady-state thermal analysis.

Transient thermal analysis is used to estimate the temperature at any point as a function of time. For transient thermal analysis, we need to specify the initial temperature, solution time, and time increment.

Steady-state thermal analysis is to estimate the temperature at any point in a steady state, which is a constant, and not a function of time.

The procedure for thermal analysis is presented and explained here.

Step 1: Create a thermal analysis study and select the type of analysis (steady or transient).

First, create a thermal analysis study as shown in Fig. 6.3. (1) Click the "Simulation" tab in the toolbars, (2) Click "New Study" and the study dialog window will appear; (3) Select the "Thermal" module; (4) Enter the name of a study; (5) Click the check mark to complete the creation of a new study.

We need to specify the type of thermal analysis as shown in Fig. 6.4. (1) Right-click the tab of the project name in the simulation study tree. (2) Select the "Properties" tab. The "Thermal" dialog window will appear. (3) If the thermal analysis is a transient thermal analysis, select "Transient", and enter the total time and time increment. (4) If the thermal analysis is a steady state thermal analysis, select "Steady state". (5) Finally, click "OK" to complete the definition of the thermal analysis.

In this example, we will run a steady thermal analysis.

Step 2: Define material properties.

We need to assign a type of material to each component. For thermal analysis, some material properties will need to be defined and customized. For example, most analyses on computer chips will use custom material properties. We can use the descriptions in Ch 3.6 to define a custom material. The custom material properties for thermal analysis should include elastic modulus, Poisson's ratio, thermal conductivity, thermal expansion coefficient, specific heat,

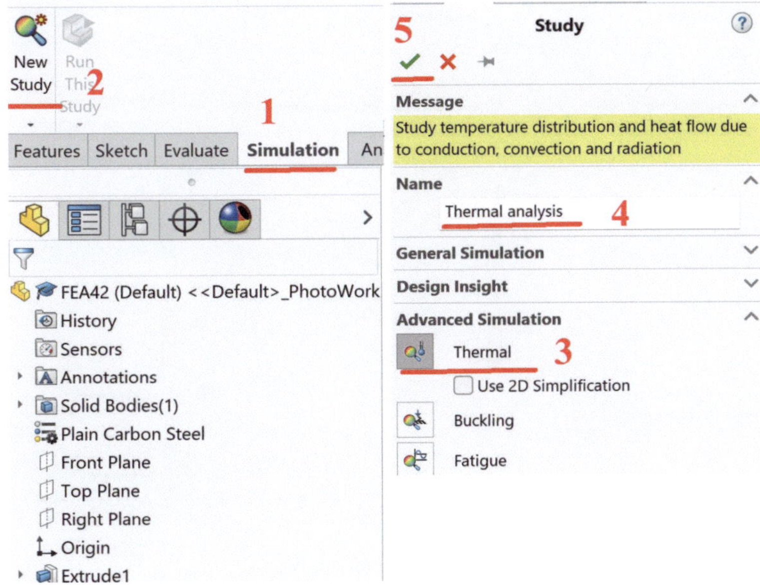

Fig. 6.3   Create a thermal analysis study

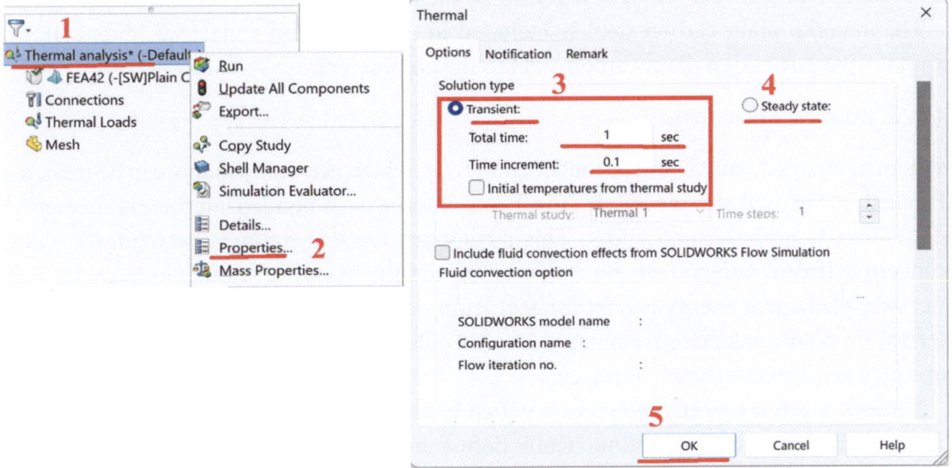

Fig. 6.4   Define the type of thermal analysis

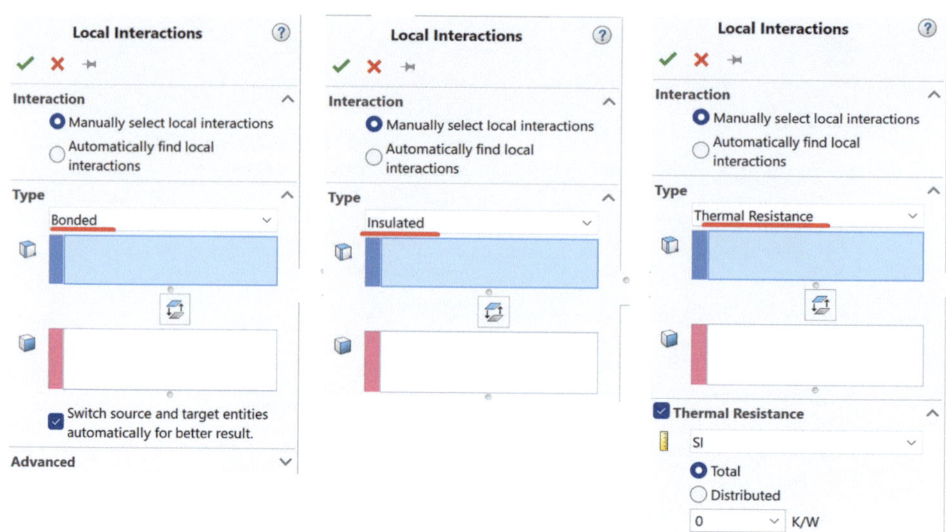

Fig. 6.5 Three types of contact interfaces for thermal analysis

and mass density. For further thermal stress analysis, additional material properties such as yield strength or ultimate strength will also be included.

The material plain carbon steel is included in the SolidWorks material library. In this example, we just assign plain carbon steel to the component.

Step 3: Define connections.

In thermal analysis, contacting or touching interfaces between components can be treated as "Bonded", "Thermal resistance", or "Insulated" interfaces. A bonded interface is an interface where there is no heat energy loss. This means that the temperatures between the shared points of different components on the interface are the same. An insulated interface is an interface where heat energy cannot transfer from one component to another component. A thermal resistance interface is an interface where there is some heat energy loss due to tiny gaps between two contacting surfaces.

There is a default global interaction which is automatically generated for an assembly. This global interaction will automatically define any touching interface as a bonded interface. We typically do not make any changes to the global interaction. We can use the tool "Local Interactions" to manually define a type of thermal contact, which will overwrite the definitions in the default global interaction.

Right-click the "Connections" tab in the simulation study tree and select "Local Interactions". The "Local Interactions" dialog window will appear. Then we can select one type of thermal contact (Bonded, Insulated, or Thermal Resistance) to define the interface as shown in Fig. 6.5.

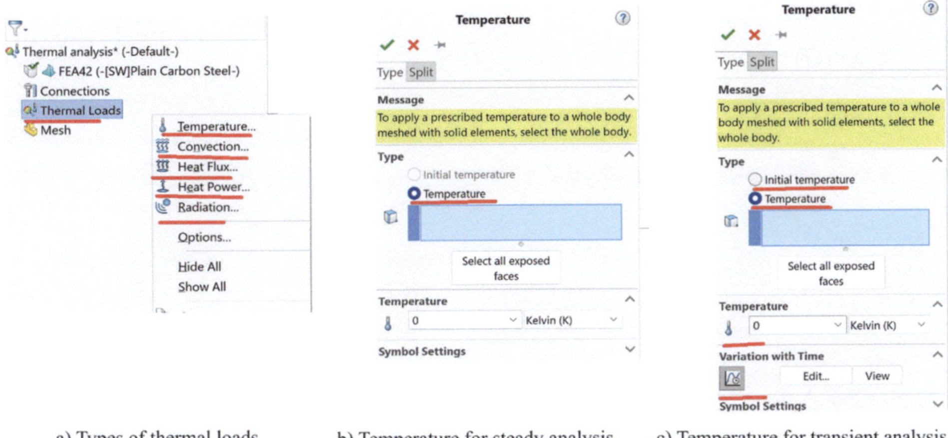

a) Types of thermal loads   b) Temperature for steady analysis   c) Temperature for transient analysis

Fig. 6.6   Types of thermal loads and temperature as a thermal load

This example is a thermal analysis for one component. We don't need to specify any connections.

Step 4: Thermal loads.

Thermal loads include temperature, convection, heat flux, heat power, and radiation as shown in Fig. 6.6a.

Isolation is a type of thermal load. Heat energy cannot be transferred in or out on the surface of a component through convection. When a surface is not selected for convection, this surface is defined as an isolated surface.

Temperature is a type of thermal load that can be applied on a point, a surface, or a component. Right-click the "Thermal loads" tab and Select "Temperatures". The "Temperature" dialog window will appear. For a steady state thermal analysis as shown in Fig. 6.6b, the option "Temperature" is a thermal load and a constant, which can be applied on a point, a surface, or a component.

For a transient thermal analysis as shown in Fig. 6.6c, the option "Initial temperature" is a constant defining an initial condition and is applied to a component. The option "Temperature" is a thermal load applied on a point, a surface, or a component. It can be a constant or a function of time. If temperature is a function of time, we click "Edit" to bring out a table where we can define temperature vs. time.

Convection is a type of thermal load. It is also a boundary condition. Right-click the "Thermal loads" tab and select "Convections". For a steady thermal analysis, the "Convection" dialogue window is shown in Fig. 6.7a. Convection can be applied on the selected surfaces. The convection coefficient can be a constant or a function of temperature. When

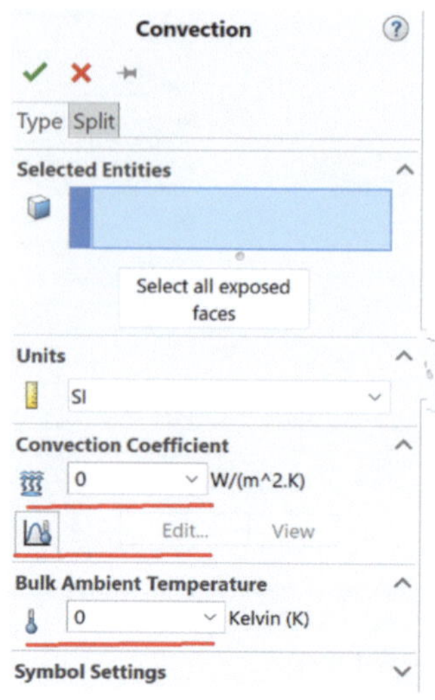
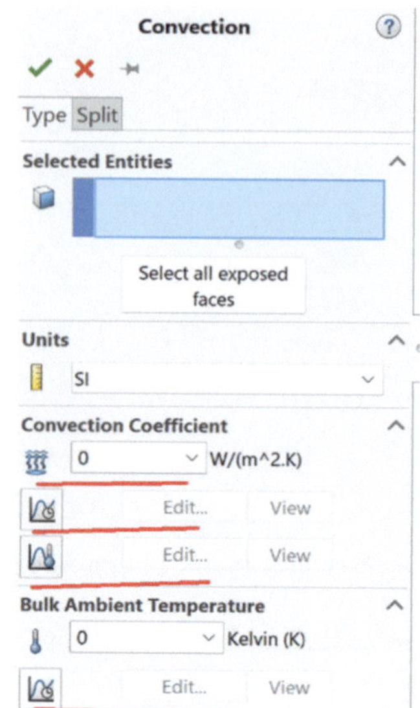

a) Convection for steady analysis   b) Convention for transient analysis

Fig. 6.7 Convection as a boundary condition

we click "Edit", we can use a table to define the convection coefficient vs. temperature. The ambient temperature is a constant.

For a transient thermal analysis, the convection dialogue is shown in Fig. 6.7b. The convection can be applied on the selected surfaces. The convection coefficient can be a constant, a function of temperature, or a function of time. The ambient temperature can be a constant or a function of time. When we click a corresponding tab "Edit" in the dialogue window, we can define the corresponding functions.

Heat flux is the rate at which heat is transferred per unit area, per unit time, through an object or substance, usually from a hotter to a colder area. Heat flux is a thermal load. Right-click the "Thermal loads" tab and select "Heat flux". For a steady thermal analysis, the "Heat Flux" dialogue window is shown in Fig. 6.8a. Heat flux is applied on the selected surfaces. It can be a constant or a function of temperature.

For transient thermal analysis, the "Heat Flux" dialogue window is shown in Fig. 6.8b. Heat flux is applied on selected surfaces. It can be a constant, a function of temperature, or a function of time.

Heat power is a thermal load. Right-click the "Thermal loads" tab and select "Heat power". For a steady thermal analysis, the "Heat Power" dialogue window is shown in Fig. 6.9a. Heat power can be applied on a point, an edge, a surface, and a component. It can be a constant or a function of temperature.

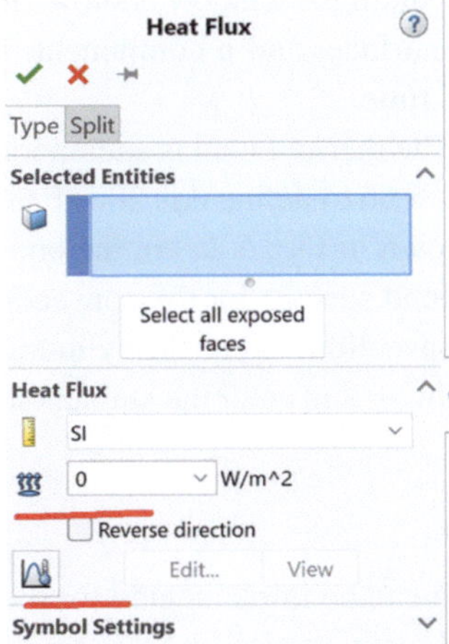

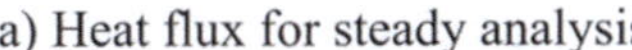

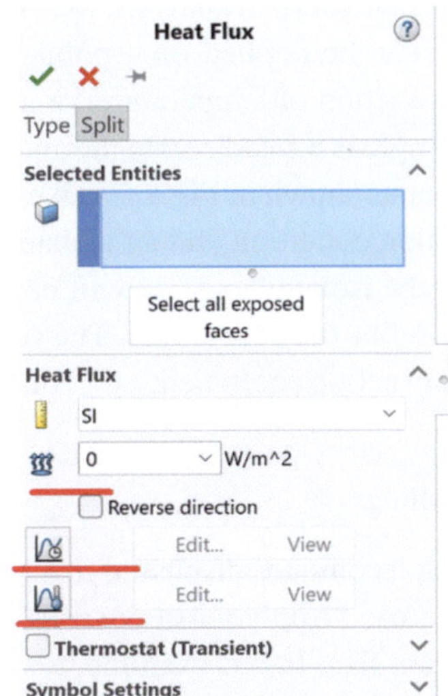

a) Heat flux for steady analysis   b) Heat flux for transient analysis

Fig. 6.8   Heat flux as a thermal load

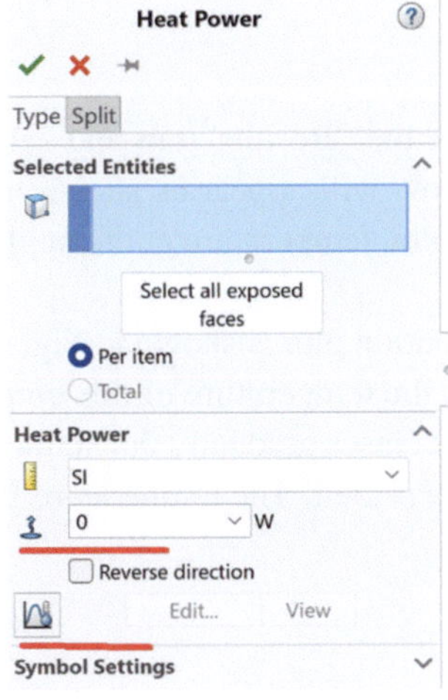

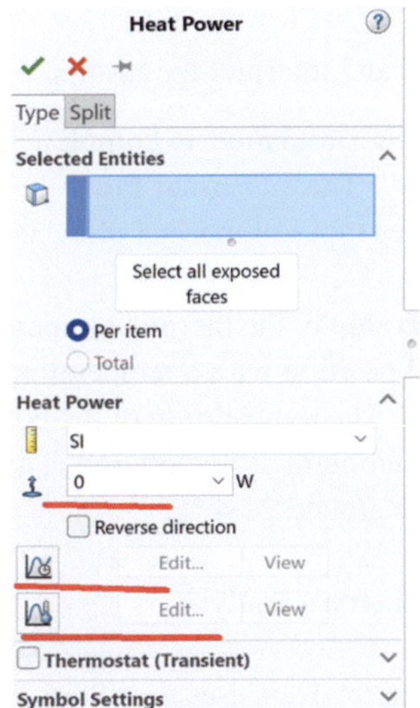

a) Heat power for steady analysis   b) Heat power for transient analysis

Fig. 6.9   Heat power as a thermal load

For a transient thermal analysis, the "Heat Power" dialogue window is shown in Fig. 6.9b. Heat power can be applied on a point, an edge, a surface, and a component. It can be a constant, a function of temperature, or a function of time.

This example is a steady-state thermal analysis. The thermal load is a temperature at one end of the bar as shown in Fig. 6.2a. We can use Fig. 6.6b to define this 200 °F temperature. The convection condition and an isolated end as shown in Fig. 6.2a are the boundary conditions. For the isolated end, we will not select this end surface for the convection. We use Fig. 6.7a to define the convention. The surface for convection will be the cylindrical surface. Enter the convection coefficient of $1.16E-5$ BTU/s-in^2-F and enter the ambient temperature of 80 °F.

Step 5: Meshing.

The meshing techniques discussed in Chaps. 3 and 4 can be implemented for thermal analysis. We can use "Apply mesh controls" for a local element size and standard mesh for a global element size. In this example, we will use a standard mesh and a global element size of 0.2".

Step 6: Run Simulation.

Right-click the tab of the project name and select "Run" to run the FEA simulation. We can also click the toolbar "Run this study" to run the FEA simulation.

Step 7: View and interpret the results.

After the FEA simulation is completed, right-click the "Results" tab and select "Define Thermal plot". The "Thermal Plot" dialogue window will appear as shown in Fig. 6.10. There are three types of thermal plots: temperature plot, temperature gradient plot, and heat flux plot.

For this example, the thermal temperature distribution plot is shown in Fig. 6.11. At the thermal load position with a temperature of 200 °F, the temperature of this component has its maximum. The temperature of the component decreases gradually due to the convection and has its minimum at the left end. This result is expected. The temperature increases on the left end is around 21.4 °F.

(b) **Thermal stress analysis**

Thermal stress analysis uses the "Static" module in SolidWorks Simulation. The procedure for thermal stress analysis is the same as the procedure for static analysis, except that the load in thermal stress analysis is a temperature distribution. The techniques discussed in Chaps. 3 and 4 can be directly implemented for thermal stress analysis.

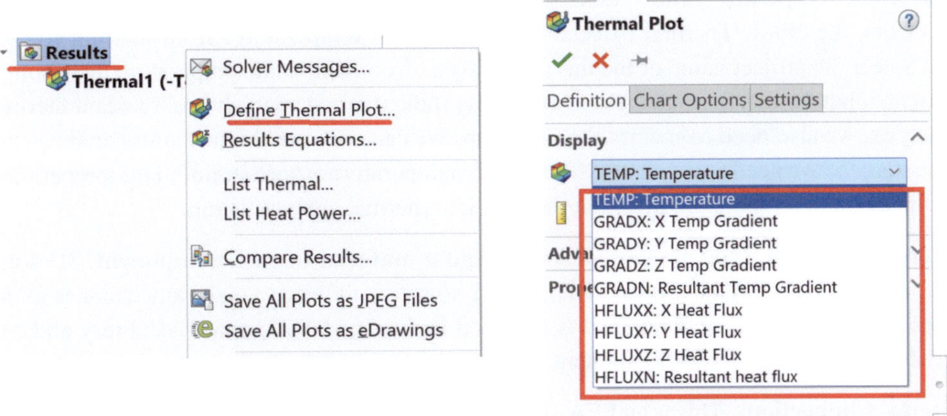

Fig. 6.10 Define a thermal plot

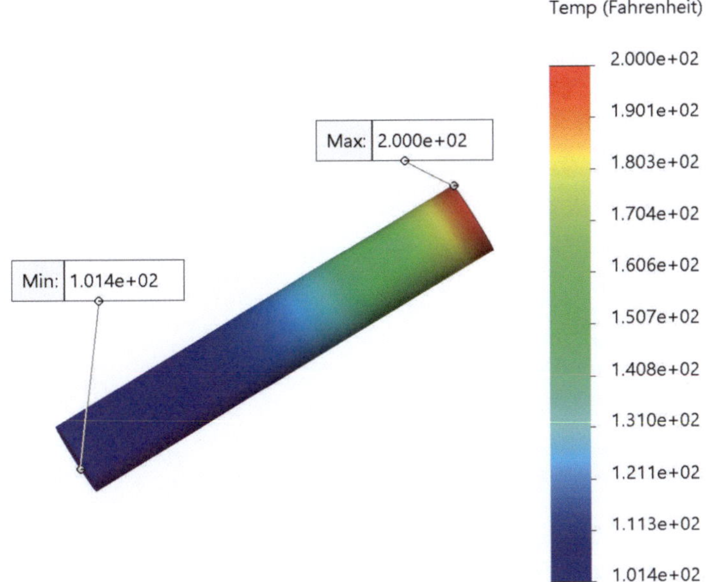

Fig. 6.11 Temperature distribution plot

Step 1: **Pre-processing**. The model FEA42 is ready for FEA simulation.

Step 2: Set up a project. This is a static stress analysis. After we set up a static stress analysis project, we need to specify where the temperature distribution will be input. After the project of a thermal stress analysis is created, right-click the tab of the project name, and select "Properties". The "Static" Property Manager will appear as shown in Fig. 6.12. (1) Click the "Flow/Thermal Effects" tab. (2) Select "Temperatures from thermal study". (3) Select the project name of the thermal analysis from which the temperature distribution will be input for the thermal stress analysis. (4) If the thermal analysis is a transient thermal analysis, we also need to specify the time step. We can only run thermal stress analysis per one step. (5) We need to specify the "Reference temperature at zero strain". This temperature is the ambient temperature at which the transient thermal analysis is run.

Step 3: **Select the type of element and assign a material to each component**. 3D solid elements with high-quality mesh will be used, so we don't need to make any changes to the type of element. Plain carbon steel is included in the SolidWorks material library and can be directly assigned to the component.

Step 4: **Connections**. This is an FEA simulation on a component. There are no connections.

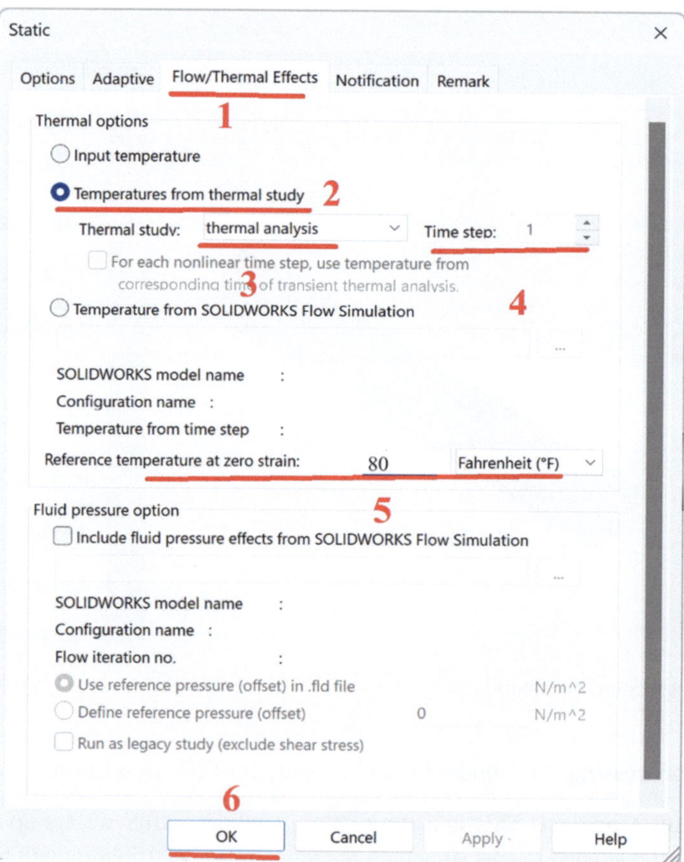

Fig. 6.12 Input temperature as a thermal load

Step 5: Fixtures. Right-click the "Fixture" tab and select "Fixture geometry" to apply the fixed end on the left end of the component. Right-click the "Fixture" tab and select "Advanced Fixture". Then select "On Flat faces" on the right end of the component and specify "No normal displacement".

Step 6: Loads. The thermal load has been specified in Step 2 as shown in Fig. 6.12. Component stress can be the combination of stress due to external forces and stress due to thermal temperatures. But in this example, thermal load is the only load.

Step 7: Meshing. We will still use the global standard element size of 0.2″. Since the left end of the component is fixed, there might be high stress induced due to the temperature change. We will the "Use mesh control" option on this fixed surface to set a local element size of $0.2''/5 = 0.04''$.

Step 8: Run Simulation. Right-click the project name and select "Run". This will start the FEA simulation.

Step 9: Post-processing.

The Von Mises stress plot of this component due to thermal load is displayed in Fig. 6.12a. The factor of safety with red areas under the value of 1.5 is displayed in Fig. 6.13b.

(2) **Interpretations and discussions**

Based on Fig. 6.11 the temperature change at the left end of the component is about 21.4 °F with a temperature increase from the ambient temperature of 80 °F to 101.4 °F. The temperature change at the right end of the component is 120 °F, which is the maximum temperature change. But from Fig. 6.13a, the maximum stress happens at the left end of the component.

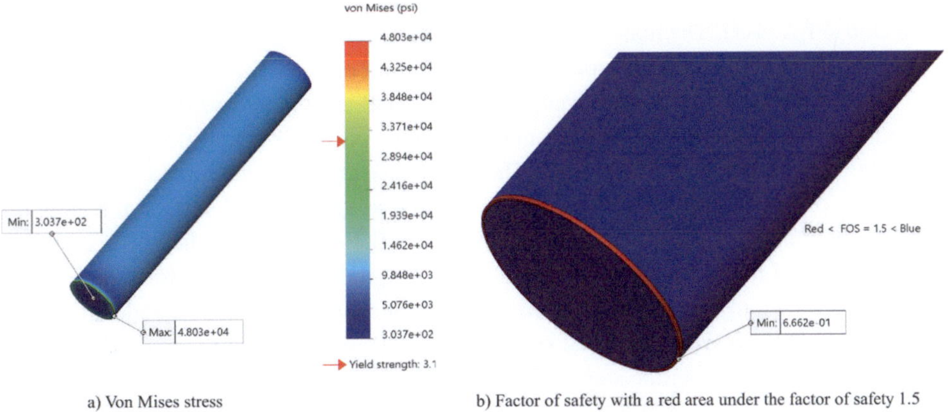

a) Von Mises stress   b) Factor of safety with a red area under the factor of safety 1.5

Fig. 6.13 Von Mises stress plot and factor of safety plot

Table 6.1 Material properties of a chip

Property	Value (SI units)	Property	Value (SI units)
Elastic modulus (EX)	4.1e11 (N/m^2)	Yield strength	4.2E8 (N/m^2)
Poisson's ratio (NUXY)	0.3	Thermal conductivity (KX)	200 (W/(m.k)
Mass density (DENS)	1250 (kg/m^3)	Specific heat (C)	670 J/(Kg.K)
Compression strength	6.6E8 (N/m^2)	Thermal expansion coefficient (ALPX)	1E$-$6 (K^{-1})
Tensile strength	4.9E8 (N/m^2)		

Table 6.2 Heat power as a function of time

Time in second	% of 0.4 W	Time in second	% of 0.4 W
0	0	90	0.99997
30	0.9682	120	1
60	0.99899	900	1

The explanation is that the left end of the component is fixed, and its material is not allowed to freely expand or contract, while the right end of the component is only restrained along the normal direction, allowing its material to freely expand or contract along the radial direction.

From Fig. 6.13b, the minimum factor of safety on the left end of the component is only 0.67. If we assume that the factor of safety of this component is 1.5, the component will fail along the edge of the left end surface.

This example tells us that when there is a significant temperature change in an area, the fixed support in this area is not a good choice.

Example 2: An assembly for transient thermal analysis (FEA43-ASM). Download, unzip, and open FEA43-ASM, which consists of two components: a substrate (FEA34-1) and a chip (FEA34-2). The material of the substrate is ceramic porcelain. The chip material properties are defined using a custom material and listed in Table 6.1. The interface between the base and the chip for thermal analysis and thermal stress analysis can be treated as a bonded interface. The chip generates a maximum heat power of 0.4 Watts. The heat power is a function of time listed in Table 6.2. The heat is dissipated by the forced air-convection from every outer surface of the substrate and chip (except the fixed surface for mounting chip-base assembly). The convection heat transfer coefficient (Film coefficient) is 25 W/(m^2 K) and the bulk (ambient) temperature is 300 K. The fixture conditions for thermal stress analysis are shown in Fig. 6.14. When creating the mesh, use 0.5 mm as the global element size for thermal analysis and thermal stress analysis. Run a transient thermal analysis with a total time of 900 s and a time step of 30 s. Generate a temperature distribution plot at step 30 and a temperature response of any point.

Fig. 6.14 Fixture for thermal
stress analysis

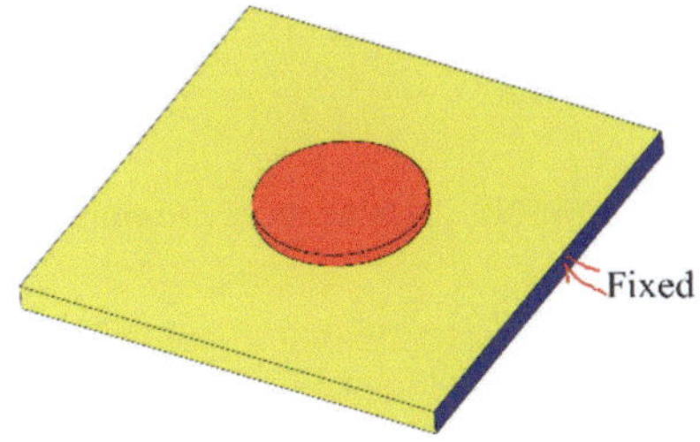

Solution

(1) **Transient thermal analysis**

Step 1: Create a thermal analysis study and select type of analysis

First, follow Fig. 6.3 to create a thermal study by using the "Thermal" module. Then follow Fig. 6.4 to define a transient thermal analysis with a total time of 900 s and a time step of 30 s as shown in Fig. 6.15.

Step 2: Define material properties

The ceramic porcelain material for the substrate (FEA43-1) is included in the SolidWorks material library. It is in the category "Other Non-metals". However, the chip material properties need to be defined using a custom material. We can follow Figs. 3.45 and 3.46 in Ch3.6 to create a custom material as shown in Fig. 6.16.

Step 3: Define Connections.

Since the bonded interface has been defined in the global interactions, we don't need to specify it again.

Step 4: Thermal loads.

In this example, we need to specify the initial temperature for all the components. We also need to specify the convection. These are boundary conditions. Finally, we need to specify the heat power of the chip which is a function of time. We can follow the descriptions of step 4 in example 1 of this chapter.

When we define the initial temperature for transient thermal analysis, we must select the components. After the "Temperature" dialogue window is shown and the initial temperature is selected, we can then select components for the feature manager tree as shown in Fig. 6.17.

In this example, we could select each outer surface for a convention. However, we can check "Select all exposed faces" and then unselect the fixed surface as shown in Fig. 6.18. Then, input the convection coefficient of 25 W/(m^2 K) and the bulk (ambient) temperature of 300 K as shown in Fig. 6.18.

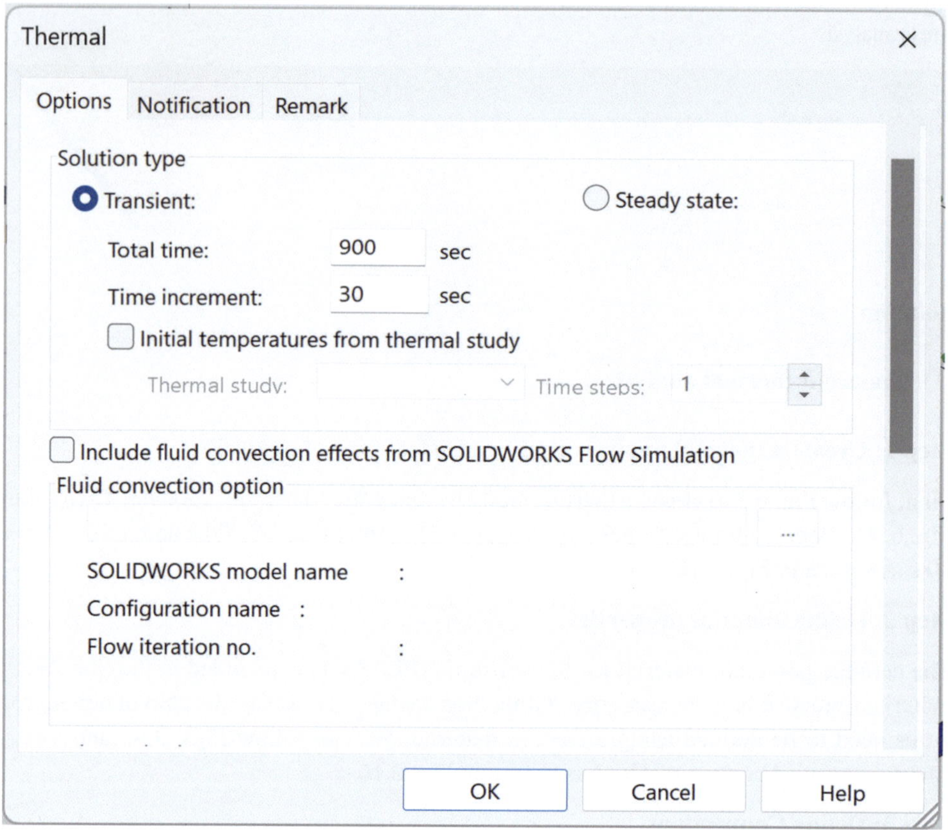

Fig. 6.15 The settings for a transient thermal analysis

In this example, the heat power is applied to the chip component. We need to select the component FEA43-2 in the feature manager tree as shown in Fig. 6.19. Since the heat power is a function of time. We select "function of time" and then click "Edit". This will bring up a table for a function of time. To add a new row to the table, double-click the first column of the table.

Step 5: Meshing. We will use 0.5 mm as the global element size for meshing.

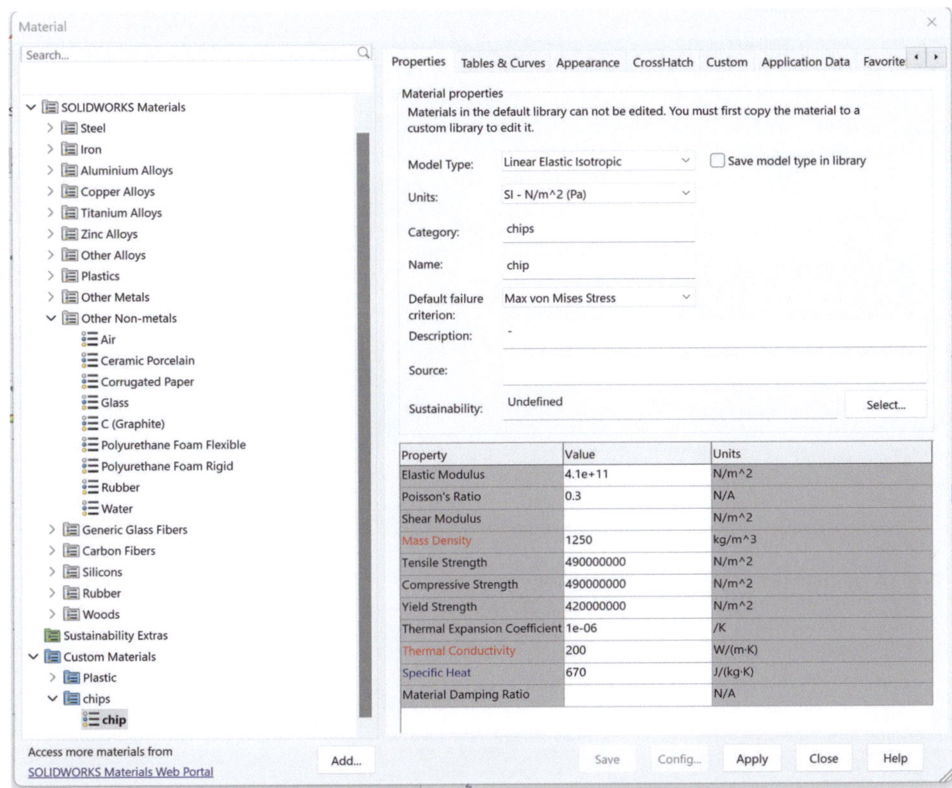

Fig. 6.16   A custom material chip

Step 6: Run Simulation. Right-click the tab of the project name and select "Run" to run the FEA simulation. We can also click "Run this study" on the toolbar to run the FEA simulation.

Step 7: View and interpret the results.

For a transient thermal analysis, we can show the temperature distribution plot at each time step. The temperature distribution at step 30 is shown in Fig. 6.20. We can get a response at any point, which is the temperature of a point vs. time. After the temperature distribution plot has been created, right-click the tab of this plot and select "List selected". This will bring up the "Probe" dialogue window as shown in Fig. 6.21. Follow this procedure to create a response to a point. (1) Select "On selected entities"; (2) Select a point on the temperature plot; (3) Click "Update", which will update the result for the selected point; (4) Select "Response". This will generate the response on the selected point. One response is shown in Fig. 6.22.

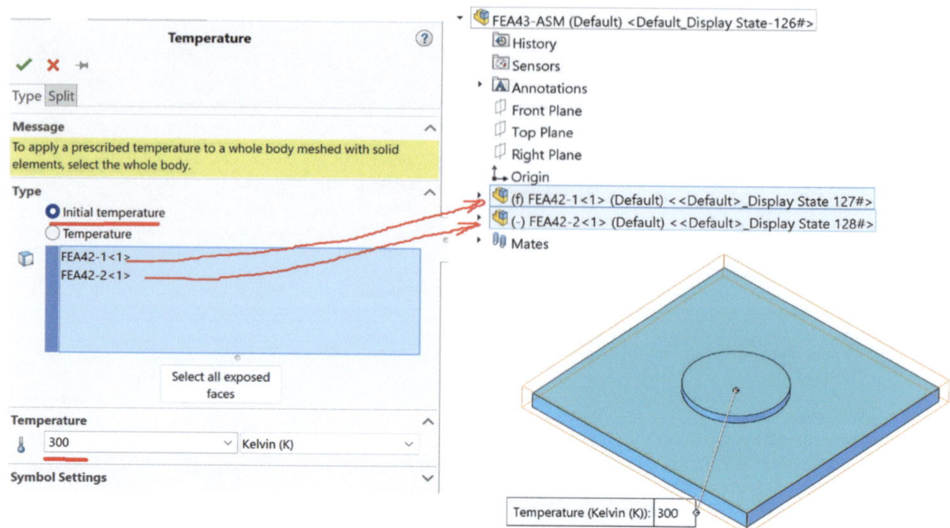

Fig. 6.17 Define the initial temperature for components

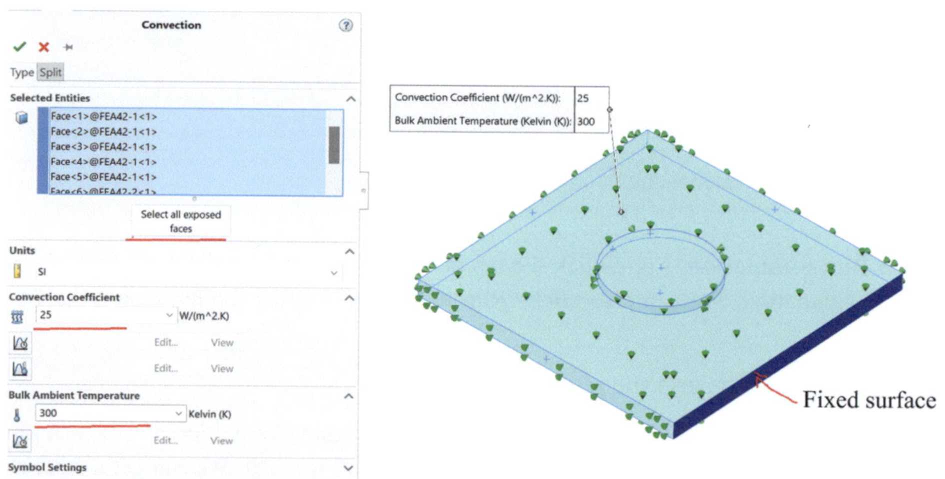

Fig. 6.18 Defining a convection

(2) **Interpretations and discussions**

The heat power in this example reaches its maximum at the time 120 s as shown in Table 6.2. But from Fig. 6.22, the temperature of that point reaches its steady state at around 360 s (the 12th time step). After that, the temperature of this point is constant.

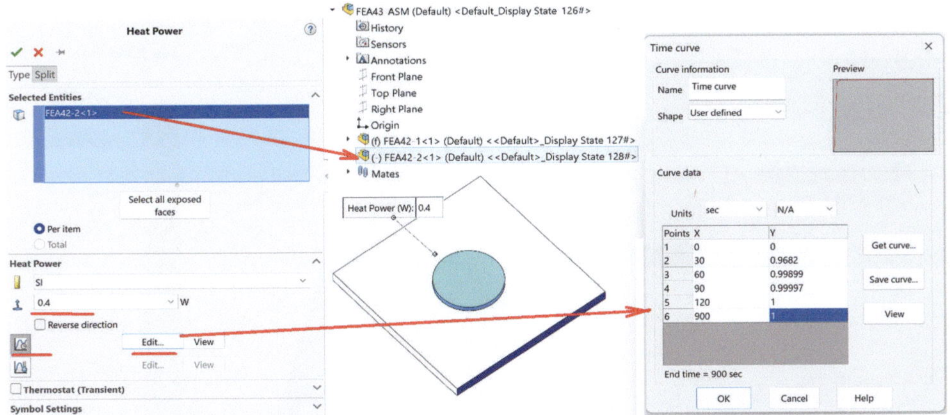

Fig. 6.19   Define heat power as a function of time

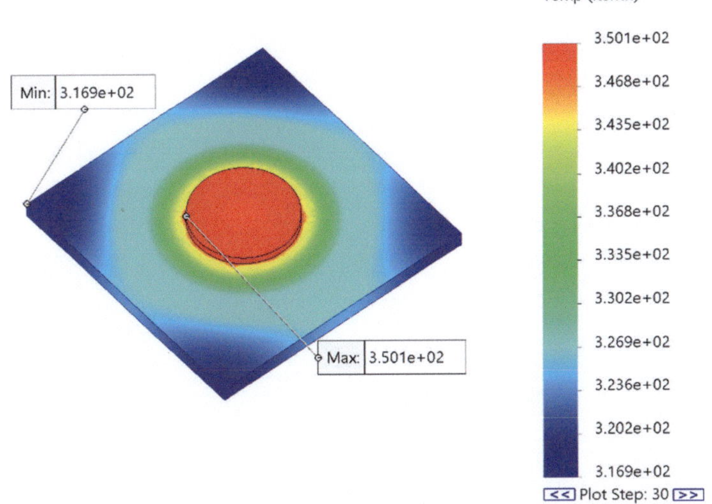

Fig. 6.20   The temperature distribution plot at the time step 30

After the transient thermal analysis is completed, we can run a thermal stress analysis by using the thermal temperature distribution of any step. Follow the procedure shown in Example 1 and we can complete this thermal stress analysis.

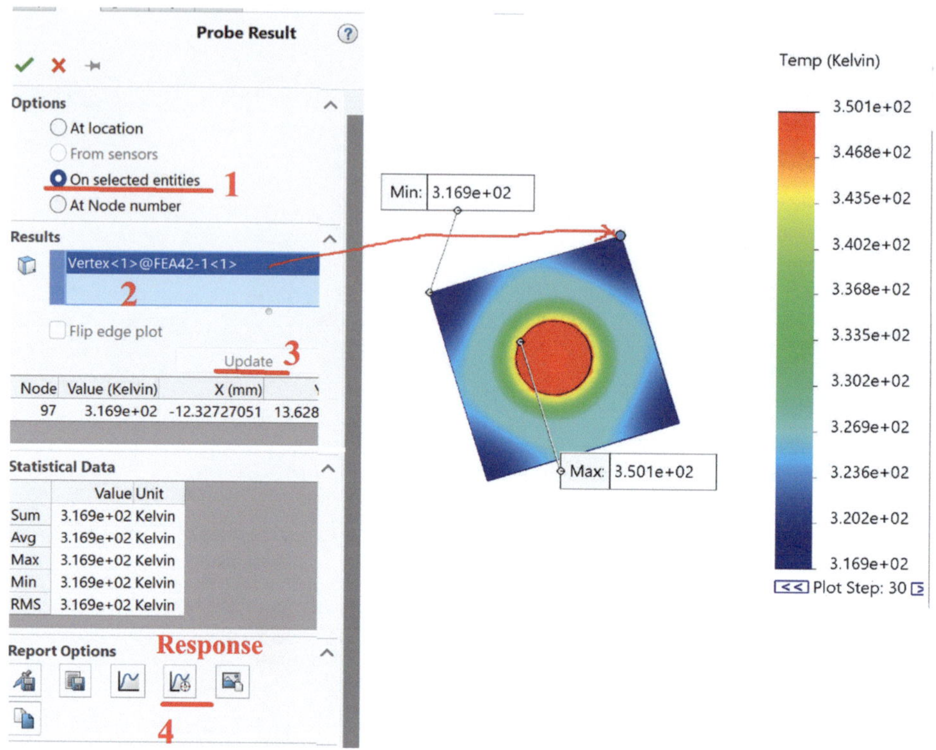

Fig. 6.21   Create a response curve

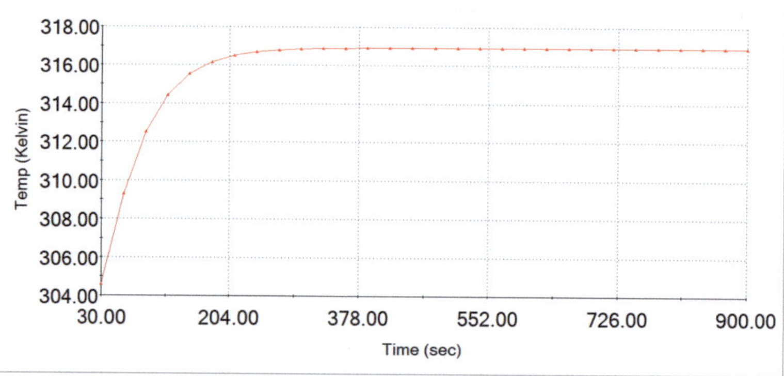

Fig. 6.22   A response at a point

Exercises

6-1 Run a transient thermal analysis of Example 1 (FEA42) of this chapter. The total time
     is 300 seconds, and the time step is 15 seconds. Show a temperature distribution plot
     and a response of any point. Then run thermal stress analysis by using the temperature
     distribution at the time step 20. Show the von Mises stress distribution and a factor of
     safety plot with red areas under the value of 1.5. (Note: Download FEA42 and then
     use this to create an assembly that contains only one part (FEA42) for a transient
     thermal analysis. In this way, the initial temperature can be defined.)

6-2 Run a steady-state thermal analysis of Example 2 (FEA43-ASM) of this chapter.
     Use the heat power as a constant 0.4 W. Create a temperature distribution plot and
     compare the temperatures at four corners of the substrate with the data obtained from
     a transient thermal analysis of Example 2. Then run a thermal stress analysis. Create
     a Von Mises stress plot and determine the reaction force on the support.

References

1. T.L. Bergman, et. al., (2018) Fundamentals of Heat and Mass Transfer, 8th ed., Wiley, Hoboken,
    NJ
2. SolidWorks Web Help, https://help.solidworks.com

Fatigue Analysis by FEA Simulation

7

Abstract

Fatigue failure is the single largest cause of failures in metal components. Fatigue is a very complicated issue. Fatigue analysis by FEA simulation is a worthy approach for fatigue design. This chapter discusses and explains how to use SolidWorks Simulation to estimate fatigue damage of metal components. First, the fundamentals of metal fatigue are presented, which cover cyclic stress, its descriptions, the definition of fatigue, fatigue damage mechanism, fatigue test, typical S–N curves, modification factors, fatigue strength reduction factor, and the famous Miers' rule. The purpose of these is to gain a basic understanding of fatigue and be ready to run fatigue analysis by FEA simulation. Then, for fatigue damage analysis in SolidWorks Simulation, the "Static" module is used to run static stress analysis first. Typically, the absolute maximum load is used to run the static analysis to check whether components are safe. This simulation will provide the maximum stress at every point when a cyclic load is estimated. Finally, another "Fatigue" module is used to estimate the fatigue damage. Several examples with step-by-step instructions are provided to demonstrate how to simulate the fatigue damage.

7.1   Introduction

Many mechanical components are made of metal materials. If metal components fail, they typically fail due to fatigue. Fatigue failure is the single largest cause of failures in metal components, estimated to cause 90% of all metallic failures. Fatigue is a very complicated

---

Supplementary Information The online version contains supplementary material available at https://doi.org/10.1007/978-3-031-64132-9_7.

issue and fatigue theories are still under development. This chapter will discuss how to use SolidWorks Simulation to estimate fatigue damage of a metal component.

This chapter contains two sections. The brief descriptions are listed here.

Ch7.2 Fundamentals of metal fatigue: We will only discuss the fundamentals of metal fatigue. The purpose of this section is to gain a basic understanding of fatigue and be ready to run fatigue analysis by FEA simulation.

Ch7.3 SolidWorks Simulation for fatigue analysis: Two examples will be used to show how to estimate fatigue damage of a metal component through FEA simulation.

7.2   Fundamentals of Metal Fatigue

This section will discuss the fundamentals of fatigue theory such as cyclic stress (loads), the definition of fatigue, fatigue damage, the fatigue tests, the S–N curves, modification factors, and the Miner's rule for cumulative fatigue damage.

Cyclic stress:

Cyclic stress or fluctuating stress is defined as the repetitive and continuous fluctuation of stress at a point of a component. Mechanical devices or systems always have at least one moving component. Due to the repeated functions or stop-start process or mechanical vibration, mechanical components are typically subjected to cyclic stress. A typical example of cyclic stress is the stress of a rotating shaft under a bending moment.

A schematic of cyclic stress with a constant stress amplitude is depicted in Fig. 7.1. The maximum stress σ_{max} and the minimum stress σ_{min} of cyclic stress are the maximum and minimum values of the cyclic stresses. The stress amplitude σ_a and the mean stress σ_m of the cyclic stress are defined as:

$$\sigma_a = \frac{\sigma_{max} - \sigma_{min}}{2} \tag{7.1}$$

$$\sigma_m = \frac{\sigma_{max} + \sigma_{min}}{2} \tag{7.2}$$

The stress ratio, or load ratio S_r is defined as the ratio of the minimum stress σ_{min} to the maximum stress σ_{max}.

$$S_r = \frac{\sigma_{min}}{\sigma_{max}} \tag{7.3}$$

A cycle is defined as a period of a point to the adjacent tsame point. For example, a cycle is a period from the current maximum stress to the adjacent maximum stress as shown in Fig. 7.1. One cause for metal fatigue is due to a large number of cycles.

A cyclic stress with a constant amplitude is typically described by the stress amplitude σ_a, the mean stress σ_m, and the number of cycles n. One specific cyclic stress is $S_r =$

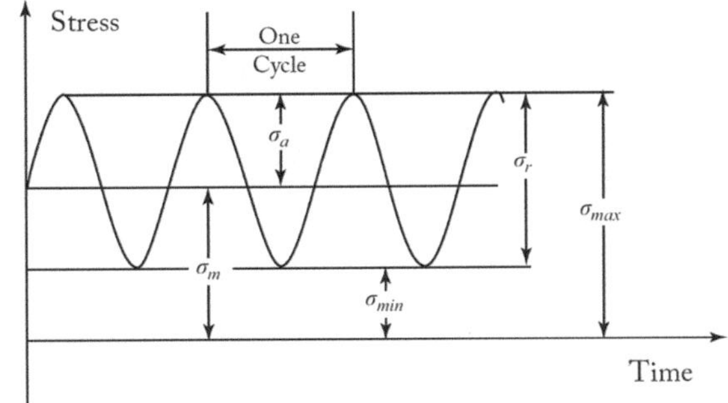

Fig. 7.1 A schematic of cyclic stress with a constant stress amplitude

$-1, \sigma_a = \sigma_{max} = -\sigma_{min}$ and $\sigma_m = 0$, which is called fully reversed cyclic stress or completely reverse cyclic stress.

Metal fatigue:

The failure mode of metal components under cyclic stress is completely different from the static failure under static load. The failure mode of metal components under cyclic stress is fatigue.

Metal fatigue is defined as "Failure under a cyclic stress, which never reaches a level sufficient to cause failure of components in a single application." Fatigue failure of a metal component under cyclic loading is due to cumulated fatigue damage. Fatigue damage is the weakening of a metal material due to a gradual crack propagation of inherent existing microscopic cracks or defects inside or on the surface of the metal component under repeated cyclic loading. Without a crack inside or on the surface of a metal component, there is no fatigue. If the magnitude of cyclic loading is not big enough to generate crack propagation, there will be no fatigue.

Fatigue damage mechanism:

The fatigue damage mechanism can be described and explained by its four features: Crack initiation, crack propagation, fracture due to static loading, and irreversible fatigue damage that is gradually accumulated [1].

(1) Crack initiation. There are always many randomly distributed defects inside a component, such as voids and dislocations, and on the surface of a component such as manufacturing scratches. Figure 7.2a shows a magnified microscopic crack on the surface of a component under a fully reversed cyclic bending moment. Assume that the nominal normal stress due to the bending stress is 20 ksi, and the material yield strength is 60 ksi. The bending moment tries to open the microscopic crack "A", which will have very high stress because of the sharp tip of the microscopic crack "A". Assume that the stress concentration factor in this situation is 3.5. So, the maximum stress at the tip of the microscopic crack "A" will be larger than the material

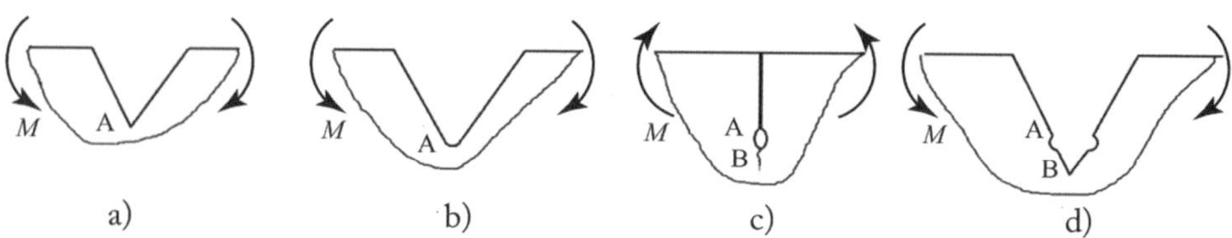

Fig. 7.2 Schematic of a gradual crack propagation [1]

yield strength. The material at the tip of crack "A" will start yielding and has some plastic deformation, as shown in Fig. 7.2b. When the material at the tip is yielding, the sharp tip of crack "A" will become a dull tip, and the stress concentration factor decreases. Let's assume that the stress concentration factor becomes 2.5. Now the maximum stress at the dull tip of the crack "A" is less than the material yield strength. Therefore, the yielding at the tip of the crack "A" will stop. If external loading is not a cyclic loading, the effect of the microscopic crack "A" on the component is negligible and can be ignored because the plastic deformation at the tip of the crack "A" is at a microscopic level.

(2) Crack propagation. When the component is under a fully reversed cyclic loading, the reversed bending moment now tries to close the microscopic crack "A", as shown in Fig. 7.2c. It is well known that when the dull tip of the microscopic crack "A" undergoes "strain hardening" due to the yielding, the material around the dull tip area will be brittle. After one or a few repetitions of such "open" and "close" actions, a new microscopic crack "B" will be generated beneath the crack "A". This result is a crack propagation, as shown in Fig. 7.2c. When the component is under a fully reversed cyclic loading, the microscopic crack "B" will be opened again as shown in Fig. 7.2d, which is the same situation as shown in Fig. 7.2a. In this stage, the crack could continue to grow because of continuously applied cyclic loading.

(3) Fracture due to static loading. Eventually, a crack will reach a critical size and the effective cross-sectional area of the component is so reduced that the actual stress on the effective area under a normal service cyclic load will be larger than the ultimate material strength, causing the component to rupture due to static loading.

(4) Fatigue damage is irreversible and will gradually accumulate. The fatigue damage is due to crack propagation. When the cyclic loading stops, the propagated microscopic cracks still exist. Therefore, the fatigue damage is irreversible and will gradually accumulate due to the continuous cyclic loading.

Fatigue test and standard specimen:

Material fatigue properties are obtained from fatigue tests. The most traditional and widely used method for fatigue analysis is the stress-life method, in which cyclic stress with a constant stress amplitude is continuously applied on a fatigue test specimen while the

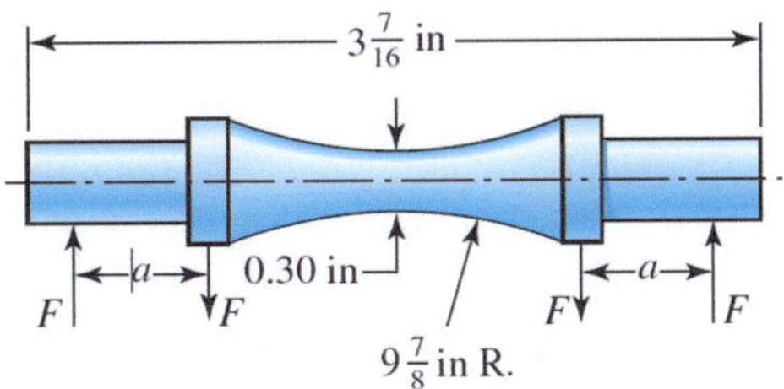

Fig. 7.3 A standard fatigue test specimen for rotating fatigue test [2]

number of cyclic stress cycles is counted until the fatigue test specimen fails. A standard fatigue test specimen for rotating fatigue tests is shown in Fig. 7.3. This specimen has a polished surface finish.

The S–N curves:

For rotating fatigue tests, the cyclic stress is a fully reversed cyclic stress. One rotating fatigue test will obtain one set of fatigue test results (S_f, N). Where S_f is called the fatigue strength, which is equal to the test stress amplitude of a fully reversed cyclic stress. N is called fatigue life at the fatigue strength S_f, which is the number of cyclic stress at failure. When a group of rotating fatigue tests are run at different cyclic stress amplitudes, the chart of plotted test results (S_f, N) can be obtained and is called the S–N curve as shown in Fig. 7.4.

Typically, the horizontal axis is in the logarithmic scale of the figure life N. The vertical axis is in the logarithmic scale of the fatigue strength S_f. The S–N curve presents the material fatigue properties and is used for fatigue design. On the S–N curve, any point will have one coupled fatigue strength data (S_f, N). The physical meanings of (S_f, N) are explained as the following.

Fig. 7.4 Schematics of a S–N curve

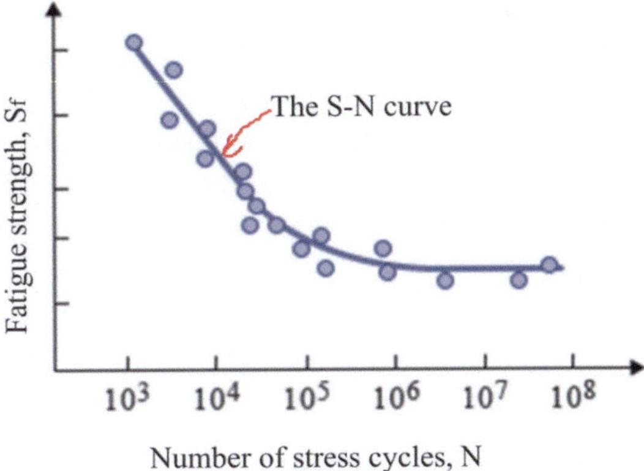

- S_f is the fatigue strength at the given fatigue life N. The design's physical meaning is that if the fatigue life is specified as N, the component will not fail when the actual cyclic stress amplitude is less than the fatigue strength S_f.
- N is the fatigue life under the cyclic stress amplitude equal to S_f. The design's physical meaning is that if the value of actual cyclic stress amplitude is equal to S_f, the component will not fail when the actual number of cyclic stress cycles is less than the fatigue life N.

Modification factors:

Material S–N curves are typically obtained from rotating fatigue tests on standard specimens. There are some significant differences between actual components and standard fatigue specimens such as geometric dimensions, type of cyclic stress, surface finish, and so on. Before material S–N curves can be used for mechanical component design, they must be modified to consider the differences between standard specimens and actual components. An example of such modification factors is Martin's modification factors [2]. More detailed information about them can be found in Chap. 6 Fatigue failure resulting from variable loading of the reference [2].

The Miner' rule:

The Miner's rule is a linear damage cumulative rule and one of the frequently used cumulative damage equations for failures caused by fatigue. The Miner's rule is expressed as:

$$\delta = \sum_{i=1}^{k} \frac{n_i}{N_i} = \begin{cases} \geq c & \text{Fatigue Failure} \\ < c & \text{safe} \end{cases} \tag{7.4}$$

Where δ is cumulated fatigue damage, which is a relative measurement index; n_i is the number of cycles at the cyclic stress level σ_i; N_i is the fatigue life from the component S–N curve at the cyclic stress level $S_f = \sigma_i$; c is the critical fatigue damage at fatigue failure and is determined by experiments. It is usually found in the range of $0.7 < c < 2.2$ with an average value of 1. k is the number of different cyclic stress levels.

The Miner's rule for fatigue design is commonly expressed as:

$$\delta = \sum_{i=1}^{k} \frac{n_i}{N_i} = \begin{cases} \geq 1 & \text{Fatigue Failure} \\ < 1 & \text{safe} \end{cases} \tag{7.5}$$

Example 1: Fatigue Damage

From a component S–N curve obtained from a fully reversed bending stress, three sets of fatigue strength properties are: $S_{f1} = 15(\text{ksi})$, $N_1 = 700,000(\text{cycles})$; $S_{f2} =$

27(ksi), $N_2 = 140,000$(cycles); $S_{f3} = 30$(ksi), $N_3 = 60,000$(cycles). The component is subjected to cyclic stress load which contains three different fully reversed bending stress levels: $\sigma_{a1} = 15$(ksi), $n_1 = 320,000$(cycles); $\sigma_{a2} = 27$(ksi), $n_2 = 40,000$(cycles); $\sigma_{a3} = 30$(ksi), $n_3 = 12,000$(cycles). Determine the cumulative fatigue damage. Is the component safe for fatigue??

Solution

Per the Miner's rule, the cumulative fatigue damage is:

$$\delta = \sum_{i=1}^{3} \frac{n_i}{N_i} = \frac{140000}{700000} + \frac{40000}{140000} + \frac{12000}{60000} = 0.686 = 68.6\%$$

Since the cumulative fatigue damage is 0.686, the component only consumes 68.6% of fatigue life and still has a remaining 21.4% of fatigue life. So, the component is safe for fatigue.◄

7.3 SolidWorks Simulation for Fatigue Analysis

One of the main failure modes for mechanical metal components is due to cumulative fatigue damage. There are lots of different fatigue theories. But most of them can only provide some estimation of fatigue life and the status of fatigue damage because fatigue damage is a very complicated phenomenon and fatigue analysis is still under development. SolidWorks simulation provides a module to estimate fatigue damage. It is worth trying it for fatigue analysis. This section will explain how to run fatigue analysis by SolidWorks Simulation.

In SolidWorks Simulation, fatigue analysis by FEA simulation needs to use two modules as shown in Fig. 7.5.

- The first module is "Static", which is used to get stress information for cyclic stress. Typically, the maximum value of the cyclic load is used for static analysis. This will create the absolute value of maximum stress at each point. This can also check whether a component is safe under this maximum static load. The techniques and skills discussed in Chaps. 3 and 4 can be directly implemented.
- The second module is "Fatigue" under constant amplitude event, which uses the Miner rule to estimate the cumulative fatigue damage at any point of a component. When the cumulative fatigue damage is less than 100%, the component is safe during the specified design life. If the cumulative fatigue damage is more than 100%, the component is said to be a failure because it will not have the specified design life.

Fig. 7.5 Two modules for fatigue analysis in SolidWorks Simulation

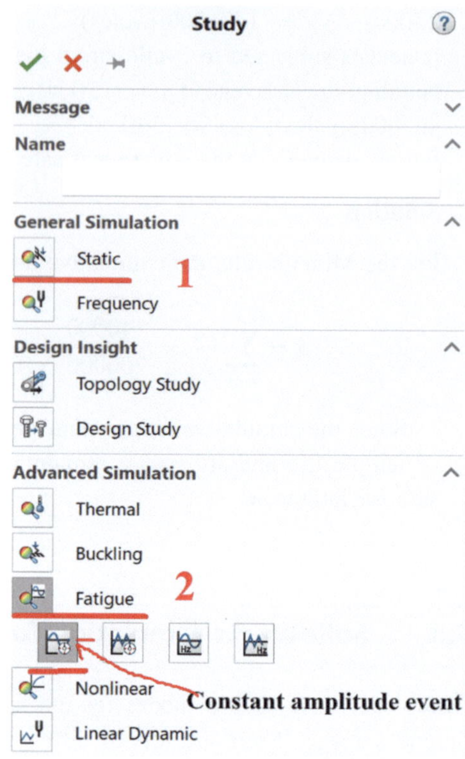

Now, two examples will be used to explain and demonstrate how to run fatigue analysis by SolidWorks simulation.

Example 2: Fatigue Analysis Under a Single Level of Cyclic Load (FEA44)

Download and open FEA44, which is an L-shape bar. The material of this machined component is AISI 1020 steel. Its top end is fixed and cyclic load with a minimum load of −200 lb and the maximum load of 300 lab with cycles of 75,000 (cycles) is applied on its right end as shown in Fig. 7.6. The required factor for the static analysis is 2.0. The factor of safety for fatigue analysis is 1.25.

(a) Run static analysis to show the Von Mises stress and the factor of safety distribution plots.
(b) Run fatigue analysis to show the fatigue damage plot.

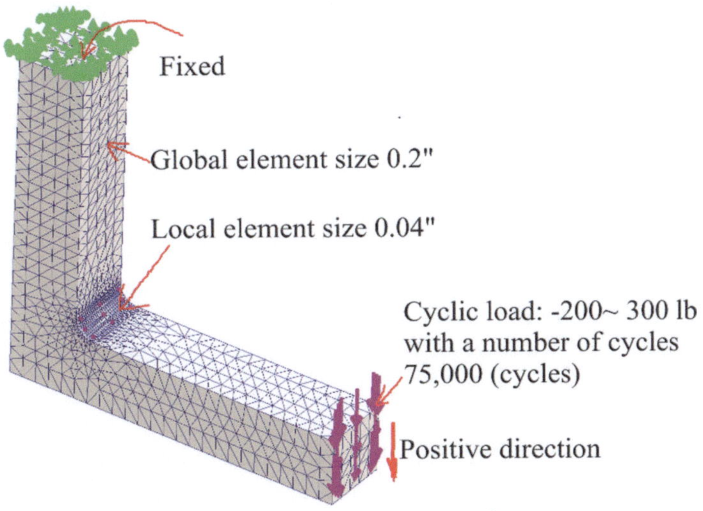

Fig. 7.6 The simulation settings of this example

Solution

(1) **FEA simulation**
(a) **The static analysis**

The main purpose of this static analysis is to provide information to calculate the maximum or minimum cyclic stress of any point for fatigue analysis. Since it is always assumed that stress is in an elastic region, the stress can be scaled up and down by a scale factor in fatigue analysis, which will be discussed later. So, any load value can be used for the static analysis. However, a component must be safe not only for fatigue cyclic loads but also for static loads. Therefore, the maximum value of the load is typically used to run a static analysis and check whether a component is safe for static load. If a component fails for static load, it is not necessary to run a fatigue analysis anymore.

In this example, the maximum value of the load of the cyclic load (-200 lb~ $+$ 300 lb) is $+300$ lb. This 300 lb load value will be used to run the static analysis.

Step 1: **Pre-processing**. The model FEA44 is ready for FEA simulation.

Step 2: **Set up a project**. Use the "Static" module to create a study for static analysis.

Step 3: **Select the type of element and assign a material to each component**. The solid element with high-quality mesh will be used for meshing. The material AISI 1020 steel is included in the SolidWorks materials library. No changes for the type of element are required and AISI 1020 can be directly assigned to the component.

Step 4: **Connections**. This is an FEA simulation on a component. There is no connection.

Step 5: Fixtures. Right-click the tab "Fixtures" in the simulation study tree and select "Fixed geometry" to apply a fixed fixture on the top end of the L-shaped bar.

Step 6: Loads. Right-click the tab "External loads" in the simulation study tree and select "Force". In the force Property Manager, apply 300 lb shear force as shown in Fig. 7.6.

Step 7: Meshing. Use the standard mesh with a global element size of 0.2″. Use "Use mesh control" in the fillet area with a local element size of 0.04″.

Step 8: Run Simulation. Right-click the tab of the study name in the simulation study tree and select "Run" or click "Run this study" in the toolbars to start the FEA simulation run.

Step 9: Post-processing.

The Von Mises stress distribution plot is displayed in Fig. 7.7a. The factor of safety distribution plot is shown in Fig. 7.7b. The factor of safety is 3.54, which is bigger than the required factor of safety 2.0. So, this component is safe under the static load.

(b) **Fatigue analysis by FEA Simulation**

The fatigue analysis discussed here is based on cyclic stress events with constant stress amplitudes. This fatigue analysis will use the maximum stress obtained from static analysis and the stress ratio (load ratio) to get information for cyclic stress and then use Miner's rule to estimate fatigue damage. The procedure for fatigue analysis through FEA simulation is presented and explained here.

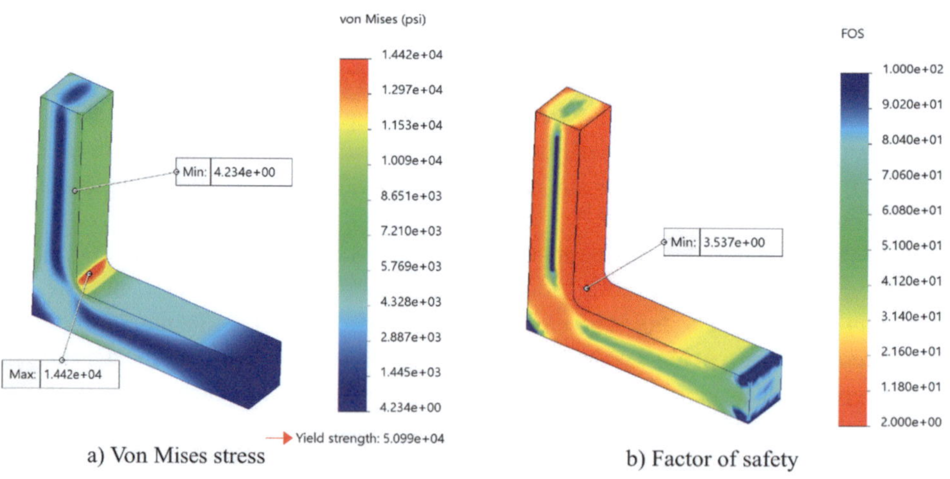

a) Von Mises stress   b) Factor of safety

Fig. 7.7 Von Mises stress and the factor of safety distribution plots

Step 1: Calculate the fatigue strength reduction factor K_f

SolidWorks simulation uses fatigue strength reduction factor K_f to account for differences in fatigue test specimens for the material S–N curve and the actual component. The fatigue strength reduction factor K_f is between 0 and 1. One approach for calculating the fatigue strength reduction factor K_f is:

$$K_f = \frac{k_a k_b k_c}{n_f} \qquad (7.6)$$

where, k_a is surface condition modification factor; k_b is size modification factor; k_c is load modification factor; and n_f is the fatigue factor of safety. The reference [2] provides formulas or tables to determine k_a, k_b and k_c.

One special note is that each simulation of a fatigue analysis will need to input a fatigue strength reduction factor. Different components of an assembly will have different fatigue strength reduction factor K_f. Even on the same component due to different geometric sizes at different locations, the K_f will have different values. So, this K_f is the value on the weakest location of the component. Therefore, fatigue analysis on an assembly is not performed. Fatigue analysis on each component will be performed by using the corresponding fatigue strength reduction factor.

In this example, the surface finish of the component is a "machined" surface. The cyclic load condition is bending. The weakest cross-section is a $1''$ by $1''$ square. By using reference [2]:

$$k_a = 0.908; \quad k_b = 1; \quad k_c = 0.899; \quad n_f = 1.25$$

Per Eq. (7.6), $K_f = 0.653$.

Step 2: Create a study for fatigue analysis

Use the "Fatigue" module under constant amplitude events as shown in Fig. 7.5 to create a study for fatigue analysis.

Step 3: Input the fatigue strength reduction factor and some settings

After a study of fatigue analysis has been created, input the fatigue strength reduction factor and select the correct settings. Right-click the tab of the study name and select "Properties". The fatigue analysis window will appear as shown in Fig. 7.8. Input the fatigue strength reduction factor, which is less than 1. There are several options for fatigue analysis. Since no solid experiment data proves which option is most favorable, any option could be selected. But generally, "No interaction", "Equivalent stress (von Mises), and "Goodman" is selected as shown in Fig. 7.8.

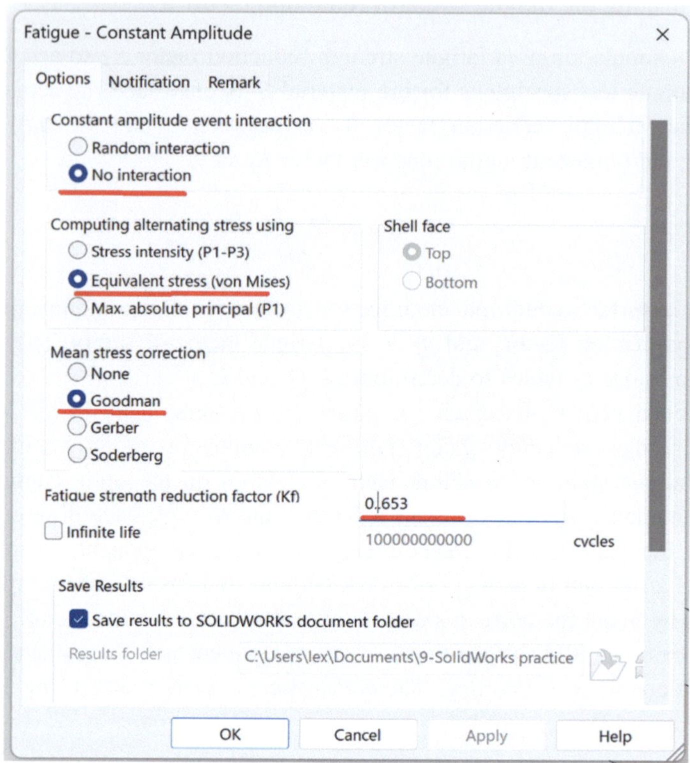

Fig. 7.8 Properties of this fatigue analysis

Step 4: Add cyclic load events

This step will specify all parameters for cyclic stresses. Right-click the tab "Loading (—Constant Amplitude—)" in the simulation study tree and select "Add events". The add event window will appear as shown in Fig. 7.9. Three groups of information as shown in the following for cyclic stress need to be inputted.

(1) Type of cyclic load (stress) or stress (load ratio) LR and number of cycles. In SolidWorks Simulation, cyclic stress is grouped into three cases: fully reversed, LR = −1; general cyclic stress, LR; and zero-based, LR = 0; as shown in Fig. 7.10. The zero-based cyclic stress includes released tension and released compression. For a general cyclic stress, the load ratio LR needs to be inputted.

$$LR = \frac{F_{min}}{F_{max}} \tag{7.7}$$

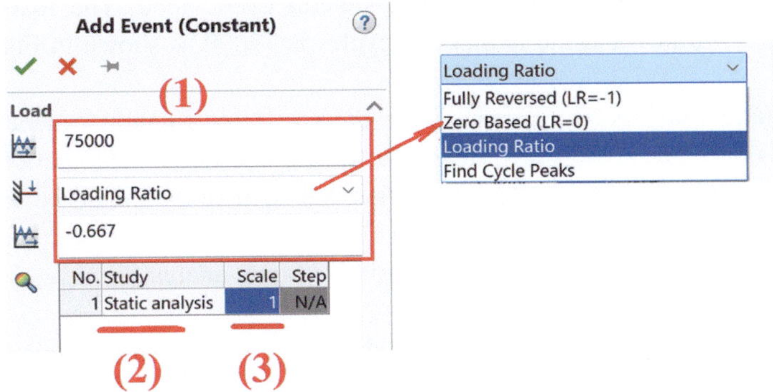

Fig. 7.9 Add load event

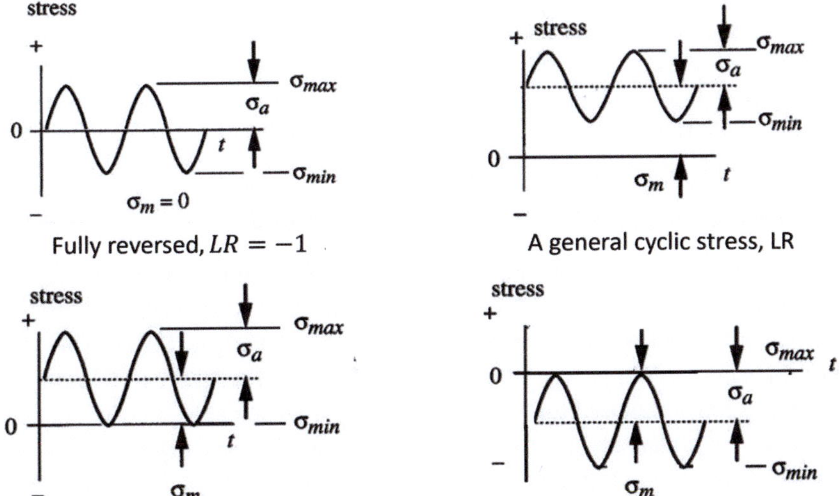

Fig. 7.10 Four types of cyclic stresses

where, LR is the load ratio, which can be positive or negative; F_{min} and F_{max} are the minimum and maximum load in the cyclic load, which can be a force, bending moment, or torque. The load ratio is equal to the stress ratio when components work in a material elastic region.

In this example, the cyclic load is a general cyclic load. The load ratio $LR = -200/300 = -0.667$ and the number of cycles is 75000 as shown in Fig. 7.9.

(2) The static study project, from which the stress information can be retrieved. If there is only one static study, the static study name will be automatically retrieved. But when there are several static studies, the right study should be selected.

In this example, the study for static analysis is "Static analysis".

(3) The value of scale, which is used to scale up or down the stress value. The equation for scale is:

$$Scale = \frac{|\text{maximum load}|}{|F_{static}|} \tag{7.8}$$

where $|\text{maximum load}|$ is the maximum value of load in this cyclic load. $|F_{static}|$ is the load used in the static analysis. Scale is a positive value. For example, if the static load of 500 lb is used in static analysis, the cyclic load is $-400 \sim 200$ lb, the scale will be $|-400|/500 = 0.8$.

In this example, the scale will be $|300|/300 = 1$.

Step 5: Define the material S–N curve for fatigue analysis

After all cyclic load events are added, the tab for parts will appear. This step is to provide material fatigue strength data, that is, the material S–N curve.

Right-click the tab of a part and select "Apply/Edit Fatigue Data". The material Property Manager will appear as shown in Fig. 7.11. The program will automatically lock the type of material used for static analysis, (1) If an actual S–N curve is available, select "Define" and then input data for the S–N curve. (2) If the material S–N curve is not available, the program provides an approach to create an approximate S–N curve that is based on material elastic modulus. Based on the type of steel, the option of "Based on ASME Austenitic Steel Curves" such as for stainless steel, or "Based on ASME Carbon steel curves" such as for plain carbon steel can be selected.

In this example, "Based on ASME Carbon steel curves" for AISI 1020 steel is selected.

Step 6: Run fatigue analysis. Right-click the tab of fatigue analysis study name and select "Run" or click "Run this study" in the toolbars to run the simulation.

Step 7: Post-processing

Right-click the tab "Results" and select "Define fatigue plot". The fatigue damage and the fatigue life of this component are shown in Fig. 7.12a and b, respectively. The fatigue life plot is only available for a single-level cyclic load.

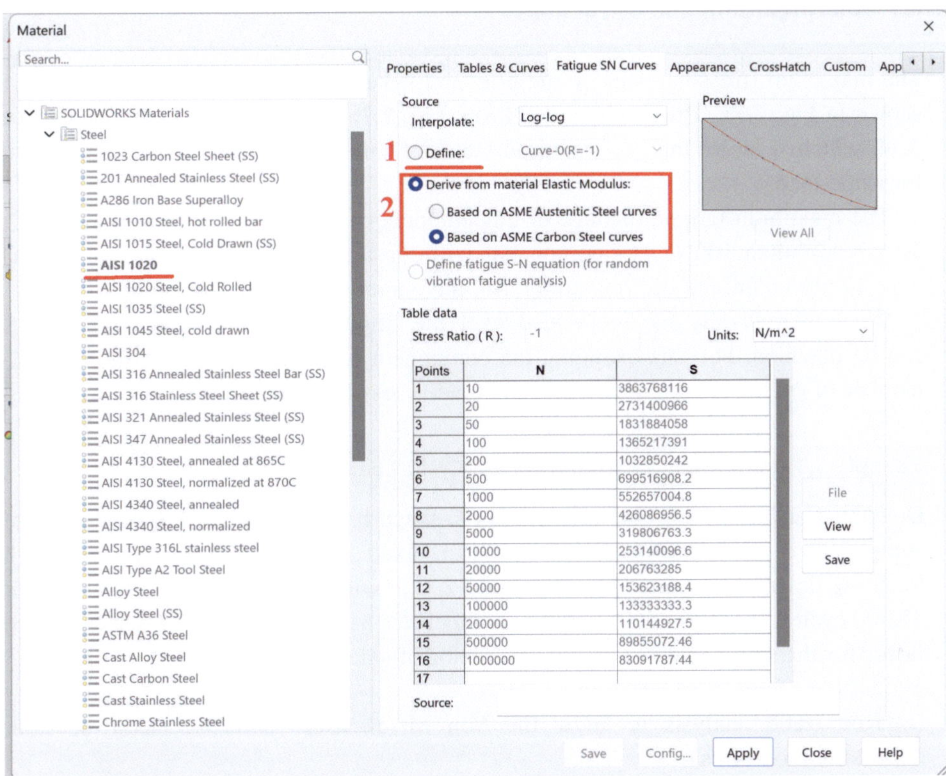

Fig. 7.11 Define material S–N curve

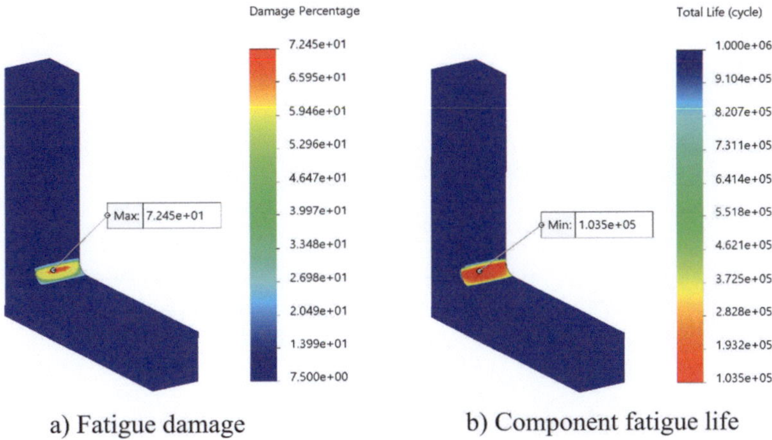

a) Fatigue damage   b) Component fatigue life

Fig. 7.12 Fatigue damage and fatigue life plots

(2) **Interpretations and discussions**

The maximum stress happens on the fillet area where there is a stress concentration as shown in Fig. 7.7a. This is expected. From Fig. 7.7b, the minimum factor of safety is 3.54, which is larger than the required factor of safety 2.0. So, the component is safe for static load.

The maximum fatigue damage of this component is 72.5% as shown in Fig. 7.12a. This is less than 100%. The component will consume 72.5% of fatigue life and still have 17.5% of fatigue life remaining. So, the component is safe for fatigue.

When there is only one cyclic load level, the fatigue life plot as shown in Fig. 7.12b can be obtained. For this example, the component will remain safe for fatigue if the number of cycles of the specified cyclic load is less than 103,500 (cycles).◄

Example 3: Fatigue Analysis Under Several Cyclic Loads (FEA44)

Download and open FEA44, which is an L-shape bar. The material of this machined component is AISI 1020 steel. Its top end is fixed. During its design life, the cyclic loads are described by three levels: $-200 \sim 200$ lb with 500,000 cycles; $0 \sim 300$ lb with 45,000 cycles, and $100 \sim 500$ lb with 20,000 cycles as shown in Fig. 7.13. The required factor for the static analysis is 2.0. The factor of safety for fatigue analysis is 1.25.

(a) Run static analysis to show the Von Mises stress and the factor of safety distribution plots
(b) Run fatigue analysis to show the fatigue damage plot.

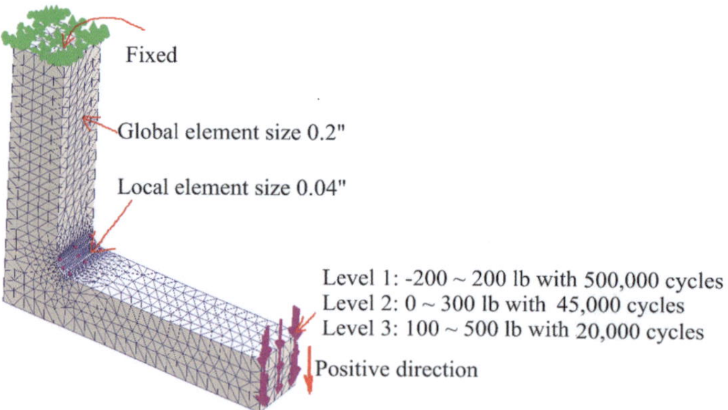

Fig. 7.13 Simulation settings for example 3

Solution

This example is very similar to Example 2 except that this example has three different levels of cyclic loads.

(1) **FEA simulation**
(a) **The static analysis**

Step 1: **Pre-processing**. The model FEA42 is ready for FEA simulation.

Step 2: Set up a project. Use the module "Static" to create a study for static analysis.

Step 3: **Select the type of element and assign a material to each component**. The solid element with high-quality mesh will be used for meshing. The type of element does not need to be changed. The material AISI 1020 steel is included in the SolidWorks materials library and can be directly assigned to the component.

Step 4: **Connections**. This is an FEA simulation on a component. There is no connection.

Step 5: Fixture. Right-click the tab "Fixtures" in the simulation study tree and select "Fixed geometry" to apply a fixed fixture on the top end of the L-shape bar.

Step 6: Loads. The maximum value of cyclic loads in this example will be 500 lb from all three levels of cyclic loads. Right-click the tab "External loads" in the simulation study tree and select "Force". In the force Property Manager window, apply 500 lb shear force as shown in Fig. 7.13.

Step 7: Meshing. Use the standard mesh with a global element size of $0.2''$. Use "Use mesh control" in the fillet area with a local element size of 0.04".

Step 8: Run Simulation. Right-click the tab of the study name in the simulation study tree and select "Run" or click "Run this study" to start the FEA simulation run.

Step 9: Post-processing. The Von Mises stress and the factor of safety distribution plots are displayed in Fig. 7.14a and b, respectively. The factor of safety is 2.12, which is bigger than 2. So, the component is safe for static loads.

(b) **Fatigue analysis**

Step 1: Calculate the fatigue strength reduction Factor K_f
The situation of Example 3 is similar to Example 2 and the same fatigue strength reduction factor $K_f = 0.653$ is used.

Step 2: Create a study for fatigue analysis. Use the "Fatigue" module under constant amplitude events as shown in Fig. 7.5 to create a study for fatigue analysis.

Step 3: Input the fatigue strength reduction factor and some settings. The properties of this example for fatigue analysis are the same as shown in Fig. 7.8.

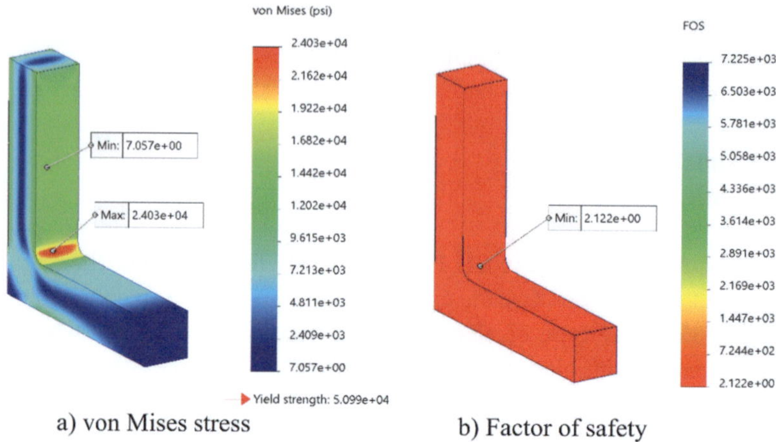

a) von Mises stress   b) Factor of safety

Fig. 7.14 The von Mises stress and the factor of safety distribution plots

Step 4: Add cyclic load events. In this example, three load events will be added. Cyclic stress information will be retrieved from the same static study. However, the types of cyclic loads and load ratios will be different.

- level 1 is a fully reversed cyclic load with 500,000 cycles. The scale is $200/500 = 0.4$.
- Level 2 is a zero-based cyclic load with 45,000 cycles. The scale is $300/500 = 0.6$.
- Level 3 is a general cyclic load with a load ratio $LR = 100/500 = 0.2$ and 20,000 cycles. The scale is $500/500 = 1$.

The setting for these three cyclic load events is shown in Fig. 7.15.

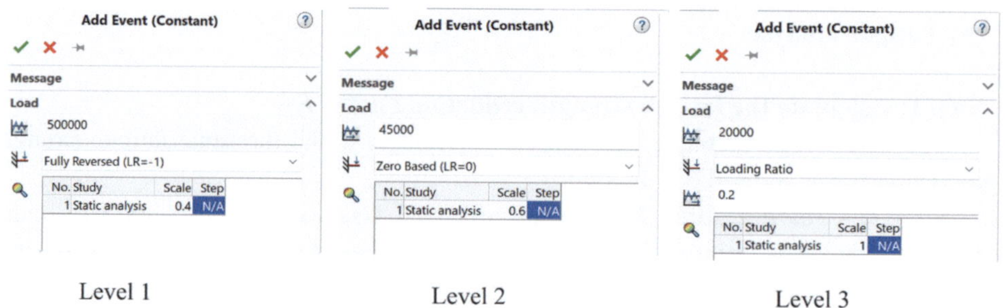

Level 1   Level 2   Level 3

Fig. 7.15 Setting for three cyclic load events

Fig. 7.16 The fatigue damage
plot of example 3

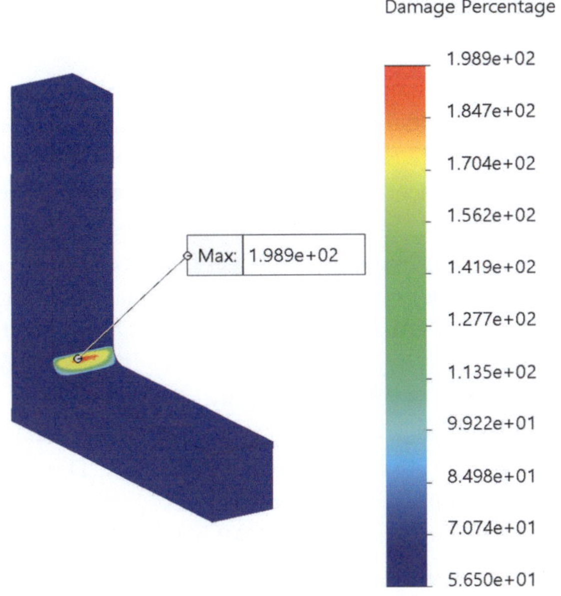

Step 5: Define the material S–N curve for fatigue analysis. In this example, use "Derive from Material Elastic Modulus" and select "Based on ASME Carbon steel curves" for AISI 1020 steel as shown in Fig. 7.11.

Step 6: Run fatigue analysis. Right-click the tab of fatigue analysis study name and select "Run" or click "Run this study" in the toolbars to run the simulation.

Step 7: Post-process

When a component is subjected to several cyclic loads during its service life, only the fatigue damage plot as shown in Fig. 7.16 can be generated.

2. **Interpretations and discussions**

The maximum stress happens on the fillet area where there is a stress concentration as shown in Fig. 7.13a. This is as expected. From Fig. 7.13b, the minimum factor of safety is 2.12, which is larger than the required factor of safety 2.0. So, the component is safe for static load.

The maximum fatigue damage of this component is 198.9% as shown in Fig. 7.16. This is more than 100%. So, the component will fail due to specified cyclic loads.◄

Exercises

7-1 From a component S–N curve obtained from a fully reversed bending stress, three sets of fatigue strength properties are: $S_{f1} = 15$ (ksi), $N_1 = 700,000$ (cycles); $S_{f2} = 27$ (ksi), $N_2 = 140,000$ (cycles); $S_{f3} = 30$ (ksi), $N_3 = 60,000$ (cycles). The component is subjected to a completely reversed bending stress with stress amplitude stress $\sigma_a = 30$ (ksi) and the number of cycles $n = 45,000$(cycles). Determine the cumulative fatigue damage. Is the component safe for fatigue?

7-2 From a component S–N curve obtained from a fully reversed bending stress, three sets of fatigue strength properties are: $S_{f1} = 15$(ksi), $N_1 = 700,000$(cycles); $S_{f2} = 27$(ksi), $N_2 = 140,000$(cycles); $S_{f3} = 30$(ksi), $N_3 = 60,000$(cycles). The component is subjected to fully reversed bending stresses, which contain three different levels: $\sigma_{a1} = 15$(ksi), $n_1 = 320,000$(cycles); $\sigma_{a2} = 27$(ksi), $n_2 = 65,000$(cycles); $\sigma_{a3} = 30$(ksi), $n_3 = 8,000$(cycles). Determine the cumulative fatigue damage. Is the component safe for fatigue??

7-3 Download and open FEA2, which is a machined component made of Alloy steel. It is fixed at one end and is subjected to cyclic load $0 \sim 2000$ lb with 12,000 (cycles) as shown in Fig. 7.17. The factors of safety of static and fatigue analysis are 2.0 and 1.25, respectively. Determine the cumulative fatigue damage. Is the component safe for static load? Is the component safe for fatigue?

7-4 Download and open FEA2, which is a machined component made of Alloy steel. It is fixed at one end and is subjected to two levels of cyclic loads. Level 1: $-600 \sim 600$ lb with 200,000 (cycles) and Level 2: $0 \sim 2000$ lb with 5,000 (cycles) as shown in Fig. 7.18. The factors of safety of static and fatigue analysis are 2.0 and 1.25, respectively. Determine the cumulative fatigue damage. Is the component safe for static load? Is the component safe for fatigue?

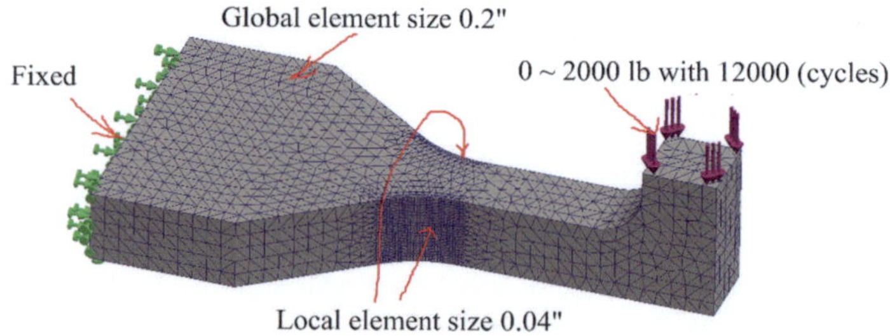

Fig. 7.17 The setting for Exercise 7.3

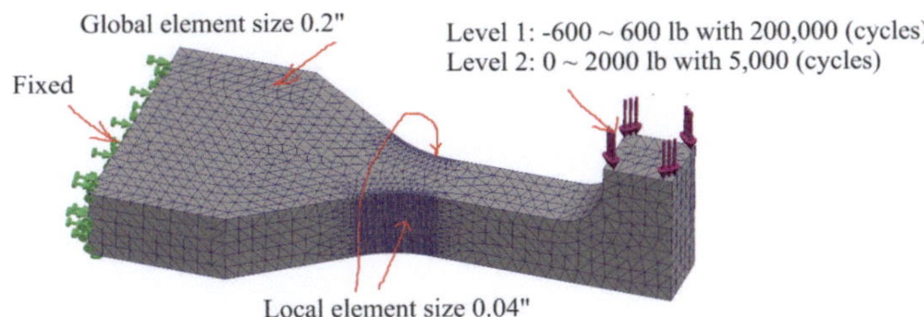

Fig. 7.18 The settings for Exercise 7.4

References

1. Xiaobin Le (2019) Reliability-based mechanical design, Volume 2, components under cyclic load and dimension design with required reliability, Morgan & Claypool, San Rafael, California
2. Richard G. Budynas, J. Keith Nisbett (2015) Shrigley's mechanical engineering design, McGraw-Hill Education, New York
3. SolidWorks Web Help, https://help.solidworks.com